PROCESS CONTROL INSTRUMENTATION TECHNOLOGY

Fourth Edition

Curtis Johnson

REGENTS/PRENTICE HALL
Englewood Cliffs, New Jersey 07632

Library of Congress Cataloging-in-Publication Data

Johnson, Curtis D.
   Process control instrumentation technology / Curtis D. Johnson.—
4th ed.
     p. cm.
   Includes bibliographical references and index.
   ISBN 0-13-721150-3
   1. Process control. 2. Engineering instruments. I. Title.
TS156.8.J63 1993
670.42—dc20   92-12781
                                             CIP

Acquisitions: Holly Hodder
Editorial/production supervision: Tally Morgan, Wordcrafters Editorial Services, Inc.
Cover design: Laura Zerardi
Manufacturing buyer: Ed O'Dougherty
Prepress buyer: Ilene Levy
Cover photo: Tom Tracy/©Stock Market

© 1977, 1982, 1988 by John Wiley & Sons, Inc.; © 1993 by Regents/Prentice-Hall, Inc.
A Division of Simon & Schuster
Englewood Cliffs, NJ 07632

Printed in the United States of America

10 9 8 7 6 5 4 3 2 1

ISBN 0-13-721150-3

Prentice-Hall International (UK) Limited, *London*
Prentice-Hall of Australia Pty. Limited, *Sydney*
Prentice-Hall Canada Inc., *Toronto*
Prentice-Hall Hispanoamericana, S.A., *Mexico*
Prentice-Hall of India Private Limited, *New Delhi*
Prentice-Hall of Japan, Inc., *Tokyo*
Simon & Schuster Asia Pte. Ltd., *Singapore*
Editora Prentice-Hall do Brasil, Ltda., *Rio de Janeiro*

CONTENTS

3 DIGITAL SIGNAL CONDITIONING, 98

4 THERMAL SENSORS, 135

Contents

11 DIGITAL CONTROLLERS, 428

12 CONTROL LOOP CHARACTERISTICS, 480

PREFACE

The advancement of both knowledge and technique has resulted in the development of specialists in process control. Today, it is possible to delineate process-control activities into three categories, each with specific requirements with regard to training. The process-control system engineer is concerned with the design of an overall process-control system that will provide the regulation specified by the manufacturing process experts. This activity requires a good theoretical understanding of stability, mode characteristic, and general process-control loop dynamic characteristics. The process-control technologist or applied engineer is responsible for the design of specific elements required to implement the overall design specified for the system. This involves a good, design-level understanding of the measurement, electronic, and pneumatic features inherent in process-control loops. The process-control technician or mechanic is responsible for the final assembly and testing of the elements of the process-control loop and the system as a whole. This requires a good working knowledge of the electronics, pneumatics, and mechanical features of process-control loop installations.

This book was developed primarily to fulfill the instructional needs of the last two categories, that is, the applied process engineer, technologist, and advanced technician. Many fine textbooks exist that cover the stability and design criteria of process-control systems, usually involving the use of Laplace transform methods. Often these books do not cover the system elements per se, such as measurement methods, signal conditioning, and final control elements. Indeed, there has been a serious lack of texts that cover process-control elements with less than the mathematical rigor of Laplace transforms and yet enough math to allow the design of loop elements to satisfy specifications. This volume has been prepared to fill this need. Its overall objective is to provide instructional material for a general understanding of process-control characteristics such as elements, modes, and stability along with detailed knowledge of measurement technique, control mode implementation, and final control element functions. In keeping with modern trends, the digital aspect of process-control technology is stressed.

The student background is assumed to consist of that acquired by an advanced sophomore in a two-year engineering/technology program or a junior in a four-year program. Consequently, a basic knowledge of physics and analog and digital electronics is assumed along with math through college algebra, although some knowledge of calculus would be beneficial.

The goal of this fourth edition is the same as that of the original text in 1977: to provide a quality book which can be used to teach students about the practical aspects of process control. Esoteric concepts of Laplace transforms, Z transforms, and other theoretical control system topics have not been included. Instead, this book presents the reader with essential information on how to put together a real control system and how to tune that control system to provide stable regulation of the controlled variable. None of the revisions to this edition lose sight of that goal.

In addition to including more problems and cleaning up errors and omissions pointed out since the previous edition, this edition updates some of the terminology. For example, the American National Standards Institute (ANSI) and the Instrumental Society of America (ISA) have recommended that the word *transducer* be replaced with *sensor* for the initial measurement device. A number of other similar changes have been made throughout.

I acknowledge Lois Jones for a great deal of help in corrections and for moral support and encouragement during the long revision process. Mason Rittman of Texas A&M University cannot be thanked enough for the help he provided in identifying errors and making constructive suggestions. Of course, the Prentice Hall staff and the reviewers they solicited were essential and appreciated.

Curtis D. Johnson

CHAPTER 1

INTRODUCTION TO PROCESS CONTROL

INSTRUCTIONAL OBJECTIVES

This chapter will consider the overall process-control loop, its function, and its description. After you have read this chapter, you should be able to

1. Draw a block diagram of a process-control loop with a description of each element.
2. List three typical controlled variables.
3. Describe three criteria used to evaluate the response of a process-control loop.
4. Define analog signal processing.
5. Describe the two types of digital process control.
6. Define accuracy, hysteresis, and sensitivity.
7. List the SI units of measure for length, time, mass, and electric current.
8. Convert a physical quantity from SI to English units and vice versa.
9. Define the types of measurement time response.

1.1 INTRODUCTION

The progression of human existence from a primitive state to the present complex technological world was paced by learning new and improved methods to control the environment. Simply stated, the term *control* means methods to force parameters in the environment to have specific values. This can be as simple as making

1

the temperature in a room stay at 21°C or as complex as manufacturing an integrated circuit or guiding a spacecraft to Jupiter. In general, all of the elements necessary to accomplish the control objective are described by the term *control system*.

The purpose of this book is to examine the elements and methods of control system operation used in industry to control industrial processes (hence the term *process control*). This first chapter will present an overall view of process-control technology and its elements. In addition, a number of important definitions and general topics will be presented. The remaining chapters will study the elements of process control in more detail.

1.2 CONTROL SYSTEMS

The basic strategy by which a control system operates is quite logical and natural. In fact, the same strategy is employed in living organisms to maintain temperature, fluid flow rate, and a host of other biological functions. This is natural process control.

The technology of artificial control was first developed with a human as an integral part of the control action. When we learned how to use machines, electronics, and computers to replace the human function, the term *automatic control* came into use.

1.2.1 Process-Control Principles

In process control, the basic objective is to regulate the value of some quantity. To regulate means to maintain that quantity at some desired value regardless of external influences. The desired value is called the *reference value* or *setpoint*.

The following paragraphs illustrate the development of a control system for a specific process-control example and introduce some of the terms and expressions used.

The process

Figure 1.1 shows the process to be used for this discussion. Liquid is flowing into a tank at some rate Q_{in} and out of the tank at some rate Q_{out}. The liquid in the tank has some height or level h. It is known that the flow rate out varies as the square root of the height, so the higher the level the faster the liquid flows out.

If the output flow rate is not exactly equal to the input flow rate, the tank will either empty, if $Q_{out} > Q_{in}$, or overflow, if $Q_{out} < Q_{in}$.

This process has a property called *self-regulation*. This means that for some input flow rate, the liquid height will rise until it reaches a height for which the output flow rate matches the input flow rate. This is somewhat misleading, however, because it is not regulated since a change in input flow rate will cause the height to change.

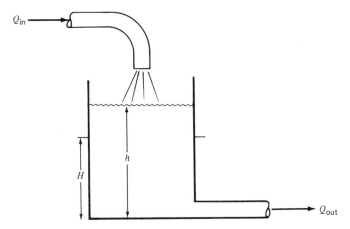

Figure 1.1   The objective is to regulate the level of liquid in the tank, h, to the value H.

The objective is to regulate the height at a specific value, the setpoint H. The height, or level, is called the *controlled variable*.

Human-aided control

Figure 1.2 shows a modification of the tank system to allow artificial control of the level by a human. A sight tube S has been added so the human can see what the level in the tank is compared to the setpoint value H, which has been marked

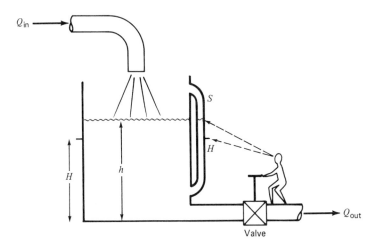

Figure 1.2   A human can regulate the level using a sight tube, S, to compare the level, h, to the objective, H, and adjust a valve to change the level.

on the tube. In addition, a valve has been added so the output flow rate can be changed by the human. The output flow rate is called the *manipulated variable* or *controlling variable*.

Now the height can be regulated apart from the input flow rate using the following strategy: The human measures the height in the sight tube and compares the value to the setpoint. If the measured value is larger, the human opens the valve a little to let the flow out increase, and thus the level lowers toward the setpoint. If the measured value is smaller than the setpoint, the human closes the valve a little to decrease the flow out and allow the level to rise toward the setpoint.

By a succession of incremental opening and closing of the valve, the human can bring the level to the setpoint value H and maintain it there by continuous monitoring of the sight tube and adjustment of the valve. The height is regulated.

Automatic control

To provide automatic control, the system is modified as shown in Figure 1.3 so machines, electronics, or computers replace the operations of the human. An instrument, called a *sensor*, is added that is able to measure the value of the level and convert it into a proportional signal s. This signal is provided as input to a machine, electronic circuit, or computer, called the *controller*. This performs the function of the human in evaluating the measurement and providing an output signal u to change the valve setting via an *actuator* connected to the valve by a mechanical linkage.

When automatic control is applied to systems like the one shown in Figure 1.3, which are designed to regulate the value of some variable to a setpoint, it is called *process control*.

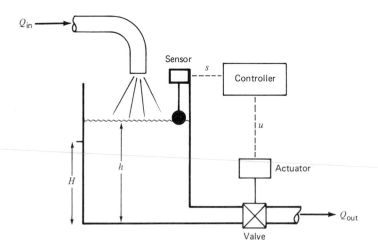

Figure 1.3 An automatic level-control system replaces the human by a controller and uses a sensor to measure the level.

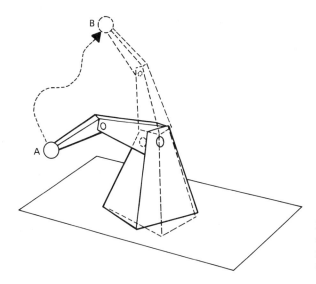

Figure 1.4 Servomechanism-type control systems are used to move a robot arm from point A to point B in a controlled fashion.

1.2.2 Servomechanisms

Another type of control system in common use, which has a slightly different objective from process control, is called a *servomechanism*. In this case the objective is to force some parameter to vary in a specific manner. This may be called a tracking control system. Instead of regulating a variable value to a setpoint, the servomechanism forces the controlled variable value to follow variation of the reference value.

For example, in an industrial robot arm like the one shown in Figure 1.4, servomechanisms force the robot arm to follow a path from point A to point B. This is done by controlling the speed of motors driving the arm and the angles of the arm parts.

The strategy for servomechanisms is very similar to process-control systems, but the dynamic differences between regulation and tracking result in differences in design and operation of the control system. This text is directed toward process-control technology.

1.3 PROCESS CONTROL BLOCK DIAGRAM

To provide a practical, working description of process control, it is useful to describe the elements and operations involved in more generic terms. Such a description should be independent of a particular application (such as the example presented in the previous section) and thus be applicable to *all* control situations. A model may be constructed using blocks to represent each distinctive element. The characteristics of control operation then may be developed from a consideration of the properties and interfacing of these elements. Numerous models have been employed in the history of process-control description; we will use one that seems most appropriate for a description of modern and developing technology of process control.

1.3.1 Identification of Elements

The elements of a process-control system are the various separate parts of the system. The following paragraphs define the basic elements of a process-control system and relate them to the example presented in the previous section.

Process

The flow of liquid in and out of the tank, the tank itself, and the liquid all constitute a process to be placed under control with respect to the fluid level. In general, a process can consist of a complex assembly of phenomena that relate to some manufacturing sequence. Many variables may be involved in such a process, and it may be desirable to control all these variables at the same time. There are *single-variable* processes, in which only one variable is to be controlled, as well as *multivariable* processes, in which many variables, perhaps interrelated, may require regulation.

Measurement

Clearly, to effect control of a variable in a process, we must have information on the variable itself. Such information is found by *measurement* of the variable. In general, a measurement refers to the transduction of the variable into some corresponding *analog* of the variable, such as a pneumatic pressure, an electrical voltage, or current. A sensor is a device that performs the initial measurement and energy conversion of a variable into analogous electrical or pneumatic information. Further transformation or *signal conditioning* may be required to complete the measurement function. The result of the measurement is a transformation of the variable into some proportional information in a useful form required by the other elements in the process-control operation.

In the system shown in Figure 1.3, the controlled variable is the level of liquid in the tank. The measurement is performed by some sensor which provides a signal s to the controller. In the case of Figure 1.2, the sensor is the sight tube showing the level to the human operator as an actual level in the tank.

The sensor is also called a transducer. The word *sensor* is preferred for the initial measurement device, however, because *transducer* represents a device which converts any signal from one form to another. Thus, for example, a device which converts a voltage into a proportional current would be a transducer. In other words, all sensors are transducers but not all transducers are sensors.

Error detector

In Figure 1.2, the human looked at the difference between the actual level h and the setpoint level H and deduced an error. This error has both a magnitude and polarity. For the automatic control system of Figure 1.3, this same kind of error determination must be made before any control action can be taken by the controller. Although the error detector is often a part of the controller device, it is important to keep a clear distinction between the two.

Controller

The next step in the process-control sequence is to examine the error and determine what action, if any, should be taken. This part of the control system has many names; however, *controller* is the most common. The evaluation may be performed by an operator (as in the previous example), by electronic signal processing, by pneumatic signal processing, or by a computer. Computer use is growing rapidly in the field of process control because it is easily adapted to the decision-making operations and because of its inherent capacity to handle control of multivariable systems. The controller requires an input of both a *measured indication* of the controlled variable and a representation of the *reference value* of the variable, expressed in the same terms as the measured value. The reference value of the variable is referred to as the *setpoint*. Evaluation consists of a determination of action required to bring the controlled variable to the setpoint value.

Control element

The final element in the process-control operation is the device that exerts a direct influence on the process, that is, that provides those required changes in the controlled variable to bring it to the setpoint. This element accepts an input from the controller, which is then transformed into some proportional operation performed on the process. In our previous example, the control element is the valve that adjusts the outflow of fluid from the tank. This element is also referred to as the *final control element*.

1.3.2 Block Diagram

Each element in a process-control system is represented in a block diagram as a separate step. Figure 1.5 is a block diagram constructed from the elements defined in the previous section. The controlled variable in the process is denoted by c in this diagram, and the measured representation of the controlled variable is labeled b. The controlled variable setpoint is labeled r for reference.

The error detector is a *subtracting–summing point* that outputs an *error signal* $e = r - b$ to the controller for comparison and action.

To illustrate further, the block diagram in Figure 1.6 shows a typical flow-

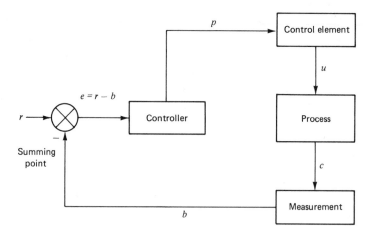

Figure 1.5 This block diagram of a control loop defines all the basic elements and signals involved.

control system and its representation by a block diagram. In this example, the controlled variable is the flow rate that is converted to electric current as an analog.

The purpose of a block diagram approach is to allow the process-control system to be analyzed as the interaction of smaller and simpler subsystems. If the characteristics of each element of the system can be determined, then the characteristics of the assembled system can be established by an analytical marriage of these subsystems. The historical development of the system approach in technology was dictated by this practical aspect: first, to specify the characteristics desired of a total system, and then, to delegate the development of subsystems that provide the overall criteria.

It becomes evident that the specification of a process-control system to regulate a variable c, within specified limits and with specified time responses, determines the characteristics the measurement system must possess. This same set of system specifications is reflected in the design of the controller and control element.

From this concept, we conclude that the analysis of a process-control system requires an understanding of the overall system behavior and the reflection of this behavior in the properties of the system elements. Most people find that an understanding of the parts leads to a better understanding of the whole. We will proceed with this assumption as a guiding concept.

The loop

Notice in Figure 1.5 that the signal flow forms a complete circuit from process through measurement, error detector, controller, and final control element. This is called a *loop*, and in general we speak of a process-control loop. In most cases

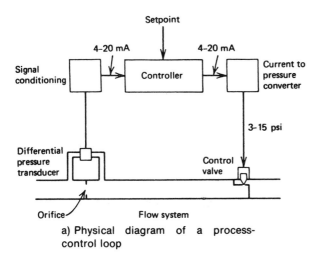

a) Physical diagram of a process-control loop

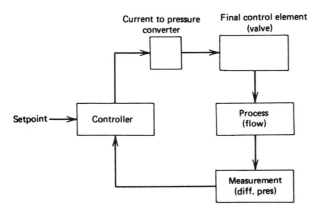

b) Block diagram of a process-control loop

Figure 1.6 The physical diagram of a control loop and its corresponding block diagram look similar. Note the use of current and pressure transmission signals.

this is called a *feedback loop* because we determine an error and feed back a correction to the process.

1.4 CONTROL SYSTEM EVALUATION

A process-control system is used to regulate the value of some process variable. When such a system is in use, it is natural to ask, "How well is it working?" This is not an easy question to answer, because it is possible to adjust a control system to provide different kinds of response to errors. This section discusses some methods for evaluating how well the system is working.

The variable used to measure the performance of the control system is the error, which is the difference between the constant setpoint or reference value r and the controlled variable $c(t)$.

$$e(t) = r - c(t) \tag{1.1}$$

Since the value of the controlled variable may vary in time, so may the error. (Note that in a servomechanism, the value of r may be forced to vary in time also.)

Control system objective

In principle, the objective of a control system is to make the error in Equation (1.1) exactly zero. But the control system only responds to errors (i.e., when an error occurs the control system takes action to drive it to zero). Conversely, if the error was zero and stayed zero, the control system was doing nothing and was not needed in the first place. Therefore, this objective can never be perfectly achieved, and there will always be some error. The question of evaluation becomes one of how large the error is and how it varies in time.

A more practical objective statement is that the control system should (1) be stable and (2) provide the best possible regulation of the controlled variable. Let's consider what this objective statement means.

1.4.1 Stability

A control system can *cause* a process variable to become unstable. Consider a process with some variable like level, flow, temperature, etc., which is unregulated but varies in some drifting fashion. A control system hooked up to regulate the variable may actually cause an unstable variation. Figure 1.7 shows this. The controlled variable is regulated when the system is turned on but then suddenly begins to exhibit growing oscillations. This is an instability.

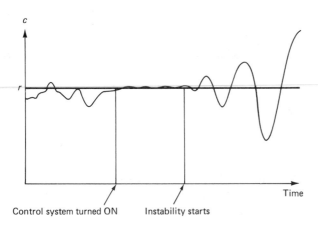

Control system turned ON   Instability starts

Figure 1.7 If a control system is improperly adjusted, it may appear to have regulated the controlled variable and then, suddenly, become unstable.

The first objective statement simply means that the control system must be designed and adjusted so the system is stable. Typically, as the control system is adjusted to give better control, the likelihood of instability also increases.

1.4.2 Regulation

The objective of best possible regulation simply means that the steady-state error should be a minimum. Generally, when a control system is specified there will be some allowable deviation, $\pm \Delta c$, about the setpoint. This means that variations of the variable within this band are expected and acceptable. External influences which tend to cause drifts of the value beyond the allowable deviation are corrected by the control system.

For example, a process-control technologist might be asked to design and implement a control system to regulate temperature at 150°C within ± 2°C. This means the setpoint is to be 150°C but the temperature may be allowed to vary in the range of 148 to 152°C.

1.4.3 Transient Regulation

What happens to the value of the controlled variable when some sudden event occurs which would otherwise cause a variation? For example, there could be a setpoint change. Suppose the setpoint in the aforementioned temperature case were suddenly changed to 160°C. Transient regulation specifies how the control system reacts to bring the temperature to this new setpoint.

Another type of transient influence is a sudden change of some other process variable. The controlled variable depends on other process variables. If one of them suddenly changes value, the controlled variable may be driven to change also, so the control system acts to minimize the effect. This is called *transient response*.

1.4.4 Evaluation Criteria

The question of how well the control system is working is answered by (1) assuring stability and (2) evaluating the response to setpoint changes and transient effects against certain standard criteria. There are many criteria for gauging the response. In general, the term *tuning* is used to indicate how a process-control loop is adjusted to provide the best control. This topic is covered in more detail in Chapter 12.

Damped response

One type of criteria requires that the controlled variable exhibit a response such as that shown in Figure 1.8 for excitations of both setpoint changes and transient effects. Note that the error is of only one polarity (i.e., it never oscillates about

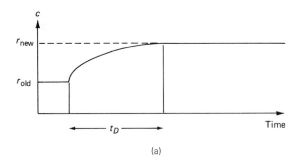

(a)

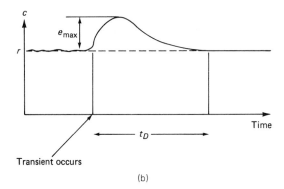

Transient occurs

(b)

Figure 1.8 A system with overdamped response will react to a change in setpoint or to a transient as shown in (a) and (b), respectively.

the setpoint). For this case, measures of quality are the duration t_D of the excursion and, for the transient, the maximum error e_{max} for a given input. The duration is the time for exceeding the allowable error to regaining the allowable error.

Different tuning will provide different values of e_{max} and t_D for the same excitation. It is up to the process designers to decide if the best control is larger duration with smaller peak error, or vice versa, or something in between.

Cyclic response

Another type of criteria applies to those cases in which the response to a setpoint change or transient is as shown in Figure 1.9. Note that the controlled variable oscillates about the setpoint. In this case, the parameters of interest are the maximum error, e_{max} and the duration t_D, also called the settling time. The duration is measured from the time the allowable error is first exceeded to the time when it falls within the allowable error and stays.

The nature of the response is modified by adjusting the control loop parameters, which is called *tuning*. There may be large maximum error but short duration or long duration with small maximum error, and everything in between.

A number of standard cyclic tuning criteria are used. Two common types

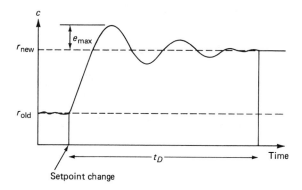

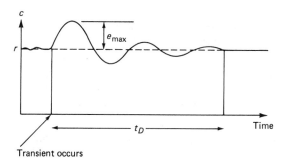

Figure 1.9 A system with underdamped response will react to a change in setpoint or to a transient with oscillations as shown in (a) and (b).

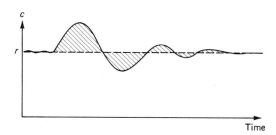

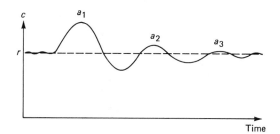

Figure 1.10 Two criteria for judging the quality of control system response are the minimum area and quarter amplitude.

are minimum area and quarter amplitude. In minimum area, the tuning is adjusted until the net area under the error-time curve is a minimum, for the same degree of excitation (setpoint change or transient). Figure 1.10a shows the area as a shaded part of the curve. Analytically, this is given by

$$A = \int |e(t)| \, dt = \text{minimum} \tag{1.2}$$

The quarter-amplitude criteria, shown in Figure 1.10b, specifies that the amplitude of each peak of the cyclic response must be a quarter of the preceding peak. Thus, $a_2 = a_1/4$, $a_3 = a_2/4$, and so on.

1.5 ANALOG AND DIGITAL PROCESSING

The evolution of process control has seen the infusion of electronics technology into almost every facet because of low cost, reliability, miniaturization, and ease of interface. It is natural that the further development of digital electronics and associated computer technology has brought about the rapid introduction of digital techniques in the field of process control. Some aspects of process control, such as the initial transduction of a controlled variable into electrical information, will probably always be of analog nature. It is inevitable, however, with the continued development of computers, miniaturized digital electronics, and associated technology that the evaluation and controller phase of process control may be digitally performed. Let us contrast the two methods of processing.

1.5.1 Analog Processing

Consider the process-control loop shown in Figure 1.11, where a process temperature is to be controlled. A thermistor, whose resistance R is proportional to and an analog of temperature, is used to measure the temperature. The resistance change is converted to a voltage V via some unspecified signal conditioning. This voltage is compared to some reference voltage (setpoint) by a differential amplifier (evaluation), the output of which activates either a heater or cooler. The range of allowed temperature is determined by the differential amplifier swing necessary to trip either relay. In this loop the temperature is represented by a proportional electrical signal. We then say that the electrical signal is an analog of the temperature. In the case of the thermistor, the resistance is an analog of the temperature.

In a similar fashion, in a pneumatic system, we may have a fluid *pressure* that is an analog of a variable. The analog relationship between processing signals and dynamic variables need not be linear and in many cases is not. The significant factor here is that the processing signal is a *smooth* and unique *representation* of the dynamic variable. In Figure 1.12 we see two examples of analog proportionalities between a variable c and an analog signal b. In case 1, we have a linear

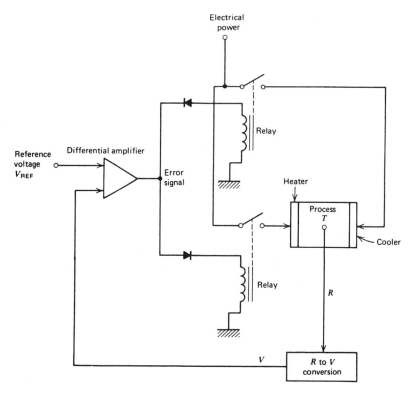

Figure 1.11   An analog process-control loop for regulation of temperature. Note that all signals in the loop have some analog to temperature.

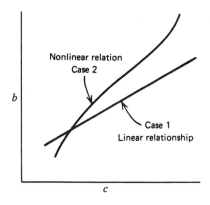

Figure 1.12   The relationship between a variable c and its measured equivalent b may be linear as in case 1, or nonlinear as in case 2.

relationship and, in case 2, a nonlinear relationship, but both are still analog representations of c.

1.5.2 Digital Processing

In digital processing, all information carried in the process-control loop is encoded into a signal that is binary in nature. Binary refers to a numbering system with a base of 2, that is, 0 and 1 are the only possible counting states.

The electrical signals on a wire are represented either by a binary "0," denoted also by low (L) and corresponding in transistor-transistor logic (TTL) to about 0 volts, or a binary "1," denoted by high (H) and corresponding in TTL to about $+5$ volts. Thus, in TTL, electrical voltages are only 0 or 5 volts and hence cannot be an analog representation of the dynamic variable. The value of the dynamic variable is represented as some *encoding* of the binary levels. The encoding itself is a correspondence between a set of binary numbers and the analog signal to be encoded. The set of binary numbers is commonly referred to as a word, which may contain many binary counts called bits.

Binary/decimal encoding

Let us consider the encoding between a 4-bit binary word and a decimal voltage signal. Such a coding can be simply the representation of numbers between the two systems, as shown in Table 1.1. Thus, if we wished to encode a 5-volt signal, where each bit corresponds to 1 volt, we would represent this in a 4-bit binary word as "**0101**".

Signal transmission

The encoding of a signal into a binary word implies that a sequence of 0 and 1 levels corresponds to the value of the signal. There are two methods by which a digitally encoded signal can be transmitted through the process-control loop. One

TABLE 1.1   Decimal-Binary Encoding

Voltage	Binary Word
0	0000
1	0001
2	0010
3	0011
4	0100
5	0101
6	0110
7	0111
8	1000
9	1001
10	1010

method, referred to as the *parallel transmission mode*, provides a separate wire for each binary number in the word. Thus, a 4-bit word requires 4 wires, an 8-bit word 8 wires, and so on. The alternate method is *serial transmission mode* of binary numbers over a single wire, where the binary levels are provided in a time sequence over the wire. These two methods are illustrated in Figure 1.13 for the digital encoding of the 5-volt level. In the parallel mode, the levels can remain set for any time on the lines, but in the serial mode, the levels appear only as a pulse train and must be interpreted in turn as they appear, introducing timing requirements.

Analog and digital converters

The sensor is most often a device that converts a variable to an electrical or pneumatic analog. Furthermore, the final control element is typically a device that converts a controller analog signal to some effect on the controlled variable in the process. If digital processing is to be employed in the process-control loop, we must have a means of converting between the analog and digital representations of the variable and controller outputs. The analog-to-digital converter (ADC) performs the function of conversion of an analog input to a digitally encoded signal. These devices are designed to output a digital word coded to a specified range of signal inputs. Thus, an ADC might accept a 0–10 volt input encoded into a 4-bit word. A 4-bit word can represent 15 spans; hence, each bit represents

$$\frac{10 \text{ volts}}{15} = 0.666 \text{ volts/state or volts/bit}$$

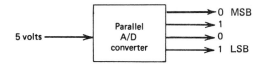

a) Parallel transmission mode

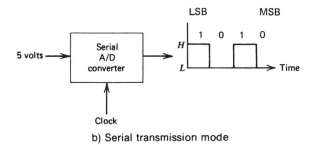

b) Serial transmission mode

Figure 1.13   Digital information may be transmitted on parallel lines, one for each bit, or serially, where the information appears on a single line in a time sequence. (a) Parallel transmission mode; (b) serial transmission mode.

If more voltage resolution is necessary, we may use a word with more bits. A 6-bit word would have 63 spans and

$$\frac{10 \text{ volts}}{63} = 0.159 \text{ volts/bit}$$

The digital-to-analog converter (DAC) provides the reverse action of converting the digitally encoded word into an appropriate analog output; that is, it decodes. Here again, each bit, by design, will correspond to a certain level of output, and thus the resolution, or the smallest increment, is the level of one state change. The example of ADC from voltage is only one example of the type of analog signal that can be used. Other converters may use analog signals of frequency variation, current, or even resistance as the primary analog signal.

There are two approaches to digital processing in industrial-control situations. One involves using digital logic circuits or computers to supervise analog process-control loops and the other direct digital control of a dynamic variable.

Supervisory digital control

The expression *supervisory digital control* is used to describe a situation in which digital techniques are employed to supervise analog control loops. Generally, in these cases, a computer is employed to adjust the setpoint of an analog controller. This approach is employed to take advantage of the well-known and trusted performance of analog control, while at the same time using the various advantages of computer supervision. In particular, the computer can examine many variables and solve complicated control equations to determine and then set optimum setpoints of several analog loops. This is particularly important when interaction exists between variables, such as changing the temperature setpoint and causing the pressure to vary. The block diagram of a supervisory system is shown in Figure 1.14. In general, the logic or computer supervisor will input the measured quantity and programming and output a calculated setpoint. In the analog loop described in Section 1.5.1, a hybrid conversion could be developed by using ADCs and DACs to provide temperature input and reference voltage output to the analog loop, as illustrated in Figure 1.15.

Direct digital control

The method of process control described by the term *direct digital control* (DDC) applies to those cases in which digital logic circuits or a computer are an integral part of the loop. In Figure 1.16 we see a block diagram of this approach to processing. Essentially, the evaluation and controller function is taken over by digital logic circuits or programming of a computer. We distinguish these two approaches to DDC by calling logic circuit methods *hardware program* controlling and computer methods *software program* controlling. Figure 1.17 shows how the temperature-control problem defined in Section 1.5.1 is implemented by DDC. The control function, setpoint, and deviation about the nominal are all defined by the

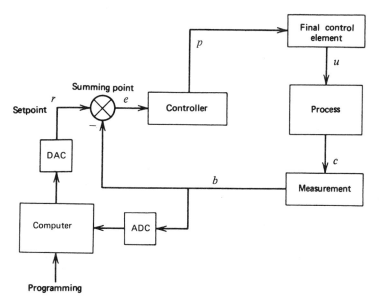

Figure 1.14 Implementation of the process-control loop using a computer in a supervisory capacity.

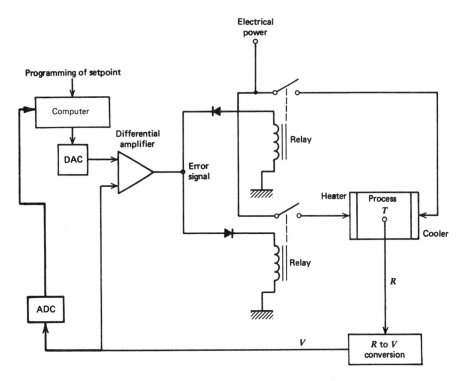

Figure 1.15 An example of supervisory digital control in the case of the temperature-control problem of Section 1.5.1.

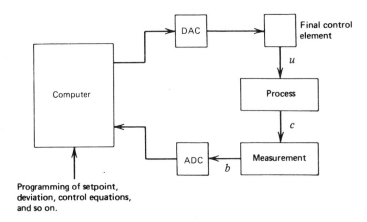

Figure 1.16 Block diagram of a process-control loop which is under computer-based, direct digital control (DDC).

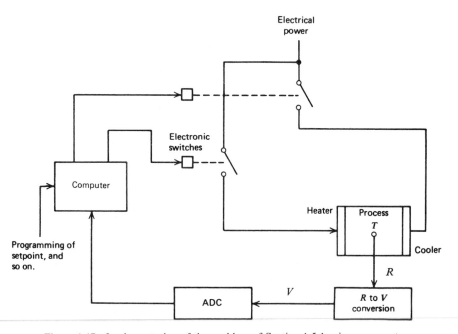

Figure 1.17 Implementation of the problem of Section 1.5.1 using a computer.

program. Direct digital control has the capacity to control multivariable processes with interaction between elements. This approach is gaining in acceptance as the reliability of computers improves and backup methods are being developed to avoid process shutdown because of a computer failure. The development of the so-called computer on a chip in the form of a single integrated circuit (IC) microprocessor has given considerable impetus to the use of DDC.

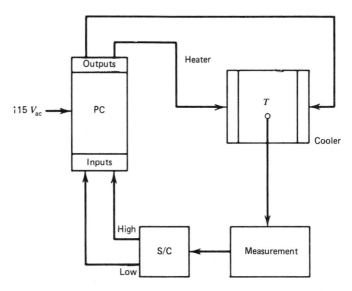

Figure 1.18   Use of a programmable to provide control for the problem of Section 1.5.1.

Programmable controllers

A special type of computer-based controller is for systems that have both input and output with only two states: ON and OFF. These systems are called *discrete-state systems*. Originally, programmable controllers were relay circuits that provided the necessary sequencing of ON/OFF output devices based on ON/OFF inputs. Examples of such inputs include limit switches, push-buttons, and thermal trip switches such as thermal overloads on motors. The ON/OFF outputs typically are used to drive motors, solenoids, and lights.

Most new programmable controllers have replaced the relays with a digital computer with the necessary interface circuits for input and output. Figure 1.18 illustrates how a programmable controller would be connected to provide ON/OFF control for the temperature-control problem of Section 1.5.1. The measurement system must be provided with some type of signal conditioning that outputs two signals for a high- and low-temperature limit. Programmable controllers will be studied in detail in Chapter 8.

1.6 UNITS, STANDARDS, AND DEFINITIONS

As in any other technological discipline, the field of process control has many sets of units, standards, and definitions to describe its characteristics. Some of these are a result of historical use, some are for convenience, and some are very confusing. As the discipline grew, there were efforts to standardize terms so that professional workers in process control could effectively communicate among themselves and with specialists in other disciplines. In this section, we summarize the present state of affairs relative to the common units, standards, and definitions.

1.6.1 Units

To assure precise technical communication among individuals employed in technological disciplines, it is essential to use a well-defined set of units of measurement. The metric system of units provides such communication and has been adopted by most technical disciplines. In process control, a particular set of metric units is used (which was developed by an international conference) called the International System (SI, Système International D'Unités). Because much technical work in the United States is still done in the English system of units, it is necessary to perform transformations between these systems.

International system of units

The international system of units is maintained by an international agreement for worldwide standardization. The system is based on seven well-defined base units and two supplementary dimensionless units. Everything else falls into the category of defined units, meaning defined in terms of the seven base and two supplementary.

Quantity	Unit	Symbol
BASE		
Length	Meter	m
Mass	Kilogram	kg
Time	Second	s
Electric current	Ampere	A
Temperature	Kelvin	K
Amount of substance	Mole	mol
Luminous intensity	Candela	cd
SUPPLEMENTARY		
Plane angle	Radian	rad
Solid angle	Steradian	sr

All other SI units can be derived from these 9 units, although in some cases a special name is assigned to the derived quantity. Thus, a force is measured by the newton (N), where $1 \text{ N} = 1 \text{ kg m/s}^2$; energy is measured by the joule (J) or watt-second (W-s), given by $1 \text{ J} = 1 \text{ kg m}^2/\text{s}^2$; and so on as shown in Appendix 1.

Other units

Although the SI system will be used in this text, other units remain in common use in some technical areas. The reader therefore should be able to identify and translate between the SI system and other systems. The centimeter-gram-second system (CGS) and English system also are given in Appendix 1. The following examples illustrate some typical translations of units.

Example 1.1

Express a pressure of $p = 2.1 \times 10^3$ dyne/cm^2 in pascals. 1 Pa = 1 N/m^2.

Solution   From Appendix 1, we find 10^2 cm = 1 m and 10^5 dyne = 1 newton; thus,

$$p = (2.1 \times 10^3 \text{ dyne/cm}^2) \left(10^2 \, \frac{\text{cm}}{\text{m}} \right)^2 \left(\frac{1 \text{ N}}{10^5 \text{ dyne}} \right)$$

$$p = \mathbf{210} \text{ pascals}$$

Example 1.2

Find the number of feet in 5.7 m.

Solution   Reference to the table of conversions in Appendix 1 shows that 1 m = 39.37 in; therefore

$$(5.7 \text{ m}) \left(39.37 \, \frac{\text{in}}{\text{m}} \right) \left(\frac{1 \text{ ft}}{12 \text{ in}} \right) = \mathbf{18.7} \text{ ft}$$

Example 1.3

Express 6 ft in meters.

Solution   Using 39.37 in/m

$$(6 \text{ ft})(12 \text{ in/ft}) \left(\frac{1 \text{ m}}{39.37 \text{ in}} \right) = \mathbf{1.829} \text{ m}$$

Example 1.4

Find the mass in kg of a 2-lb object.

Solution   We first find the mass in slugs

$$m = \frac{2 \text{ lb}}{32.17 \text{ ft/s}^2}$$

$$m = 0.062 \text{ slugs}$$

Where

$$1 \text{ slug} = 1 \text{ lb ft/s}^2$$

Then, from Appendix 1, we have

$$m = (0.062 \text{ slugs}) \left(14.59 \, \frac{\text{kg}}{\text{slug}} \right)$$

$$m = \mathbf{0.905} \text{ kg}$$

Metric prefixes

With the wide variation of variable magnitudes that occurs in industry, there is a need to abbreviate large and small numbers. Scientific notation allows the expression of such numbers through powers of 10. A set of standard metric prefixes has been adopted by the Institute of Electrical and Electronic Engineers (IEEE) (Std. No. 268A) to express these powers of 10, which are employed to simplify the expression of large or small numbers. These prefixes are given in Appendix 1.

Example 1.5

Express 0.0000215 s and 3,781,000,000 W using decimal prefixes.

Solution   We first express quantities in scientific notation and then find the appropriate decimal prefix from Appendix 1.

$$0.0000215 \text{ s} = 21.5 \times 10^{-6} \text{ s} = \textbf{21.5 } \boldsymbol{\mu}\textbf{s}$$

and

$$3,781,000,000 \text{ W} = 3.781 \times 10^{9} \text{ W} = \textbf{3.781 GW}$$

1.6.2 Standard Signals

Certain standard signal levels are used in process control to provide an easier interface of control loop elements as well as for other reasons. The most common standard concerns the manner of signal transmission, such as from the measurement element to the process controller.

Current signal transmission

When a control loop has been implemented using analog electrical signals, it is most common to transmit the analog signal as a current level. Whenever a process-control loop is designed, some operating range is specified for the controlled variable to be regulated. It may be necessary to define a setpoint anywhere within this range. One standard specifies that the signal conditioning be such that a 4–20 mA current range on the signal transmission wires represents the specified range of the variable. There are three significant points regarding the use of current transmission to represent the controlled variable.

   1. *Load impedance*   By using a current to carry analog information about the variable, we avoid errors introduced by attaching different loads to the transmitting circuit. Thus, changing lead resistance or inserting a series resistance in the leads will not change the current delivery. Generally, the transmitting circuits are designed to work into any load from 0 Ω to about 1000 Ω.

   2. *Interchangeability*   By using a specified current range to represent the variable range, we provide for interchangeability of the controller in the process-control loop. Once the dynamic variable range has been translated to a fixed current range, then all control can be based on a setpoint and deviation as some percentage of this range. Thus, a controller only sees a 4–20 mA signal, for example, and it does not matter what specific dynamic variable this represents.

3. *Measurement/power supply*   Generally speaking, the current signal lines are also the power delivery lines to energize the transducer and local signal conditioning. Thus, only two wires are necessary to connect a transducer and signal conditioning measurement system to the rest of the loop. The signal conditioning is designed so that the circuit draws more or less current from the power source in proportion to the value of the dynamic variable. This is shown in Figure 1.19, where the controller signal is taken as some voltage drop across a series resistor in the line.

Pneumatic signal transmission

The final control element in a process-control loop is a pneumatic device. In some cases, an *entire* process-control loop may be pneumatic. In either instance, the standard transmission signal is a pressure level of a range 3–15 *pounds per square inch* (*psi*). Often, a current-to-pressure converter is employed to scale the 4–20 mA signal to a 3–15 psi signal. The 3–15 psi standard remains very common in the process-control industry despite the desired conversion to SI.

In SI, the proper derived unit of pressure is the pascal (Pa). As conversion to SI occurs throughout the world, the appropriate range for pneumatic signal pressure will be 20 kPa to 100 kPa.

1.6.3 Definitions

In this section, several terms commonly applied to process control are defined. Effective communication demands a common understanding of specialized words and expressions. This has led to the establishment of standard definitions of commonly used terms.

Error

The algebraic difference between the *indicated* value and the *actual* value of a measured variable is called the *error*. Thus, error can be either negative or pos-

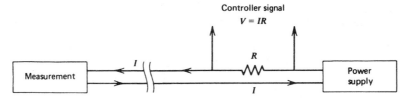

Figure 1.19 A two-wire system can be used for both power leads to a measurement system and also carry information on the measurement through a current.

itive. This deviation represents an *uncertainty* in knowledge of the actual value of a variable on the basis of a measurement.

Accuracy

This term is used to specify the maximum overall error to be expected from a device, such as measurement of a variable. Accuracy usually is expressed as the *inaccuracy* and can appear in several forms:

1. Measured variable, as the accuracy is ±2°C in some temperature measurement. Thus, there would be an uncertainty of ±2°C in any value of temperature measured.
2. Percentage of the instrument full-scale (FS) reading. Thus, an accuracy of ±0.5% FS in a 5-volt full-scale range meter would mean the inaccuracy or uncertainty in any measurement is ±0.025 volts.
3. Percentage of instrument span, that is, percentage of the range of instrument measurement capability. Thus, for a device measuring ±3% of span for a 20–50 psi range of pressure, the accuracy would be (±0.03) (50 − 20) = ±0.9 psi.
4. Percentage of the actual reading. Thus, for a ±2% of reading voltmeter, we would have an inaccuracy of ±0.04 volts for a reading of 2 volts.

Example 1.6

A temperature transducer has a span of 20°–250°C. A measurement results in a value of 55°C for the temperature. Specify the error if the accuracy is (a) ±0.5% FS, (b) ±0.75% of span, and (c) ±0.8% of reading. What is the possible temperature in each case?

Solution   Using the given definitions, we find

(a) Error = (±0.005)(250°C) = **±1.25°C**. Thus, the actual temperature is in the range 53.75–56.25°C.

(b) Error = (±0.0075)(250 − 20)°C = ±1.725°C. Thus, the actual temperature is in the range 53.275–56.725°C.

(c) Error = (±0.008)(55°C) = ±0.44°C. Thus, the temperature is in the range 54.56°C–55.44°C.

Transfer function

This term describes the relationship between the input and output of any element in the process-control loop. Often, a static (time-independent) transfer function is specified separately from the dynamic (time-dependent) transfer function. Thus, if a pressure transducer produces an output of 2 volts for every 1 psi pressure input, the transfer function would be 2 volts/psi. If this were only the static transfer function, we might also need to specify the time response of the transducer. It also is possible that this transfer function is only valid over some range of the input pressure. In most cases, the transfer function itself has an uncertainty associated with it.

Example 1.7

A temperature transducer has a transfer function of 5 mV/°C with an accuracy of ±1%. Find the possible range of the transfer function.

Solution   The transfer function range will be $(\pm 0.01)(5 \text{ mV/°C}) = \pm 0.05 \text{ mV/°C}$. Thus, the range is 4.95–5.05 mV/°C.

Example 1.8

Suppose a reading of 27.5 mV results from the transducer used in Example 1.7. Find the temperature that could provide this reading.

Solution   Because the range of transfer function is 4.95–5.05 mV/°C, the possible temperature range that could be inferred from a reading of 27.5 mV is

$$(27.5 \text{ mV}) \left(\frac{1}{4.95 \text{ mV/°C}} \right) = \mathbf{5.56°C}$$

$$(27.5 \text{ mV}) \left(\frac{1}{5.05 \text{ mV/°C}} \right) = \mathbf{5.45°C}$$

Thus, we can be certain only that the temperature is between 5.45°C and 5.56°C.

The application of digital processing has necessitated an accuracy definition compatible with digital signals. In this regard we are most concerned with the error involved in the digital representation of analog information. Thus, the accuracy is quoted as the percentage deviation of the analog variable per bit of the digital signal. As an example, an A/D converter may be specified as 0.635 volts per bit ±1%. This means that a bit will be set for an input voltage change of 0.635 ± 0.006 volts or 0.629 to 0.641 V.

System accuracy

Often, one must consider the overall accuracy of *many* elements in a process-control loop to represent a process variable. Generally, the best way to do this is to express the accuracy of each element in terms of the transfer functions. For example, suppose we have a process with two transfer functions that act on the dynamic variable to produce an output voltage as shown in Figure 1.20. We can describe the output as

$$V \pm \Delta V = (K \pm \Delta K)(G \pm \Delta G) \, C \qquad (1.3)$$

where

$$V = \text{output voltage}$$
$$\pm \Delta V = \text{uncertainty in output voltage}$$

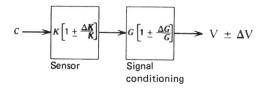

Figure 1.20   Each element of a process-control loop contributes uncertainty to a measurement through uncertainties in its transfer functions.

$$K, G = \text{nominal transfer functions}$$
$$\Delta K, \Delta G = \text{uncertainties in transfer functions}$$
$$C = \text{dynamic variable}$$

From Equation (1.3), we can find the output uncertainty to be

$$\Delta V = \pm GC\Delta K \pm KC\Delta G \pm \Delta K \Delta GC \qquad (1.4)$$

Equation (1.4) can be further simplified by noting that the nominal output is $V = KGC$ and by ignoring second-order errors

$$\frac{\Delta V}{V} = \pm\frac{\Delta K}{K} \pm \frac{\Delta G}{G} \qquad (1.5)$$

where

$$\frac{\Delta V}{V} = \text{fractional uncertainty in } V$$

$$\frac{\Delta K}{K'}, \frac{\Delta G}{G} = \text{fractional uncertainties in transfer functions}$$

We can best interpret Equation (1.5) as stating that the worst-case accuracy would be the sum of the uncertainties of each transfer function.

From a statistical point of view, it is more realistic to express the overall system accuracy as the root-mean-square (rms) of the individual element accuracy. This will give a system accuracy better than worst case but more likely to reflect the actual response. This is found from the relation

$$\left[\frac{\Delta V}{V}\right]_{\text{rms}} = \pm\sqrt{\left(\frac{\Delta K}{K}\right)^2 + \left(\frac{\Delta G}{G}\right)^2}$$

Example 1.9

Find the system accuracy of a flow process if the transducer transfer function is 10 mV/(m^3/s) $\pm$ 1.5% and the signal conditioning system transfer function is 2 mA/mV $\pm$ 0.5%.

Solution   Here we have a direct application of

$$\frac{\Delta V}{V} = \pm\left[\frac{\Delta K}{K} + \frac{\Delta G}{G}\right]$$

$$\frac{\Delta V}{V} = \pm[0.015 + 0.005] \tag{1.5}$$

$$\frac{\Delta V}{V} = \pm 0.02 = \pm 2\%$$

so that the net transfer function is 20 mA/(m³/s) $\pm$ 2%. If we use the more statistically correct rms approach, the system accuracy would be

$$\left[\frac{\Delta V}{V}\right]_{\mathrm{rms}} = \pm\sqrt{(0.015)^2 + (0.005)^2}$$

$$= \pm 0.0158$$

So the accuracy is about $\pm 1.6\%$.

Sensitivity

Sensitivity is a measure of the change in output of an instrument for a change in input. Generally speaking, *high* sensitivity is desirable in an instrument because a large change in output for a small change in input implies that a measurement may be taken easily. Sensitivity must be evaluated together with other parameters, such as *linearity* of output to input, *range*, and *accuracy*. The value of the sensitivity is generally indicated by the transfer function. Thus, when a temperature transducer outputs 5 mV per degree Celsius, the sensitivity is 5 mV/°C.

Hysteresis and reproducibility

Frequently, an instrument will not have the same output value for a given input in repeated trials. Such variation can be due to inherent uncertainties that imply a limit on *reproducibility* of the device. This variation is random from measurement to measurement and is not predictable. A similar effect is related to the history of a particular measurement taken with an instrument. In this case, a different reading results for a specific input, depending on whether the input value is approached from higher or lower values. This effect, called *hysteresis*, is shown in Figure 1.21, where the output of an instrument has been plotted against input. We see that if the input parameter is varied from low to high, curve *A* gives values of the output. If the input parameter is decreasing, then curve *B* relates input to output. Hysteresis usually is specified as a percentage of full-scale maximum deviation between the two curves. This effect is predictable if measurement values are always approached from one direction, because hysteresis will not cause measurement errors.

Resolution

Inherent in many measurement devices is a minimum measurable value of the input variable. Such a specification is called the *resolution* of the device. This

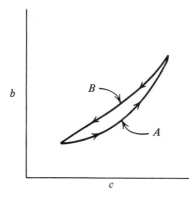

Figure 1.21 Hysteresis is a predictable error resulting from differences in the transfer functions when a reading is taken from above or below the value to be measured.

characteristic of the instrument only can be changed by redesign. A good example is a wire-wound potentiometer where the slider moves across windings to vary resistance. If one turn of the winding represents a change of ΔR ohms, then the potentiometer cannot provide a resistance change *less* than ΔR. We say that the potentiometer *resolution* is ΔR. This is often expressed as a percentage of the full-scale range.

Example 1.10

A force sensor measures a range of 0–150 N with a resolution of 0.1% FS. Find the smallest change in force that can be measured.

Solution   Because the resolution is 0.1% FS, we have a resolution of (0.001)(150 N) = 0.15 N, which is the smallest measurable change in force.

In some cases, the resolution of a measurement system is limited by the sensitivity of associated signal conditioning. When this occurs, the resolution can be improved by employing better conditioning.

Example 1.11

A sensor has a transfer function of 5 mV/°C. Find the required voltage resolution of the signal conditioning if a temperature resolution of 0.2°C is required.

Solution   A temperature change of 0.2°C will result in a voltage change of

$$\left(5\,\frac{mV}{°C}\right)(0.2°C) = \textbf{1.0 mV}$$

Thus, the voltage system must be able to resolve 1.0 mV.

In analog systems the resolution of the system is usually determined by the smallest measurable change in the analog output signal of the measurement system. In digital systems, the resolution is a well-defined quantity that is simply the change in dynamic variable represented by a *one-bit change* in the binary word output. In these cases, resolution can be improved only by a different coding of the analog information or adding more bits to the word. (This will be discussed further in Section 3.4.)

Linearity

In both sensor and signal conditioning, output is represented in some functional relationship to the input. The only stipulation is that this relationship be unique; that is, that for each value of the input variable there exists one unique value of the output variable. For simplicity of design, a linear relationship between input and output is highly desirable. Indeed, many nonlinear devices are employed for transduction simply because no other linear devices can be found! Figure 1.12 illustrates cases of both linear and nonlinear relationships. We see that linear relationships can be represented by the equation of a straight line:

$$C_M = mC + C_0 \qquad\qquad (1.6)$$

where

$$C = \text{variable to be measured}$$
$$m = \text{slope of straight line}$$
$$C_0 = \text{offset or intercept of straight line}$$
$$C_M = \text{output of measure}$$

No simple relationship such as Equation (1.6) can usually be found for the nonlinear cases, although in some cases *approximations* of a linear or quadratic nature are fitted to portions of these curves, as will be shown in Chapter 4.

Example 1.12

A sensor resistance changes linearly from 100 to 180 Ω as temperature changes from 20 to 120°C. Find a linear equation relating resistance and temperature.

Solution   Using Equation (1.6) as a guide, the desired equation would be of the form

$$R = mT + R_0$$

To find the two constants, m and R_0, we form two equations and two unknowns from the facts given,

$$100\ \Omega = (20°C)m + R_0$$

$$180\ \Omega = (140°C)m + R_0$$

Subtracting the first equation from the second gives

$$80\ \Omega = (100°C)m \quad \text{or} \quad m = 0.8\ \Omega/°C$$

Then, from the first equation we find

$$100 \ \Omega \ = \ (20°C)(0.8 \ \Omega/°C) \ + \ R_0$$

from which

$$R_0 \ = \ 84 \ \Omega$$

The equation relating resistance and temperature is

$$R \ = \ 0.8T \ + \ 84$$

One of the specifications of sensor outputs is the degree to which they are linear with the measured variable and the span over which this occurs. A measure of sensor linearity is to determine the deviation of the sensor output from a best-fit straight line over a particular range. A common specification of linearity is the maximum deviation from a straight line expressed as percent of FS.

Consider a sensor which outputs a voltage as a function of pressure from 0 to 100 psi with a linearity of 5% FS. This means that at some point on the curve of voltage versus pressure, the deviation between actual pressure and linearly indicated pressure for a given voltage deviates by 5% of 100 psi or 5 psi.

Figure 1.22 indicates this graphically. A straight line has been fitted to the slightly nonlinear sensor curve. One can either specify that for a given voltage

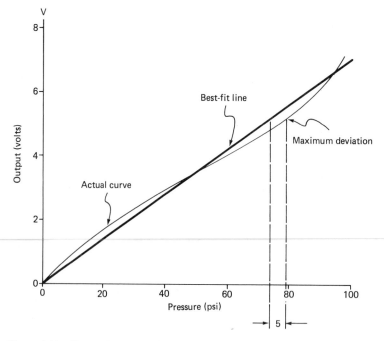

Figure 1.22 Comparison of an actual curve and its best-fit straight line, where the maximum deviation is 5% FS.

there is a deviation between actual and linearly predicted pressure or that for a given pressure there is a deviation between actual and linearly predicted voltage.

1.6.4 Process-Control Drawings

A standard form and set of symbols are used to prepare drawings of process-control systems. These drawings are much the same as a *schematic* of some circuit in electricity. In process control, the drawing is often referred to as a piping and instrumentation drawing (P&ID), although the symbols and diagram may be used in processes for which there is no piping. It is important to recognize the nature of the P&ID and the symbols. Appendix 5 presents a detailed account of the set of symbols and abbreviations. Here, we merely note the basic structure of such diagrams. Thus, Figure 1.23 shows some part of a typical P&ID. First, the P&ID contains a diagram of the *entire* process, that is, the reactants and products, as well as the instrumentation and signals that form the various process-control loops. Thus, someone with a good understanding of the P&ID will be able to see how the overall manufacturing sequence works as well as being able to recognize the individual process-control loops. The actual process flow lines, such as reactant flow and crackers on a conveyor, are shown as heavy solid lines in the diagram.

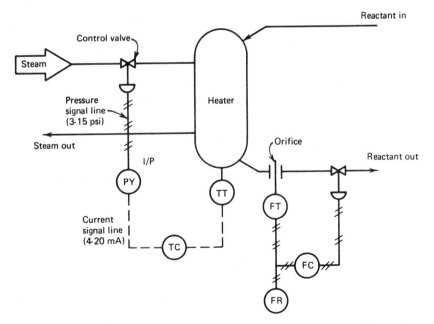

Figure 1.23 A P&ID uses special symbols and lines to show the devices and interconnections in a process-control system.

Instrument signal lines

As shown in Figure 1.23, the instrumentation signal lines are presented in a form that shows whether they are pneumatic or electric. A cross-hatched line is used to show all lines that are pneumatic, as a 3–15 psi signal, for example. The common electric current line, as 4–20 mA, is presented as a dashed line.

Instrumentation

A balloon symbol with an enclosed two- or three-letter code is used to represent the instrumentation associated with the process-control loops. Thus, the balloon in Figure 1.23 with TT enclosed is a temperature sensor and that with TC enclosed is the temperature controller associated with that loop. Special items such as control valves and in-line instruments (for example, flow meters) have special symbols, as shown in Figure 1.23 by the control valves and flow transducers.

1.7 TIME RESPONSE

The static transfer function of a process-control loop element specifies how the output is related to the input if the input is constant. An element also has a time dependence which specifies how the output changes in time when the input is changing in time. It is independent from the static transfer function. This dynamic transfer function is often simply called the *time response*. It is particularly important for sensors because they are the primary element for providing knowledge of the controlled variable value. This section will discuss the two most common types of sensor time responses.

To specify a sensor time response, a worst-case input change is applied and the output of the sensor in time is evaluated. The worst-case input is simply a step change, as illustrated in Figure 1.24a. Note that at $t = 0$ the input to the sensor is suddenly changed from an initial value c_i to a final value c_f. If the sensor were perfect, its output would be determined by the static transfer function to be b_i before $t = 0$ and b_f after $t = 0$. However, all sensors will exhibit some lag between the output and the input and some characteristic variation in time before settling on the final value.

1.7.1 First-Order Response

The simplest time response is shown in Figure 1.24b as the output change in time following a step input as in Figure 1.24a. This is called *first order* because for all sensors of this type the time response is determined by the solution of a first-order differential equation.

A general equation can be written for this response independent of the sensor, the variable being measured, or the static transfer function. The equation

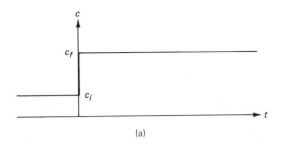

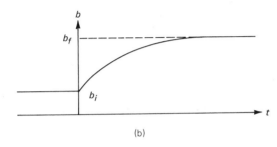

Figure 1.24 Characteristic first-order exponential time response of a sensor (b) to a step change of input (a).

gives the sensor output as a function of time following the step input (i.e., it traces the curve of Figure 1.24b in time):

$$b(t) = b_i + (b_f - b_i)\,[1 - e^{-t/\tau}] \tag{1.7}$$

where

b_i = initial sensor output from static transfer function and initial input
b_f = final sensor output from static transfer function and the final input
τ = sensor time constant

The sensor output is *in error* during the transition time of the output from b_i to b_f. The actual variable value was changed instantaneously to a new value at $t = 0$. Equation (1.7) describes transducer output very well except during the initial time period, that is, at the *start* of the response near $t = 0$. In particular, the actual transducer response generally starts the change with a *zero* slope, and Equation (1.7) predicts a finite starting slope.

Time constant

Equation (1.7) relates initial sensor output, final sensor output, and a constant that is a specification of the sensor. The significance of this quantity can be found by consideration of Equation (1.7) for the case where the initial output is zero. In this special case, the output response is

$$b(t) = b_f(1 - e^{-t/\tau}) \tag{1.8}$$

Now suppose we wish to find the value of the output exactly τ seconds after the change occurs; then

$$b(\tau) = b_f(1 - e^{-1})$$
$$b(\tau) = 0.6321\, b_f$$

(1.9)

Thus, we see that one time constant represents the time at which the output value has changed by approximately 63% of the total change.

The time constant τ is sometimes referred to as the 63% time, the *response time*, or the e-folding time. For a step change, the output response has approximately reached its final value after five time constants, since from Equation (1.8).

$$b(5\tau) = 0.993\, b_f$$

Example 1.13

A sensor measures temperature linearly with a transfer function of 33 mV/°C and has a 1.5-s time constant. Find the output 0.75 s after the input changes from 20°C to 41°C. Find the error in temperature this represents.

Solution   We first find the initial and final values of the sensor output:

$$b_I = (33 \text{ mV/°C})(20°C)$$

$$b_I = 660 \text{ mV}$$

$$b_f = (33 \text{ mV/°C})(41°C)$$

$$b_f = 1353 \text{ mV}$$

Now,

$$b(t) = b_i + (b_f - b_i)[1 - e^{-t/\tau}]$$

$$b(0.75) = 660 + (1353 - 660)[1 - e^{-0.75/1.5}]$$

(1.7)

$$b(0.75) = \textbf{932.7 mV}$$

This corresponds to a temperature of

$$T = \frac{932.7}{33 \text{ mV/°C}}$$

$$T = \textbf{28.3°C}$$

so the error is 12.7°C.

In many cases, the transducer output may be inversely related to the input. Equation (1.7) still describes the time response of the element where the final output is less than the initial output.

Real-time effects

The concept of the exponential time response and associated time constant is based on a sudden discontinuous change of the input value. In the real world such instantaneous changes occur rarely, if ever, and thus we have presented a *worst-case* situation in the time response. In general, a sensor should be able to track any changes in the physical dynamic variable in a time less than one time constant.

1.7.2 Second-Order Response

In some sensors, a step change in the input will cause the output to oscillate for a short period of time before settling down to value that corresponds to the new input. Such oscillation (and the decay of the oscillation itself) is a function of the sensor. This output transient generated by the transducer is an error and must be accounted for in any measurement involving a transducer with this behavior.

This is called a *second-order response* because for this type of sensor the time behavior is described by a second-order differential equation. It is not possible to develop a universal solution, as it is for the first-order time response. Instead we simply describe the general nature of the response.

Figure 1.25 shows a typical output curve that might be expected from a transducer having a second-order response for a discontinuous change in the input. It is impossible to describe this behavior by an analytic expression, as it is with the first-order response. However, the behavior can be described in time as

$$R(t) \propto R_0 e^{-at} \sin(2\pi f_n t) \tag{1.10}$$

where

$$R(t) = \text{the transducer output}$$
$$a = \text{output damping constant}$$
$$f_n = \text{natural frequency of the oscillation}$$
$$R_0 = \text{amplitude}$$

This equation shows the basic damped oscillation output of the device. The damping constant a and natural frequency f_n are characteristics of the transducer itself and must be considered in many applications.

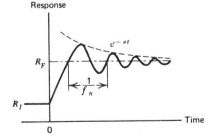

Figure 1.25 Characteristic second-order oscillatory time response of a sensor.

In general, such a transducer can be said to track the input when the input changes in a time that is *greater* than the period represented by the natural frequency. The damping constant defines the time one must wait after a disturbance at $t = 0$ for the transducer output to be a true indication of the transducer input. Thus, we see that in time of $(1/a)$ the amplitude of the oscillations would be down to e^{-1} or approximately 37%. More will be said of the effects of natural frequency and damping in the treatment of specific transducers that exhibit this behavior.

1.8 SIGNIFICANCE AND STATISTICS

Process control is vitally concerned with the value of variables, as the stated objective is to regulate the value of selected variables. It is therefore very important that the true significance of some measured value be understood. We have already seen that inherent errors may lend uncertainty to the value indicated by a measurement. In this section we need to consider another feature of measurement that may be misleading about the actual value of a variable, as well as a method to help interpret the significance of measurements.

1.8.1 Significant Figures

In any measurement we must be careful not to attach more significance to a variable value than the instrument can support. This is particularly true with the growing use of digital reading instruments and calculators with 8–12 digit readouts. Suppose, for example, that a digital instrument measures a resistance to be 125 $k\Omega$. Even if we ignore the instrument accuracy, this does not mean that the resistance is 125,000 Ω. Rather, it means that the resistance is closer to 125,000 than it is to 124,000 or 126,000 Ω. We can use the 125 $k\Omega$ number in subsequent calculations, but we cannot draw conclusions about results having more than three numbers, that is, three significant figures. The significant figures are the digits (places) actually read or known from a measurement or calculation.

Significance in measurement

When using a measuring instrument, the number of significant figures is indicated either by readability in the case of analog instruments or by the number of digits in a digital instrument. This is not to be confused with accuracy, which supplies an uncertainty to the reading itself. The following example illustrates how significant figures in measurement and accuracy are treated in the same problem.

Example 1.14

A digital Multimeter measures the current through a 12.5-$k\Omega$ resistor of 2.21 mA, using the 10-mA scale. The instrument accuracy is $\pm0.2\%$ FS. Find the voltage across the resistor and the uncertainty in the value obtained.

Solution   First, we note that the significant figures of the numbers given is 3, so no result we find can be significant to more than three digits. Then we see that the given accuracy becomes an uncertainty in the current of ± 0.02 mA. From Ohm's law we find the voltage as

$$V = IR = (2.21 \text{ mA})(12.5 \text{ k}\Omega) = 27.625 \text{ volts}$$

But in terms of significant figures we give this as $V = 27.6$ volts. The accuracy means the current could vary from 2.19 to 2.23 mA, which introduces an uncertainty of ± 0.25 volts. Thus, the complete answer is 27.6 ± 0.3 volts, because we must express the uncertainty so that our significance is not changed.

Significance in calculations

In calculations one must be careful to not obtain a result that has more significance than the numbers employed in the calculation. The answer can have no more significance than the last of the numbers used in the calculation.

Example 1.15

A transducer has a specified transfer function of 22.4 mV/°C for temperature measurement. The measured voltage is 412 mV. What is the temperature?

Solution   Using the values given, we find

$$T = (412 \text{ mV})/(22.4 \text{ mV/°C}) = 18.392857\text{°C}$$

This was found using an eight-digit calculator, but the two given values are significant to only three places. This result can be significant to only three places; the answer is **18.4°C.**

Significance in design

The reader should be aware of the difference in significant figures associated with measurement and conclusions drawn from measurement and significant figures associated with *design*. A design is a hypothetical development that makes implicit assumptions about selected values in the design. If the designer specifies a 1.1-kΩ resistor, the assmption is that it is exactly 1100 Ω. If the designer specifies that there are 4.7 volts across the resistor, then there are exactly 4.7 volts and the current can be calculated as 4.2727272 mA. Now, suppose we *measure* the resistor when the design is built and find it to be 1.1 kΩ (two significant figures) and measure the voltage and find it to be 4.7 volts (two significant figures). In this case, we report the calculated current of 4.3 mA, because we are dealing with two significant figures.

In the examples and problems in this book, we will try to maintain the distinction between design values and measurement values. Whenever a problem or example involves design, perfect values are assumed. Thus, if a design specifies a 4.7-kΩ resistor, then it is assumed that the value is exactly 4.7 kΩ (i.e., 4700.00

Ω). Whenever measurement is suggested, the figures given are assumed to be the significant figures. If a problem specifies that the measured voltage was 5.0, it is assumed that it is known to only two significant figures.

1.8.2 Statistics

Often, confidence in the value of a variable can be improved by the use of elementary statistical analysis of measurements. This is particularly true where random errors in measurement cause a distribution of readings of the value of some variable.

Arithmetic mean

If many measurements of some variable are taken, the arithmetic mean is calculated to obtain an average value for the variable. There are many instances in process control when such an average value is of interest. For example, one may wish to control the average temperature in a process. The temperature might be measured in 10 locations and averaged to give a controlled variable value for use in the control loop.

Another common application is the calibration of transducers and other process instruments. In such cases, the average gives information about the transfer function. In digital or computer process control, it is often easier to use the average value of process variables. The arithmetic mean of a set of n values, given by x_1, $x_2, x_3, \cdots , x_n$ is defined by the equation

$$\bar{x} = \frac{x_1 + x_2 + x_3 + \cdots + x_n}{n} \tag{1.11}$$

where

$$\bar{x} = \text{arithmetic mean}$$
$$n = \text{number of values to be averaged}$$
$$x_1, x_2 \cdots x_n = \text{individual values}$$

We often use the symbol Σ to represent a sum of numbers such as that used in Equation (1.11). Here we would write the equation as

$$\bar{x} = \frac{\Sigma x_i}{n} \tag{1.12}$$

where

$$\Sigma x_i = \text{symbolic for all the values } x_1, x_2, \cdots , x_n$$

Standard deviation

It often is insufficient to know the value of the arithmetic mean of a set of measurements. To interpret the measurements properly, it may be necessary to know something about how the individual values are spread out about the mean. Thus, although the mean of the set (50, 40, 30, 70) is 47.5 and the mean of the set (5, 150, 21, 14) is also 47.5, the second group of numbers is obviously far more spread out. The standard deviation is a measure of this spread. Given a set of n values $x_1, x_2, \cdots, x_n$, we first define a set of deviations by the difference between the individual values and the arithmetic mean of the values, $\bar{x}$. The deviations are

$$d_1 = x_1 - \bar{x}$$

$$d_2 = x_2 - \bar{x}$$

and so on

$$d_n = x_n - \bar{x}$$

The set of these n deviations is now used to define the standard deviation according to the equation

$$\sigma = \sqrt{\frac{d_1^2 + d_2^2 + d_3^2 + \cdots + d_n^2}{n - 1}} \tag{1.13}$$

or, using the summation symbol

$$\sigma = \sqrt{\frac{\Sigma d_i^2}{n - 1}} \tag{1.14}$$

Of course, the larger the standard deviation, the more spread out the numbers from which it is calculated.

Example 1.16

Temperature was measured in eight locations in a room, and the values obtained were 21.2°, 25.0°, 18.5°, 22.1°, 19.7°, 27.1°, 19.0°, and 20.0°C. Find the arithmetic mean of the temperature and the standard deviation.

Solution   Using Equation (1.11) we have

$$\bar{T} = \frac{21.2 + 25 + 18.5 + 22.1 + 19.7 + 27.1 + 19 + 20}{8}$$

$$\bar{T} = \textbf{21.6°C} \text{ (remember significant figures)}$$

The standard deviation is found from Equation (1.13):

$$\sigma = \sqrt{\frac{(21.2 - 21.6)^2 + (25 - 21.6)^2 + \cdots + (20 - 21.6)^2}{(8 - 1)}}$$

$$\sigma = \textbf{3.04°C}$$

Interpretation of standard deviation

A more quantitative evaluation of spreading can be made if we make certain assumptions about the set of data values used. In particular, we assume that the errors are truly random and that we have taken a large sample of readings. We then can claim that the standard deviation and data are related to a special curve called the *normal probability curve* or *bell curve*. If this is true, then

1. 68% of all readings lie within $\pm 1\sigma$ of the mean.
2. 95.5% of all readings lie within $\pm 2\sigma$ of the mean.
3. 99.7% of all readings lie within $\pm 3\sigma$ of the mean.

This gives us the added ability to make quantitative statements about how the data is spread about the mean. Thus, if one set of pressure readings has a mean of 44 psi with a standard deviation of 14 psi and another a mean of 44 psi with a standard deviation of 3 psi, we know the latter is much more peaked about the mean. In fact, 68% of all the readings in the second case lie from 41 to 47 psi, and in the first case 68% of readings lie from 30 to 58 psi.

Example 1.17

A control system was installed to regulate the weight of potato chips dumped into bags in a packaging operation. Given samples of 15 bags drawn from the operation before and after the control system was installed, evaluate the success of the system. Do this by comparing the arithmetic mean and standard deviations before and after. The bags should be 200 g.

    Samples before: 201, 205, 197, 185, 202, 207, 215, 220, 179, 201, 197, 221, 202, 200, 195

    Samples after: 197, 202, 193, 210, 207, 195, 199, 202, 193, 195, 201, 201, 200, 189, 197

Solution In the before case, we use Equations (1.12) and (1.14) to find the mean and standard deviation and get

$$\overline{W}_b = 202 \text{ g}$$

$$\sigma_b = 11 \text{ g}$$

Now the mean and standard deviation are found for the after case:

$$W_a = 199 \text{ g}$$

$$\sigma_a = 5 \text{ g}$$

Thus, we see that the control system has brought the average bag weight closer to the ideal of 200 g and that it has cut the spread by a factor of 2. In the before case, 99% of the bags weighed 202 ± 33 g, but in the case with the control system, 99% of the bags weighed in the range of 199 ± 15 g.

SUMMARY

This chapter presented an overview of process control and its elements. Subsequent chapters will examine the topics discussed in more detail and provide a more quantitative understanding.

The following list is provided to help the reader view the chapter from the key points.

1. Process control itself has been described as suitable for application to any situation where some variable is regulated to some desired value or range of values. Figure 1.5 shows a block diagram in which the elements of measurement, error detector, controller, and control element are connected to provide the required regulation.

2. Numerous criteria have been discussed that allow the evaluation of process-control loop performance of which the settling time, peak error, and minimum area are the most indicative of loop characteristics.

3. Both analog and digital processing are used in process-control applications. The current trend is to make analog measurements of the controlled variable, digitize these, and use a digital controller for evaluation. The basic technique of digital encoding allows each bit of a binary word to correspond to a certain quantity of the measured variable. The arrangement of "0" and "1" states in the word then serves as the encoding.

4. The SI system of units forms the basis of computations in this text as well as in the process-control industry in general. It also is still necessary to understand conversions to other systems, notably the English system (see Appendix 1).

5. A standard, adopted for analog process-control signals, is the 4–20 mA current range to represent the span of measurements of the dynamic variable.

6. The definitions of accuracy, resolution, and other terms used in process control are necessary and are similar to those in related fields.

7. The concept of transducer time response was introduced. The *time constant* becomes part of the dynamic properties of a transducer.

8. The use of significant figures is important to properly interpret measurements and conclusions drawn from measurements.

9. Statistics can help interpret the validity of measurements through the use of the arithmetic mean and the standard deviation.

10. The P&ID drawing and symbols were introduced as the typical representation used to display process-control systems.

PROBLEMS

Section 1.2

1.1 Explain how the basic strategy of control is employed in a room air-conditioning system. What is the controlled variable? What is the manipulated variable? Is the system self-regulating?

1.2 Is the driving of an automobile best described as a servomechanism or a process-control system? Why?

Section 1.3

1.3 Construct a block diagram of a refrigerator control system. Define each block in terms of the refrigerator components. (If you do not know the components, look them up in the library or an encyclopedia.)

Section 1.4

1.4 A process-control loop has a setpoint of 175°C and an allowable deviation of ± 5°C. A transient causes the response shown in Figure 1.26. Specify the maximum error and settling time.

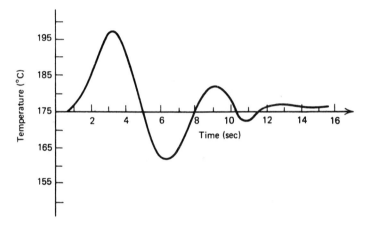

Figure 1.26   Figure for Problem 1.4.

1.5 Two different tunings of a process-control loop result in the transient responses shown in Figure 1.27. Estimate which would be preferred to satisfy the minimum area criteria.

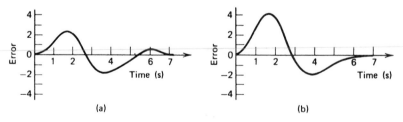

Figure 1.27   Figure for Problem 1.5.

1.6 The second cyclic transient error peak of a response test measures 4.4%. For the quarter-amplitude criteria, what error should be the third peak value?

1.7 Does the response of Figure 1.26 satisfy the quarter-amplitude criteria?

Section 1.5

1.8 An analog sensor converts flow linearly so that flow from 0 to 300 m^3/hr becomes a current from 0 to 50 mA. Calculate the current for a flow of 225 m^3/hr.

1.9 What binary word would represent the decimal number 14 if Table 1.1 were continued?

1.10 Calculate the volts per bit for an 8-bit ADC for a 0–10-volt encoding.

1.11 For the process-control system of Figure 1.11, suppose that the relays close at | 1.5 | volts and open at | 1.1 | volts. This means that as the voltage on the relay reaches ±1.5 volts, it closes but does not open again until the voltage drops to 1.1 volts (i.e., there is a deadband). The amplifier has a gain of 10, the reference is 3 V, and the sensor outputs 150 mV/°C. Calculate the temperatures at which the heater turns on and off and at which the cooler turns on and off.

Section 1.6

1.12 What is your mass in kilograms? What is your height in meters?

1.13 Atmospheric pressure is about 14.7 lb/in^2 (psi). What is this pressure in pascals?

1.14 An accelerometer is used to measure the constant acceleration of a race car that covers a quarter mile in 7.2 s.
   a. Using $x = at^2/2$ to relate distance x, acceleration a, and time t, find the acceleration in ft/s^2.
   b. Express this acceleration in m/s^2.
   c. Find the car speed, v, in m/s at the end of the quarter mile using the relation $v^2 = 2ax$.
   d. Find the car energy in joules at the end of the quarter mile if it weighs 2000 lb, where the energy $W = mv^2/2$.

1.15 An instrument has an accuracy of ±0.5% FS and measures resistance from 0 to 1500 Ω. What is the uncertainty in an indicated measurement of 397 Ω?

1.16 A sensor has a transfer function of 0.5 mV/°C and an accuracy of ±1%. If the temperature is known to be 60°C, what can be said with absolute certainty about the output voltage?

1.17 The sensor of Problem 1.16 is used with an amplifier with a gain of 15 ± 0.25 and displayed on a meter with a range of 0 to 2 volts at ±1.5% FS. What is the worst-case and rms uncertainty for the total measurement?

1.18 Using the nominal transfer function values, what is the maximum measurable temperature of the system of Problems 1.16 and 1.17?

1.19 A temperature sensor transfer function is 44.5 mV/°C. The output voltage is measured to be 8.86 volts on a three-digit voltmeter. What can you say about the value of the temperature?

1.20 A level sensor inputs a range from 4.50 to 10.6 ft and outputs a pressure range from 3 to 15 psi. Find an equation such as Equation (1.6) between level and pressure. What is the pressure for the level of 9.2 ft?

1.21 Draw Figure 1.6a in the standard P&ID symbols.

Section 1.7

1.22 A temperature sensor has a transfer function of 0.15 mV/°C and a time constant of 3.3 s. If a step change of 22°C to 50°C is applied at $t = 0$, find the output voltages at 0.5 s, 2.0 s, 3.3 s, and 9 s. What is the *indicated* temperature at these times?

1.23 A pressure sensor measures 44 psi just before a sudden change to 70 psi. The sensor measures 52 psi at a time 4.5 seconds after the change. What is the sensor time constant?

1.24 A photocell with a 35-ms time constant is used to measure light flashes. How long after a sudden dark to light flash before the cell output is 80% of the final value?

1.25 An alarm light goes ON when a pressure sensor voltage rises above 4.00 volts. The pressure sensor outputs 20 mV/kPa and has a time constant of 4.9 seconds. How long after the pressure rises suddenly from 100 kPa to 400 kPa does the light go ON?

Section 1.8

1.26 A circuit design calls for a 1.5-kΩ resistor to have 4.7 volts across its terminals. What would be the expected current? The circuit is built and the resistance is measured to be 1500 Ω and the voltage 4.7 V. What is the current through the resistor?

1.27 Flow rate was monitored for a week, and the following values were recorded as gal/min: 10.1, 12.2, 9.7, 8.8, 11.4, 12.9, 10.2, 10.5, 9.8, 11.5, 10.3, 9.3, 7.7, 10.2, 10.0, 11.3. Find the mean and the standard deviation for these data.

1.28 A manufacturer specification sheet lists the transfer function of a pressure sensor as $45 \pm 5\%$ mV/kPa with a time constant of $4 \pm 10\%$ seconds. A highly accurate test system applies a step change of pressure from 20 kPa to 100 kPa.
  a. What is the range of sensor voltage outputs initially and finally?
  b. What range of voltages would be expected to be measured 2 seconds after the step change is applied?

CHAPTER 2

ANALOG SIGNAL CONDITIONING

INSTRUCTIONAL OBJECTIVES

The purpose of this chapter is to just familiarize the reader with the basic techniques of signal conditioning in process control, not to produce experts in the subject. In view of this goal, attention has been given to only the most common techniques. After you have read this chapter, you should be able to

1. Define the common types of analog signal conditioning.
2. Design a Wheatstone bridge for resistance measurement.
3. Draw a diagram of a current balance bridge and describe its operation.
4. Design *RC* low-pass and high-pass filters for specific applications.
5. Define the operation of a silicon-controlled rectifier.
6. Design a high-input impedance op amp dc amplifier for specific gain.
7. Analyze a simple op amp circuit for its transfer characteristics.
8. Explain the purpose of compensation leads in a bridge circuit.
9. Design a voltage-to-current converter for specified voltage input and current output.
10. Define the basic linearization procedure.

2.1 INTRODUCTION

The wide variety of sensors needed to transform the wide variety of process variables in process-control systems into electrical analogs produces an equally wide variety of signal characteristics. *Signal conditioning* refers to operations performed on such signals to convert them to a form suitable for *interface* with other elements in the process-control loop. In this chapter, we are concerned only with *analog* conversions, where the conditioned output is still an analog representation of the variable. Even in applications involving digital processing, some type of analog conditioning usually is required before analog-to-digital conversion is made. Specifics of digital signal conditioning are considered in Chapter 3.

2.2 PRINCIPLES OF ANALOG SIGNAL CONDITIONING

A sensor measures a variable by converting information about that variable into a dependent signal of either electrical or pneumatic nature. To develop such transducers, we take advantage of fortuitous circumstances in nature where a dynamic variable influences some characteristic of a material. Consequently, there is little choice of the type or extent of such proportionality. For example, once we have researched nature and found that cadmium sulfide resistance varies inversely and nonlinearly with light intensity, we must then learn to employ this device for light measurement within the confines of that dependence. Analog signal conditioning provides the operations necessary to transform a sensor output into a form necessary to interface with other elements of the process-control loop. We will confine our attention to electrical transformations.

We often describe the effect of the signal conditioning by the term *transfer function*. By this term we mean the effect of the signal conditioning on the input signal. Thus, a simple voltage amplifier has a transfer function of some constant that, when multiplied by the input voltage, gives the output voltage.

It is possible to categorize signal conditioning into several general types.

2.2.1 Signal-Level Changes

The simplest method of signal conditioning is to change the level of a signal. The most common example is the necessity to either amplify or attenuate a voltage level. Generally, process-control applications result in slowly varying low-frequency signals where dc or low-frequency response amplifiers can be employed. An important factor in the selection of an amplifier is the input impedance that the amplifier offers to the sensor (or any other element that serves as an input). In process control, the signals are always representative of a process variable, and any loading effects obscure the correspondence between the measured signal and the variable value. In some cases, such as accelerometers and optical detectors, the frequency response of the amplifier is very important.

2.2.2 Linearization

As pointed out earlier, the process-control designer has little choice of the characteristics of a sensor output versus process variable. Often the dependence that exists between input and output is nonlinear. Even those devices that are approximately linear may present problems when precise measurements of the variable are required. One of the functions of analog signal conditioning is to linearize a transducer's response.

Linearization may be provided by an amplifier whose gain is a function of input voltage level to linearize the overall variation of input voltage to output voltage. An example of this linearization occurs quite frequently for a sensor where the output is exponential with respect to the dynamic variable. In Figure 2.1 we see such a case (contrived) where the voltage of a transducer is assumed to be exponential with respect to light intensity I. We may write this variation as

$$V_I = V_0\, e^{-\alpha I} \tag{2.1}$$

where

V_I = output voltage at intensity I
V_0 = zero intensity voltage
α = exponential constant
I = light intensity

To linearize this signal, we employ an amplifier whose output varies at the natural logarithm or inverse of the input

$$V_A = K \log_e (V_{IN}) \tag{2.2}$$

where

V_A = amplifier output voltage
K = calibration constant
V_{IN} = amplifier input voltage = V_I [in Equation (2.1)]

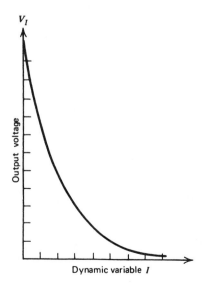

Figure 2.1   An example of a nonlinear sensor output. Here, light intensity is assumed to produce an output voltage.

By substituting Equation (2.1) into Equation (2.2) where $V_{IN} = V_I$, we find

$$V_A = K \log_e(V_0) - \alpha KI \tag{2.3}$$

where all terms have been defined.

The amplifier output *does* vary linearly with intensity but with an offset $K \log_e (V_0)$ and a scale factor of αK as shown in Figure 2.2. Further signal conditioning may be employed, if required, to eliminate the offset and provide any desired calibration of voltage versus intensity.

Other types of linearization are possible, including eliminating small variations (in response) away from linearity.

2.2.3 Conversions

Often, signal conditioning is used to convert one type of electrical variation into another. Thus, a large class of sensors provides changes of resistance with changes in a dynamic variable. In these cases, it is necessary to provide a circuit to convert this resistance change either to a voltage or current signal. This is generally accomplished by bridges when the fractional resistance change is small and/or by amplifiers whose gain varies with resistance.

Signal transmission

An important type of conversion is associated with the process-control standard of transmitting signals as 4–20 mA current levels in wire. This gives rise to the need for converting resistance and voltage levels to an appropriate current level at the transmitting end and for converting the current back to voltage at the receiving end. Of course, current transmission is used because such a signal is independent of load variations other than accidental shunt conditions that may

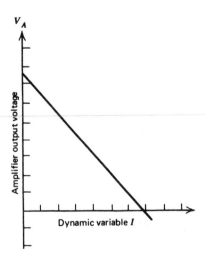

Figure 2.2 Proper signal conditioning has produced an output voltage which now varies linearly with light intensity.

draw off some current. Thus, voltage-to-current and current-to-voltage converters are often required.

Digital interface

The use of computers in process control requires conversion of analog data into a digital format by integrated circuit devices called analog-to-digital converters (ADCs). Analog signal conversion is usually required to adjust the analog measurement signal to match the input requirements of the ADC. For example, the ADC may need a voltage that varies between 0 and 5 volts, but the sensor provides a signal that varies from 30 to 80 mV. Signal conversion circuits can be developed to interface the output to the required ADC input.

2.2.4 Filtering and Impedance Matching

Two other common signal conditioning requirements are filtering and matching impedance.

Often, spurious signals of considerable strength are present in the industrial environment, such as the 60-Hz line frequency signals. Motor start transients also may cause pulses and other unwanted signals in the process-control loop. In many cases, it is necessary to use high-pass, low-pass, or notch *filters* to eliminate unwanted signals from the loop. Such filtering can be accomplished by *passive* filters using only resistors, capacitors, inductors, or *active* filters, using gain and feedback.

Impedance matching is an important element of signal conditioning when transducer internal impedance or line impedance can cause errors in measurement of a dynamic variable. Both active and passive networks are employed to provide such matching.

2.3 PASSIVE CIRCUITS

Bridge and divider circuits are two passive measurement techniques that have been extensively used for signal conditioning for many years. Although modern active circuits often replace these techniques, there are still many applications where their particular advantages make them useful.

Bridge circuits are primarily used as an accurate means of measuring changes in impedance. Such circuits are particularly useful when the fractional *changes* in impedance are *very small*.

It is quite common in the industrial environment to find signals that possess high- and/or low-frequency noise as well as the desired measurement data. For example, a transducer may convert temperature information into a dc voltage, proportional to temperature. Because of the ever-present ac power lines, however, there may be a 60-Hz noise voltage impressed on the output that makes determination of the temperature difficult. A passive circuit consisting of a resistor and capacitor often can be used to eliminate both high- and low-frequency noise without changing the desired signal information.

2.3.1 Divider Circuit

The elementary voltage divider shown in Figure 2.3 often can be used to provide conversion of resistance variation into a voltage variation. The voltage of such a divider is given by the well-known relationship

$$V_D = \frac{R_2 V_s}{R_1 + R_2} \qquad (2.4)$$

where

$$V_s = \text{supply voltage}$$
$$R_1, R_2 = \text{divider resistors}$$

Either R_1 or R_2 can be the transducer whose resistance varies with some measured variable.

It is important to consider the following issues when using a divider for conversion of resistance to voltage variation:

1. The variation of V_D with either R_1 or R_2 is nonlinear; that is, even if the resistance varies linearly with the measured variable, the divider voltage will not vary linearly.
2. The effective output impedance of the divider is the parallel combination of R_1 and R_2. This may not necessarily be high, so loading effects must be considered.
3. In a divider circuit, current flows through both resistors; that is, power will be dissipated by both, including the transducer. The power rating of both the resistor and transducer must be considered.

Example 2.1

The divider of Figure 2.3 has $R_1 = 10.0$ kΩ and $V_s = 5.00$ V. Suppose R_2 is a transducer whose resistance varies from 4.00 to 12.0 kΩ as some dynamic variable

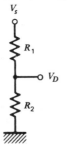

Figure 2.3 The simple voltage divider can often be used to convert resistance variation into voltage variation.

varies over a range. Then find (a) the minimum and maximum of V_D, (b) the range of output impedance, and (c) the range of power dissipated by R_2 (the transducer).

Solution   The solution to part (a) is given by Equation (2.4). For $R_2 = 4$ kΩ, we have

$$V_D = \frac{(5 \text{ V})(4 \text{ k}\Omega)}{10 \text{ k}\Omega + 4 \text{ k}\Omega} = 1.43 \text{ V}$$

For $R_2 = 12$ kΩ, the voltage is

$$V_D = \frac{(5 \text{ V})(12 \text{ k}\Omega)}{10 \text{ k}\Omega + 12 \text{ k}\Omega} = 2.73 \text{ V}$$

Thus, the voltage varies from 1.43 to 2.73 volts.

For part (b), the range of output impedance is found from the parallel combination of R_1 and R_2 for the minimum and maximum of R_2. Simple parallel resistance computation shows that this will be from 2.86 to 5.45 kΩ.

For part (c), the power dissipated by the transducer can be determined most easily from V^2/R, as the voltage across R_2 has been calculated. The power dissipated varies from 0.51 to 0.62 mW.

2.3.2 Bridge Circuits

Bridge circuits are passive networks often used to measure impedances by the technique of potential matching. In this case, a set of accurately known impedances is adjusted in value in relation to an unknown until a condition exists where the potential difference between two points in the network is zero, that is, *null*. This condition defines an equation used to find the unknown impedance in terms of the known values. A distinct advantage of such measurement techniques is that they are based on reaching the null condition, that is, zero voltage or current, as opposed to an absolute measurement. It is usually much easier to refine and improve techniques for detection of a null than for measurement of some other specific value. This leaves the accuracy of the measurement predominantly dependent on the accuracy with which the known impedances have been determined.

Wheatstone bridge

The simplest and most common bridge circuit is the dc Wheatstone bridge, as shown in Figure 2.4. This network is used in signal conditioning applications where a sensor changes resistance with process variable changes. Many modifications of this basic bridge are employed for other specific applications. In Figure 2.4 the object labeled D is a *null detector* used to compare the potentials of points a and b of the network. In most modern applications the null detector is a very high-input impedance differential amplifier. In some cases, a highly sensitive galvanometer with a relatively low impedance may be used, especially for calibration purposes and single measurement instruments.

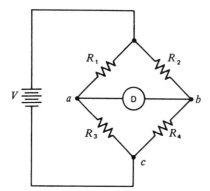

Figure 2.4 The basic dc Wheatstone bridge.

For our initial analysis, assume the null detector impedance is infinite, that is, an open circuit.

In this case the potential difference, ΔV between points a and b, is simply

$$\Delta V = V_a - V_b \tag{2.5}$$

where

V_a = potential of point a with respect to c
V_b = potential of point b with respect to c

The values of V_a and V_b now can be found by noting that V_a is just the supply voltage V divided between R_1 and R_3.

$$V_a = \frac{VR_3}{R_1 + R_3} \tag{2.6}$$

In a similar fashion, V_b is a divided voltage given by

$$V_b = \frac{VR_4}{R_2 + R_4} \tag{2.7}$$

where

V = bridge supply voltage
R_1, R_2, R_3, R_4 = bridge resistors as given in Figure 2.4

If we now combine Equations (2.5), (2.6), and (2.7), the voltage difference or voltage offset can be written

$$\Delta V = \frac{VR_3}{R_1 + R_3} - \frac{VR_4}{R_2 + R_4} \tag{2.8}$$

After some algebra, the reader can show that this equation reduces to

$$\Delta V = V \frac{R_3R_2 - R_1R_4}{(R_1 + R_3) \cdot (R_2 + R_4)} \tag{2.9}$$

Equation (2.9) shows how the difference in potential across the detector is a function of the supply voltage and the values of the resistors. Because a difference appears in the *numerator* of Equation (2.9), it is clear that a particular combination of resistors can be found that will result in zero difference and zero voltage across the detector, that is, a null. Obviously, this combination, from examination of Equation (2.9), is

$$R_3 R_2 = R_1 R_4 \qquad (2.10)$$

Equation (2.10) indicates that whenever a Wheatstone bridge is assembled and resistors are adjusted for a detector null, the resistor values must satisfy the indicated equality. It does *not* matter if the supply voltage drifts or changes; the null is maintained. Equations (2.9) and (2.10) underlie the application of Wheatstone bridges to process-control applications using high-input impedance detectors.

Example 2.2

If a Wheatstone bridge, as shown in Figure 2.4, nulls with $R_1 = 1000\ \Omega$, $R_2 = 842\ \Omega$, and $R_3 = 500\ \Omega$, find the value R_4.

Solution   Because the bridge is nulled, find R_4 using

$$R_1 R_4 = R_3 R_2$$

$$R_4 = \frac{R_3 R_2}{R_1} = \frac{(500\ \Omega)(842\ \Omega)}{1000\ \Omega} \qquad (2.10)$$

$$R_4 = \textbf{421 } \boldsymbol{\Omega}$$

Example 2.3

The resistors in a bridge are given by $R_1 = R_2 = R_3 = 120\ \Omega$ and $R_4 = 121\ \Omega$. If the supply is 10.0 volts, find the voltage offset.

Solution   Assuming the detector impedance to be very high, we find the offset from

$$\Delta V = V \frac{R_3 R_2 - R_1 R_4}{(R_1 + R_3) \cdot (R_2 + R_4)}$$

$$\Delta V = 10V \frac{(120\ \Omega)(120\ \Omega) - (120\ \Omega)(121\ \Omega)}{(120\ \Omega + 120\ \Omega) \cdot (120\ \Omega + 121\ \Omega)} \qquad (2.9)$$

$$\Delta V = \textbf{-20.8 mV}$$

Galvanometer detector

The use of a galvanometer as a null detector in the bridge circuit introduces some differences in our calculations because the detector resistance may be low and because we must determine the bridge offset as current offset. When the bridge is nulled, Equation (2.10) still defines the relationship between the resistors in the bridge arms. Equation (2.9) must be modified to allow determination of current drawn by the galvanometer when a null condition is *not* present. Perhaps the easiest way to determine this offset current is first to find the Thévenin equivalent circuit between points a and b of the bridge (as drawn in Figure 2.4 with the detector removed). The Thévenin voltage is simply the open circuit voltage difference between points a and b of the circuit. But wait! Equation (2.9) *is* the open circuit voltage, so

$$V_{Th} = V \frac{R_3 R_2 - R_1 R_4}{(R_1 + R_3)(R_2 + R_4)} \tag{2.11}$$

The Thévenin resistance is found by replacing the supply voltage by its internal resistance and calculating the resistance between terminals a and b of the network. We may assume that the internal resistance of the supply is negligible compared to the bridge arm resistances. It is left as an exercise for the reader to show that the Thévenin resistance seen at points a and b of the bridge is

$$R_{Th} = \frac{R_1 R_3}{R_1 + R_3} + \frac{R_2 R_4}{R_2 + R_4} \tag{2.12}$$

The Thévenin equivalent circuit for the bridge enables us easily to determine the current through any galvanometer with internal resistance R_G, as shown in Figure 2.5. In particular, the offset current is

$$I_G = \frac{V_{Th}}{R_{Th} + R_G} \tag{2.13}$$

Using this equation in conjunction with Equation (2.10) defines the Wheatstone bridge response whenever a galvanometer null detector is used.

Example 2.4

A bridge circuit has resistances of $R_1 = R_2 = R_3 = 2.00$ kΩ and $R_4 = 2.05$ kΩ, and a 5.00-V supply. If a galvanometer with a 50.0 Ω internal resistance is used for a null detector, find the offset current.

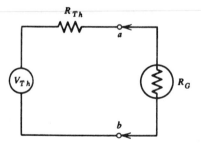

Figure 2.5 When a galvanometer is used for a null detector, it is convenient to use the Thévenin equivalent circuit of the Wheatstone bridge.

Solution   From Equation (2.11), the offset voltage is V_{Th}.

$$V_{Th} = 5V \frac{(2 \text{ k}\Omega)(2 \text{ k}\Omega) - (2 \text{ k}\Omega)(2.05 \text{ k}\Omega)}{(2 \text{ k}\Omega + 2 \text{ k}\Omega)(2 \text{ k}\Omega + 2.05 \text{ k}\Omega)} \tag{2.11}$$

$$V_{Th} = -30.9 \text{ mV}$$

We next find the bridge Thévenin resistance.

$$R_{Th} = \frac{(2 \text{ k}\Omega)(2 \text{ k}\Omega)}{(2 \text{ k}\Omega + 2 \text{ k}\Omega)} + \frac{(2 \text{ k}\Omega)(2.05 \text{ k}\Omega)}{(2 \text{ k}\Omega + 2.05 \text{ k}\Omega)} \tag{2.12}$$

$$R_{Th} = 2.01 \text{ k}\Omega \text{ '}$$

Finally, the current is

$$I_G = \frac{-30.9 \text{ mV}}{2.01 \text{ k}\Omega + 0.05 \text{ k}\Omega} \tag{2.13}$$

$$I_G = -15.0 \text{ }\mu\text{A}$$

Bridge resolution

The resolution of the bridge circuit is a function of the resolution of the null detector used to determine the bridge offset. Thus, referring primarily to the case where a voltage offset occurs, we define the resolution in resistance as that resistance change in one arm of the bridge that causes an offset voltage that is equal to the resolution of the null detector. If a null detector can measure a null to 100 μV, this sets a limit on the minimum measurable resistance change in a bridge using this detector. In general, once given the detector resolution, we may use Equation (2.9) to find the change in resistances from null that causes this offset.

Example 2.5

A bridge circuit has $R_1 = R_2 = R_3 = R_4 = 120.0 \text{ }\Omega$ resistances and a 10.0-volt supply. Clearly, the bridge is nulled, as Equation (2.10) shows. If a detector of 10.0-mV resolution is employed to detect the null, find the resolution in resistance change in R_4.

Solution   We can simply use Equation (2.8) with R_4 unspecified and find the change in R_4 that will produce a 10.0-mV offset voltage as

$$10 \text{ mV} = \frac{(120 \text{ }\Omega)(10 \text{ V})}{120 \text{ }\Omega + 120 \text{ }\Omega} - \frac{R_4(10V)}{120 \text{ }\Omega + R_4}$$

Solving for R_4, we get

$$R_4 = 119.52 \text{ }\Omega$$

so the bridge resolution is

$$\Delta R = 0.48 \text{ }\Omega$$

In this example, we see that a minimum resistance change of 0.48 Ω must occur before the detector indicates a change in offset voltage.

One may also view this as an overall *accuracy* of the instrument, because it can also be said that ΔR represents the uncertainty in any determination of resistance using the given bridge and detector.

The same arguments can be applied to a galvanometer measurement where the resolution is limited by the minimum measurable current.

Lead compensation

In many process-control applications, a bridge circuit may be located at considerable distance from the sensor whose resistance changes are to be measured. In such cases, the remaining fixed bridge resistors can be chosen to account for the resistance of leads required to connect the bridge to the sensor in providing a null. Furthermore, any measurement of resistance can be adjusted for lead resistance to determine the actual resistance. Another problem exists that is not so easily handled, however. There are many effects that can change the resistance of the long lead wires on a transient basis, such as frequency, temperature, stress, and chemical vapors. Such changes normally are picked up by the bridge response and interpreted as changes in the sensor output. This problem is reduced using *lead compensation*, where any changes in lead resistance are introduced equally into *two* (both) arms of the bridge circuit, thus causing no effective change in bridge offset. Lead compensation is shown in Figure 2.6. Here we see that R_4, which is assumed to be the sensor, has been removed to a remote location with lead wires (1), (2), and (3). Wire (3) is the power lead and has no influence on the bridge balance condition. If wire (2) changes in resistance because of spurious influences, it introduces this change into the R_4 leg of the bridge. Wire (1) is exposed to the same environment and changes by the same amount but is in the

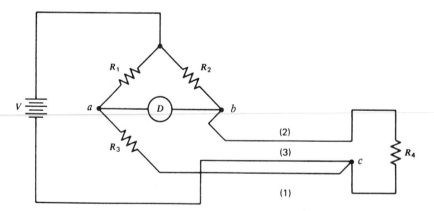

Figure 2.6 For remote sensor applications, a compensation system is used to avoid errors for lead resistance.

R_3 leg of the bridge. Effectively, both R_3 and R_4 are identically changed, and thus Equation (2.10) shows that no change in the bridge null occurs. This type of compensation is often employed where bridge circuits must be used with long leads to the active element of the bridge.

Current balance bridge

One disadvantage of the simple Wheatstone bridge is the need to obtain a null by variation of resistors in bridge arms. In the past, many process-control applications used a feedback system in which the bridge offset voltage was amplified and used to drive a motor whose shaft altered a variable resistor to renull the bridge. Such a system does not suit the modern technology of electronic processing because it is not very fast, is subject to wear, and generates electronic noise. A technique that provides for an electronic nulling of the bridge and that uses only fixed resistors (except as may be required for calibration) can be used with the bridge. This method uses a *current* to null the bridge. A closed-loop system can even be constructed that provides the bridge with a *self-nulling* ability.

The basic principle of the current balance bridge is shown in Figure 2.7. The standard Wheatstone bridge is modified by splitting one arm resistor into two, R_4 and R_5. A current I is fed into the bridge through the junction of R_4 and R_5 as shown. We now stipulate that the size of the bridge resistors is such that the current flows predominantly through R_5. This can be provided for by any of several requirements. The least restrictive is to require

$$R_4 \gg R_5 \tag{2.14}$$

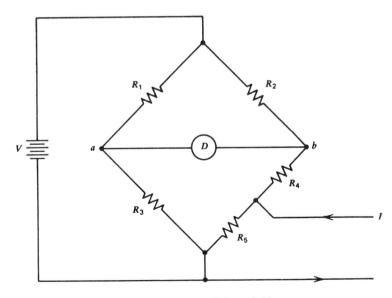

Figure 2.7 The current balance bridge.

Often, if a high-impedance null detector is used, then the restriction of Equation (2.14) becomes

$$(R_2 + R_4) \gg R_5 \tag{2.15}$$

Assuming that either conditions of Equations (2.14) or (2.15) are satisfied, the voltage at point b is the sum of the divided supply voltage plus the voltage dropped across R_5 from the current I.

$$V_b = \frac{V(R_4 + R_5)}{R_2 + R_4 + R_5} + IR_5 \tag{2.16}$$

The voltage of point a is still given by Equation (2.6). Thus, the bridge offset voltage is given by $\Delta V = V_a - V_b$, or

$$\Delta V = \frac{VR_3}{R_1 + R_3} - \frac{V(R_4 + R_5)}{R_2 + R_4 + R_5} - IR_5 \tag{2.17}$$

This equation shows that a null is reached by adjusting the magnitude and polarity of the current I until IR_5 equals the voltage difference of the first two terms. If one of the bridge resistors changes, the bridge can be renulled by changing current I. In this manner, the bridge is electronically nulled from any convenient current source. In most applications the bridge is nulled at some nominal set of resistances with zero current. Changes of a bridge resistor are detected as a bridge offset signal that is used to provide the renulling current. The action is explained in Example 2.6.

Example 2.6

A current balance bridge, as shown in Figure 2.7, has resistors $R_1 = R_2 = 10$ kΩ, $R_4 = 950$ Ω, $R_3 = 1$ kΩ, $R_5 = 50$ Ω and a high-impedance null detector. Find the current required to null the bridge if R_3 changes by 1 Ω. The supply voltage is 10 volts.

Solution   First, for the nominal resistance values given, the bridge is at a null with $I = 0$, because

$$V_a = \frac{(10V)(1 \text{ k}\Omega)}{10 \text{ k}\Omega + 1 \text{ k}\Omega} \tag{2.6}$$
$$V_a = 0.9091 \text{ volts}$$

With $I = 0$, Equation (2.16) gives

$$V_b = \frac{(10V)(950 \ \Omega + 50 \ \Omega)}{10 \text{ k}\Omega + 950 \ \Omega + 50 \ \Omega} \tag{2.16}$$
$$V_b = 0.9091 \text{ volts}$$

When R_3 increases by 1 Ω to 1001 Ω, V_a becomes

$$V_a = \frac{(10V)(1001)}{10 \text{ k}\Omega + 1001}$$

$$V_a = 0.9099 \text{ volts}$$

which shows that the voltage at b must increase by 0.0008 volts or 0.8 mV to renull the bridge. This can be provided by a current, from Equation (2.17) with $\Delta V = 0$, of $50I = 0.8$ mV.

$$\mathbf{I = 16.0\ \mu A}$$

Potential measurements using bridges

A bridge circuit is also useful to measure small potentials at a very high impedance, using either a conventional Wheatstone bridge or a current balance bridge. This type of measurement is performed by placing the potential to be measured in series with the null detector, as shown in Figure 2.8. The null detector responds to the potential between points c and b. In this case, V_b is given by Equation (2.7) and V_c by

$$V_c = V_x + V_a \tag{2.18}$$

where V_a is given by Equation (2.6), and V_x is the potential to be measured. The voltage appearing across the null detector is

$$\Delta V = V_c - V_b = V_x + V_a - V_b$$

A null condition is established when $\Delta V = 0$; furthermore, no current flows through the unknown potential when such a null is found. Thus, V_x can be measured by varying bridge resistors to provide a null with V_x in the circuit and solving for V_x using the null condition

$$V_x + \frac{R_3 V}{R_1 + R_3} - \frac{V R_4}{R_2 + R_4} = 0 \tag{2.19}$$

A similar analysis using a current balance bridge and fixed bridge resistors provides a null condition that can be solved for V_x in terms of the nulling current I.

$$V_x + \frac{R_3 V}{R_1 + R_3} - \frac{V(R_4 + R_5)}{R_2 + R_4 + R_5} - I R_5 = 0 \tag{2.20}$$

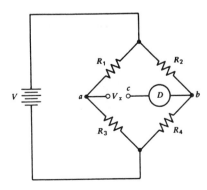

Figure 2.8 Using the basic Wheatstone bridge for potential measurement.

If the fixed resistors are chosen to null the bridge with $I = 0$ when $V_x = 0$, then the two middle terms in Equation (2.20) cancel leaving a very simple relationship between V_x and the nulling current

$$V_x - IR_5 = 0 \qquad (2.21)$$

Example 2.7

A bridge circuit for potential measurement nulls when $R_1 = R_2 = 1 \text{ k}\Omega$, $R_3 = 605$ Ω, and $R_4 = 500 \ \Omega$ with a 10-volt supply. Find the unknown potential.

Solution   Here we simply use Equation (2.19) to solve for V_x.

$$V_x + \frac{(605)(10)}{605 + 1000} - \frac{(10)(500)}{1000 + 500} = 0$$

$$V_x + 3.769 - 3.333 = 0 \qquad (2.19)$$

$$V_x = \mathbf{-0.436 \ volts}$$

Example 2.8

A current balance bridge is used for potential measurement. The fixed resistors are $R_1 = R_2 = 5 \text{ k}\Omega$, $R_3 = 1 \text{ k}\Omega$, $R_4 = 990 \ \Omega$, and $R_5 = 10 \ \Omega$ with a 10-volt supply. Find the current necessary to null the bridge if the potential is 12 mV.

Solution   First, an examination of the resistances shows that the bridge is nulled when $I = 0$ and $V_x = 0$ because, from Equation (2.20),

$$\frac{VR_3}{R_1 + R_3} = \frac{10(1 \text{ k})}{1 \text{ k} + 5 \text{ k}} = 1.667 \text{ volts}$$

and

$$\frac{V(R_4 + R_5)}{R_2 + R_4 + R_5} = \frac{10(990 + 10)}{5 \text{ k} + 990 + 10} = 1.667 \text{ volts}$$

Thus, we can use Equation (2.21)

$$12 \text{ mV} - 10I = 0$$

$$I = \mathbf{1.2 \ mA}$$

Ac bridges

The bridge concept described in this section can be applied to the matching of impedances in general as well as to resistances. In this case, the bridge is represented as in Figure 2.9 and employs an ac excitation, usually a sine wave voltage signal. The analysis of bridge behavior is basically the same as in the previous treatment, but impedances replace resistances. The bridge offset voltage then is represented as

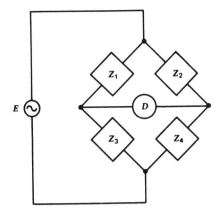

Figure 2.9 A general ac bridge circuit.

$$\Delta E = E\,\frac{Z_3 Z_2 - Z_1 Z_4}{(Z_1 + Z_3)(Z_2 + Z_4)}$$ (2.22)

$$\Delta E = \text{ac offset voltage}$$

where

$$E = \text{sine wave excitation voltage}$$
$$Z_1, Z_2, Z_3, Z_4 = \text{bridge impedances}$$

A null condition is defined as before by a zero offset voltage $\Delta E = 0$. From Equation (2.22), this condition is met if the impedances satisfy the relation

$$Z_3 Z_2 = Z_1 Z_4$$ (2.23)

This condition is analogous to Equation (2.10) for resistive bridges.

A special note is necessary concerning the achievement of a null in ac bridges. In some cases, the null detection system is phase sensitive with respect to the bridge excitation signal. In these instances, it is necessary to provide a null of both the in-phase and quadrature (90° out-of-phase) signals before Equation (2.23) applies.

Example 2.9

An ac bridge employs impedances as shown in Figure 2.10. Find the value of R_x and C_x when the bridge is nulled.

Solution Because the bridge is at null, we have

$$Z_2 Z_3 = Z_1 Z_x$$ (2.23)

or

$$R_2\left(R_3 - \frac{j}{\omega C}\right) = R_1\left(R_x - \frac{j}{\omega C_x}\right)$$

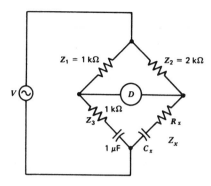

Figure 2.10 The ac bridge circuit and components for Example 2.9.

$$R_2R_3 - j\frac{R_2}{\omega C} = R_1R_x - \frac{jR_1}{\omega C_x}$$

The real and imaginary parts must be independently satisfied so that

$$R_x - \frac{R_2R_3}{R_1} = 0$$

$$R_x = \frac{(2 \text{ k}\Omega)(1 \text{ k}\Omega)}{1 \text{ k}\Omega}$$

$$R_x = 2 \text{ k}\Omega$$

and

$$C_x = C\frac{R_1}{R_2}$$

$$C_x = (1 \text{ }\mu\text{F})\frac{1 \text{ k}\Omega}{2 \text{ k}\Omega}$$

$$C_x = 0.5 \text{ }\mu\text{F}$$

2.3.3 *RC* Filters

To eliminate unwanted noise signals from measurements, it is often necessary to use circuits that block certain frequencies or bands of frequencies. These circuits are called *filters*. In many cases a simple filter can be constructed from a single resistor and a single capacitor to accomplish the desired rejection.

Low-pass *RC* filter

The simple circuit shown in Figure 2.11 is called a *low-pass RC* filter. It is called low-pass because it blocks high frequencies and passes low frequencies. It would be most desirable if a low-pass filter had a characteristic such that all signals with

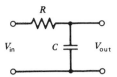

Figure 2.11   Circuit for the low-pass *RC* filter.

frequency above some critical value are simply rejected. Unfortunately, such circuits do not exist.

In the case of the *RC* low-pass filter, the variation of rejection with frequency is shown in Figure 2.12. In this graph the vertical is the ratio of output voltage to the input voltage without regard to phase. When this ratio is one, the signal is passed without effect; when it is very small or zero, the signal is effectively blocked.

The horizontal is actually the logarithm of the ratio of the input voltage signal frequency to a *critical frequency*. This critical frequency is that frequency for which the ratio of the output to the input voltage is 0.707. In terms of the resistor and capacitor, the critical frequency is given by

$$f_c = 1/(2\pi RC) \tag{2.24}$$

The voltage ratio for any signal frequency can be determined graphically from Figure 2.12 or can be computed by

$$\frac{V_{out}}{V_{in}} = \frac{1}{[1 + (f/f_c)^2]^{1/2}} \tag{2.25}$$

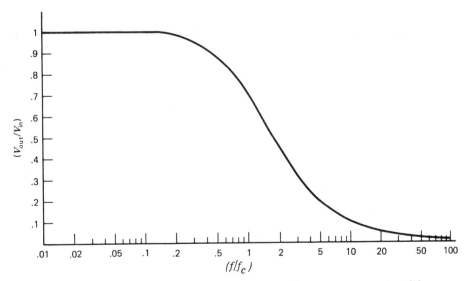

Figure 2.12   Variation of V_{out}/V_{in} as a function of frequency expressed as f/f_c and plotted on a semilog graph.

Example 2.10

A measurement signal has a frequency of <1 kHz, but there is unwanted noise at 1 MHz. Design a low-pass filter that attenuates the noise to 1%. What is the effect on the measurement signal at its maximum of 1 kHz?

Solution   Use Equation (2.25) to determine what critical frequency will give $(V_{out}/V_{in}) = 0.01$ at 1 MHz. To do this we have the relationship

$$0.01 = \frac{1}{[1 + (1 \text{ MHz}/f_c)^2]^{1/2}}$$

which gives us

$$(1 \text{ MHz}/f_c)^2 = 9999$$

or

$$f_c = 10 \text{ kHz}$$

Any resistor and capacitor combination that gives this critical frequency will be a correct solution. Let us pick a capacitor of some practical value, say, 0.01 μF; then, from Equation (2.24), the resistor is

$$R = 1/[2\pi(0.01 \ \mu\text{F})(10 \text{ kHz})] = 1.59 \text{ k}\Omega$$

To determine the effect on a signal of 1 kHz, we use the graph of Figure 2.12 or Equation (2.25) with the (now) known critical frequency of 10 kHz. The graph is used first to find $(f/f_c) = (1 \text{ kHz}/10 \text{ kHz}) = 0.1$; from the graph you can see that the rejection is very small. Equation (2.25) can be used to find the rejection more accurately, that is, $(V_{out}/V_{in}) = 0.995$. Thus, the measurement signal at maximum frequency is attenuated by only 0.5%.

High-pass *RC* filter

A high-pass filter passes high frequencies (no rejection) and blocks (rejects) low frequencies. A filter of this type can be constructed using a resistor and capacitor, as shown in the schematic of Figure 2.13. Similar to the low-pass filter, the rejection is not sharp in frequency but distributed over a range around a critical frequency. This critical frequency is defined by the *same* value Equation (2.24) as for the low-pass filter.

The graph of voltage output to input versus logarithm of frequency to critical frequency is shown in Figure 2.14. Note that the magnitude of $V_{out}/V_{in} = 0.707$ when the frequency is equal to the critical frequency.

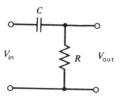

Figure 2.13 Circuit for the high-pass *RC* filter.

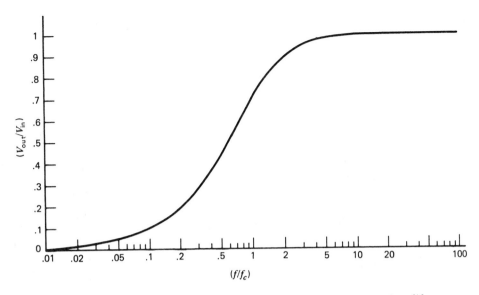

Figure 2.14   Variation of V_{out}/V_{in} as a function of frequency expressed as f/f_c and plotted on a semilog graph.

An equation for the ratio of output voltage to input voltage as a function of the frequency for the high-pass filter is found to be

$$V_{out}/V_{in} = \frac{(f/f_c)}{[1 + (f/f_c)^2]^{1/2}} \qquad (2.26)$$

Example 2.11

Pulses for a stepping motor are being transmitted at 2000 Hz. Design a filter to reduce 60-Hz noise, but reduce the pulses by no more than 3 dB.

Solution   Let us first find what voltage ratio corresponds to a 3-dB reduction. We remember that

$$P(\text{dB}) = 20 \log(V_{out}/V_{in})$$

so *down* by 3 dB means that $P = -3$. Therefore

$$(V_{out}/V_{in}) = 10^{-3/20} = 0.707$$

You probably saw that coming. The critical frequency is that frequency for which the output is attenuated by 3 dB. Thus, in this case, $f_c = 2$ kHz. The effect on 60-Hz noise is found using Equation (2.26), with $f = 60$ Hz.

$$V_{out}/V_{in} = \frac{(60/2000)}{[1 + (60/2000)^2]^{1/2}}$$

$$V_{out}/V_{in} = 0.03$$

Thus, we see that only 3% of the 60-Hz noise remains; that is, it has been reduced by 97%.

2.4 OPERATIONAL AMPLIFIERS

As discussed in Section 2.2, there are many diverse requirements for signal conditioning in process control. In Section 2.3 we considered common, passive circuits that can provide some of the required signal operations, the divider, bridge, and RC filters. Historically, the detectors used in bridge circuits consisted of tube and transistor circuits. In many other cases where impedance transformations, amplification, and other operations were required, a circuit was designed that depended on discrete electronic components. With the remarkable advances in electronics and integrated circuits (ICs), the requirement to implement designs from discrete components has given way to easier and more reliable methods of signal conditioning. Many special circuits and general-purpose amplifiers are now contained in integrated circuit (IC) packages producing a quick solution to signal conditioning problems together with small size, low power consumption, and low cost.

In general, the application of ICs requires familiarity with an available line of such devices, their specifications and limitations, before they can be applied to a specific problem. Apart from these specialized ICs, there also is a type of amplifier that finds wide application as the building block of signal conditioning applications. This device, called an operational amplifier (op amp), has been in existence for many years. It was first constructed from tubes, then from discrete transistors, and now as integrated circuits. Although many lines of op amps with diverse specifications exist from many manufacturers, they all have common characteristics of operation that can be employed in basic designs relating to any general op amp.

2.4.1 Op Amp Characteristics

Taken alone, an op amp is an exceedingly simple and apparently useless electronic amplifier. In Figure 2.15a we see the standard symbol of an op amp with the designations $(+)$ input, $(-)$ input, and the output. The $(+)$ input is also called the *noninverting* input and the $(-)$ the *inverting* input. The relation of op amp input to output is very simple indeed, as will be seen by considering an idealization of its description.

Ideal op amp

To describe the response of an ideal op amp, we label V_1 the voltage on the $(+)$ input, V_2 the voltage on the $(-)$ input terminal, and V_0 the output voltage. Ideally, if $V_1 - V_2$ is positive $(V_1 > V_2)$, then V_0 saturates positively. If $V_1 - V_2$ is

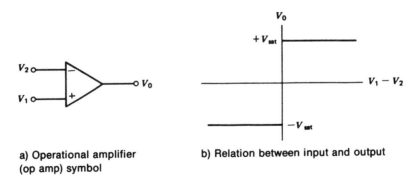

a) Operational amplifier
(op amp) symbol

b) Relation between input and output

Figure 2.15   Symbol and ideal characteristics of an op amp.

negative ($V_2 > V_1$), then V_0 saturates negatively, as shown in Figure 2.15b. The ($-$) input is called the inverting input. If the voltage on this input is more positive than that on the ($+$) input, the output saturates negatively. This ideal amplifier has infinite gain because an infinitesimal difference between V_1 and V_2 results in a saturated output.

Other characteristics of ideal op amps are (1) an infinite impedance between inputs and (2) a zero output impedance. Basically, the op amp is a device that has only two output states: $+V_{sat}$ and $-V_{sat}$. In practice, the device is always used with feedback of output to input. Such feedback permits implementation of many special relationships between input and output voltage.

Ideal inverting amplifier

To see how the op amp is used, let us consider the circuit of Figure 2.16. Resistor R_2 is used to feed back the output to the inverting input of the op amp, and R_1 connects the input voltage V_{in} to this same point. The common connection is called the summing point. We can see that with no feedback and the ($+$) grounded, $V_{in} > 0$ saturates the output negative and $V_{in} < 0$ saturates the output positive. With feedback, the output adjusts to a voltage such that

1. The summing point voltage is equal to the ($+$) op amp input level, zero in this case.

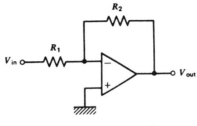

$$V_{out} = -\frac{R_2}{R_1} V_{in}$$

Figure 2.16   Inverting amplifier using an op amp.

2. No current flows through the op amp input terminals because of the assumed infinite impedance.

In this case, the sum of currents at the summing point must be zero.

$$I_1 + I_2 = 0 \tag{2.27}$$

where

I_1 = current through R_1
I_2 = current through R_2

Because the summing point potential is assumed to be zero, we have

$$\frac{V_{\text{in}}}{R_1} + \frac{V_{\text{out}}}{R_2} = 0 \tag{2.28}$$

From Equation (2.27), we can write the circuit response as

$$V_{\text{out}} = -\frac{R_2}{R_1} V_{\text{in}} \tag{2.29}$$

Thus, the circuit of Figure 2.16 is an inverting *amplifier* with gain R_2/R_1 that is shifted 180° in phase (inverted) from the input. This device is also an *attenuator* by making $R_2 < R_1$.

A similar approach may be applied to the ideal analysis of many other op amp circuits where steps (1) and (2) lead to equations such as Equations (2.27) and (2.28). We must note, however, that the inverting amplifier of Figure 2.16 has an input impedance of R_1 that, in general, may not be high. Thus, although blessed with the virtue of variable gain or attenuation, the circuit does not have inherently high-input impedance.

Nonideal effects

Analysis of op amp circuits with nonideal response is performed by considering the following parameters:

1. *Finite open loop gain*   A real op amp has finite voltage gain, as shown by the amplifier response in Figure 2.17a. The voltage gain is defined as the change in output voltage ΔV_0 produced by a change in differential input voltage $\Delta[V_1 - V_2]$.

2. *Finite input impedance*   A real op amp has an input impedance and, consequently, a finite voltage across and current through input terminals.

3. *Nonzero output impedance*   A real op amp has a nonzero output impedance, although this low output impedance is typically only a few ohms.

In most modern applications, these nonideal effects can be ignored in de-

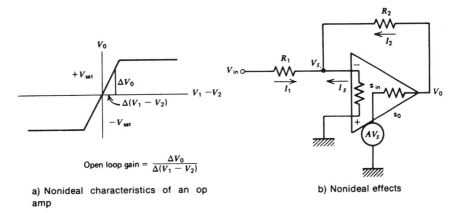

a) Nonideal characteristics of an op amp

b) Nonideal effects

Figure 2.17 Types of nonideal effects in op amp and circuit analysis.

signing op amp circuits. For example, consider the circuit of Figure 2.17b, where the finite impedances and gain of the op amp have been included. We can employ standard circuit analysis to find the relationship between input and output voltage for this circuit. Summing the currents at the summing point gives

$$I_1 + I_2 + I_s = 0$$

Then, each current can be identified in terms of the circuit parameters to give

$$\frac{V_{in} - V_s}{R_1} + \frac{V_0 - V_s}{R_2} - \frac{V_s}{z_{in}} = 0$$

Finally, V_0 can be related to the op amp gain as

$$V_0 = AV_s - \left(\frac{V_0 - V_s}{R_2}\right) z_0$$

Now, combining the equations, we find

$$V_0 = -\frac{R_2}{R_1}\left(\frac{1}{1 - \mu}\right) V_{in} \tag{2.30}$$

where

$$\mu = \frac{\left(1 + \dfrac{z_0}{R_2}\right)\left(1 + \dfrac{R_2}{R_1} + \dfrac{R_2}{z_{in}}\right)}{\left(A + \dfrac{z_0}{R_2}\right)} \tag{2.31}$$

If we assume that μ is very small compared with unity, then Equation (2.30) reduces to the ideal case given by Equation (2.29). Indeed, if typical values for an IC op amp are chosen for a case where $R_2/R_1 = 100$, we can show that $\mu \ll 1$. For example, a common, general-purpose IC op amp shows

$$A = 200,000$$

$$z_0 = 75 \ \Omega$$

$$z_{in} = 2 \ M\Omega$$

If we use a feedback resistance R_2 of 100 kΩ and substitute the aforementioned values into Equation (2.31), we find $\mu \simeq 0.0005$, which shows that the gain from Equation (2.30) differs from the ideal by only 0.05%. This was, of course, only one example of the many op amp circuits that are employed, but in most cases a similar analysis shows that the ideal characteristics may be assumed.

2.4.2 Op Amp Specifications

There are characteristics of op amps other than those given in the previous section that enter into design applications. These characteristics are given in the specifications for particular op amps together with the open loop gain and input and output impedance previously defined. Several of these characteristics are as follows:

- *Input offset voltage*   In many cases, the op amp output voltage may not be zero when the voltage across the input is zero. The voltage that must be applied across the input terminals to drive the output to zero is the *input offset voltage.*

- *Input offset current*   Just as a voltage offset may be required across the input to zero the output voltage, so a net current may be required between the inputs to zero the output voltage. Such a current is referred to as an *input offset current.* This is taken as the difference of the two input currents.

- *Input bias current*   This is the average of the two input currents required to drive the output voltage to zero.

- *Slew rate*   If a voltage is suddenly applied to the input of an op amp, the output will saturate to the maximum. For a step input the *slew rate* is the rate at which the output changes to the saturation value. This typically is expressed as volts per microsecond (V/μs).

- *Unity gain frequency bandwidth*   The frequency response of an op amp is typically defined by a Bode plot of open loop voltage gain versus frequency. Such a plot is very important for the design of circuits that deal with ac signals. It is beyond the scope of this text to consider the details of designs employing Bode plots. Instead, the gross frequency behavior can be seen by determination of the frequency at which the open loop gain of the op amp has become unity, thus defining the *unity gain frequency bandwidth.*

2.5 OP AMP CIRCUITS IN INSTRUMENTATION

As the op amp became familiar to the individuals working in process control and instrumentation technology, a large variety of circuits were developed with direct application to this field. In general, it is much easier to develop a circuit for a specific service using op amps than discrete components; with the development of low-cost IC op amps, it is also practical. Perhaps one of the greatest disadvantages is the requirement of a bipolar power supply for the op amp. This section presents a number of typical circuits and their basic characteristics together with a derivation of the circuit response assuming an ideal op amp.

2.5.1 Voltage Follower

Figure 2.18 shows an op amp circuit with unity gain and very high input impedance. The input impedance is essentially the input impedance of the op amp itself, which can be greater than 100 MΩ. The voltage output tracks the input over a range defined by the plus and minus saturation voltage outputs. Current output is limited to the short circuit current of the op amp, and output impedance is typically much less than 100 Ω. In many cases a manufacturer will market an op amp voltage follower whose feedback is provided internally. Such a unit is usually specifically designed for very high input impedance. The unity gain voltage follower is essentially an impedance transformer in the sense of converting a voltage at high impedance to the same voltage at low impedance.

2.5.2 Inverting Amplifier

The inverting amplifier has already been discussed in connection with our treatment of op amp characteristics. Equation (2.29) shows that this circuit inverts the input signal and may have either attenuation or gain depending on the ratio of input resistance R_1 and feedback resistance R_2. The circuit for this amplifier is shown in Figure 2.16. It is important to note that input impedance of this circuit is essentially equal to R_1, the input resistance. In general, this resistance is not large, and hence the input impedance is not large.

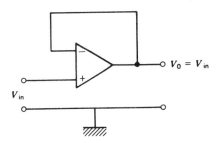

Figure 2.18   An op amp voltage follower. This circuit has a very high impedance; depending on the op amp, it may be 10^6 to 10^{11} Ω. This circuit serves as an impedance transformer.

Summing amplifier

A common modification of the inverting amplifier is an amplifier that sums or adds two or more applied voltages. This circuit is shown in Figure 2.19 for the case of summing two input voltages. The transfer function of this amplifier is given by

$$V_{out} = -\left[\frac{R_2}{R_1} V_1 + \frac{R_2}{R_3} V_2\right] \qquad (2.32)$$

The sum can be scaled by proper selection of resistors. For example, if we make $R_1 = R_2 = R_3$, then the output is simply the (inverted) sum of V_1 and V_2. The average can be found by making $R_1 = R_3$ and $R_2 = R_1/2$.

Example 2.12

Develop an op amp circuit that can provide an output voltage that is related to the input voltage by

$$V_{out} = 3.4V_{in} + 5$$

Solution There are many ways to do this. One way is to use a summing amplifier with V_{in} on one input and 5 volts on the other. The gains will be selected to be 3.4 and 1.0, respectively. The summing amplifier of Figure 2.19 is also an inverter, however, so the sign will be wrong. Thus, a second amplifier will be used with a gain of -1 to make the sign correct. The result is shown in Figure 2.20. Selection

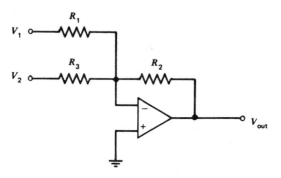

Figure 2.19 Summing amplifier.

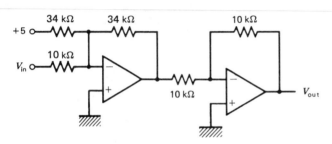

Figure 2.20 Op amp solution to Example 2.12.

of the values of the resistors is based on the general notion of keeping the currents in milliamperes.

2.5.3 Noninverting Amplifier

A noninverting amplifier may be constructed from an op amp, as shown in Figure 2.21. The gain of this circuit is found by summing the currents at the summing point S, and using the fact that the summing point voltage is V_{in} so that no voltage difference appears across the input terminals.

$$I_1 + I_2 = 0$$

where

I_1 = current through R_1
I_2 = current through R_2

But these currents can be found from Ohm's law such that this equation becomes

$$\frac{V_{in}}{R_1} + \frac{V_{in} - V_{out}}{R_2} = 0$$

Solving this equation for V_{out}, we find

$$V_{out} = \left[1 + \frac{R_2}{R_1}\right] V_{in} \tag{2.33}$$

Equation (2.33) shows that the noninverting amplifier has a gain that depends on the ratio of feedback resistor R_2 and the ground resistor R_1, but this gain can never be used for voltage attenuation. Because the input is taken directly into the noninverting input of the op amp, the input impedance is very high since it is effectively equal to the op amp input impedance.

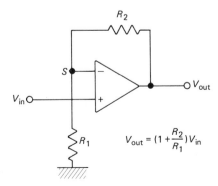

Figure 2.21 Noninverting amplifier.

Example 2.13

Design a high-impedance amplifier with a voltage gain of 42.

Solution   We use the noninverting circuit of Figure 2.21 with resistors selected from

$$V_{out} = \left[1 + \frac{R_2}{R_1} \right] V_{in}$$

$$42 = 1 + \frac{R_2}{R_1} \tag{2.33}$$

$$R_2 = 41R_1$$

so we could choose $R_1 = $ **1 kΩ**, which requires $R_2 = $ **41 kΩ**.

2.5.4 Differential Amplifier

Frequently, in the instrumentation associated with process control, differential voltage amplification is required (for example, for a bridge circuit). A differential amplifier is constructed using an op amp, as shown in Figure 2.22a. Analysis of this circuit shows that the output voltage is given by

$$V_{out} = \frac{R_2}{R_1} (V_2 - V_1) \tag{2.34}$$

This circuit has a variable gain or attenuation given by the ratio of R_2 and R_1 and responds to the difference in voltage inputs as required. It is very important that the resistors in Figure 2.22a that have the same indicated value be carefully matched to assure rejection of voltage common to both inputs. A significant disadvantage of this circuit is that the input impedance at each input terminal is not large, that is, $R_1 + R_2$ at the V_2 input and R_1 at the V_1 input. To employ this circuit when a high input impedance differential amplification is desired, voltage followers may be employed before each input, as shown in Figure 2.22b. This circuit makes a very versatile high-gain, high-input impedance differential amplifier for use in instrumentation systems.

Example 2.14

A sensor outputs a range of 20.0 to 250 mV as a variable varies over its range. Develop signal conditioning so that this becomes 0 to 5 V. The circuit must have very high input impedance.

Solution   A very logical way to approach problems of this sort is to develop an equation for the output in terms of the input, such as that shown in Example 2.12. A circuit can then be developed to provide the variation of the equation. The equation is that of a straight line; we can then write

$$V_{out} = mV_{in} + V_0$$

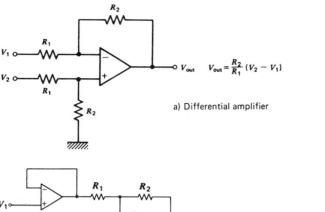

a) Differential amplifier

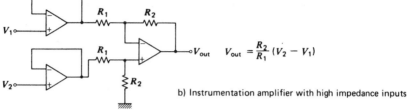

b) Instrumentation amplifier with high impedance inputs

Figure 2.22 Differential amplifiers. (a) Differential amplifier. (b) Instrumentation amplifier with high-impedance inputs.

where m is the slope of the line and represents the gain ($m > 1$) or attenuation ($m < 1$) required and V_0 is the intercept; that is, the value V_{out} would be if $V_{in} = 0$.

For the two conditions we have in this problem, form two equations to solve for m and V_0.

$$0 = m(0.02) + V_0$$

$$5 = m(0.25) + V_0$$

We get $m = 21.7$ and $V_0 = -0.434$ V using standard algebra. The equation is

$$V_{out} = 21.7V_{in} - 0.434$$

This also can be written in the form

$$V_{out} = 21.7(V_{in} - 0.02)$$

There are many, many circuits that provide this answer. Figure 2.23 shows one. Notice the voltage follower on the input that provides high input impedance and a differential amplifier with a gain of 21.7. The 0.02 volts could be provided by a divider from a well-regulated source.

2.5.5 Voltage-to-Current Converter

Because signals in process control are most often transmitted as a current, specifically 4–20 mA, it is often necessary to employ a linear voltage-to-current converter. Such a circuit must be capable of sinking a current into a number of

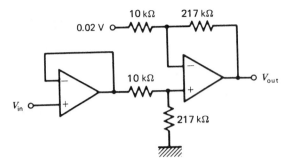

Figure 2.23 Op amp solution to Example 2.14.

different loads without changing the voltage-to-current transfer characteristics. An op amp circuit that provides this function is shown in Figure 2.24. An analysis of this circuit shows that the relationship between current and voltage is given by

$$I = -\frac{R_2}{R_1 R_3} V_{in} \tag{2.35}$$

provided that the resistances are selected so that

$$R_1(R_3 + R_5) = R_2 R_4 \tag{2.36}$$

The circuit can deliver current in either direction, as required by a particular application.

The maximum load resistance and maximum current are related and determined by the condition that the amplifier output saturates in voltage. Analysis of the circuit shows that when the op amp output voltage saturates, the maximum load resistance and maximum current are related by

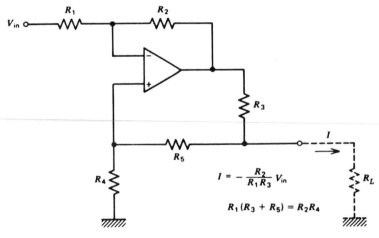

Figure 2.24 Voltage-to-current converter.

$$R_{ml} = \frac{(R_4 + R_5)\left[\dfrac{V_{sat}}{I_m} - R_3\right]}{R_3 + R_4 + R_5}$$

R_{ml} = maximum load resistance

V_{sat} = op amp saturation on voltage   (2.37)

I_m = maximum current

A study of Equation (2.37) shows that the *maximum* load resistance is always less than V_{sat}/I_m. The *minimum* load resistance is zero.

Example 2.15

A sensor outputs 0 to 1 volts. Develop a voltage-to-current convertor so that this becomes 0 to 10 mA. Specify the maximum load resistance if the op amp saturates at ± 10 V.

Solution   If we make $R_1 = R_2$ in Figure 2.24, then Equation (2.35) reduces to $I = V_{in}/R_3$. To satisfy 10 mA at 1 V, we must have

$$R_3 = 1 \text{ V}/10 \text{ mA} = 100 \ \Omega$$

Let us take $R_5 = 0$ (which is allowed) so that Equation (2.36) also specifies

$$R_3 = R_4 = 100 \ \Omega$$

This completes the voltage-to-current converter. The maximum load resistance is found from Equation (2.37).

$$R_{ml} = 100[10 \text{ V}/10 \text{ mA} - 100]/200$$

$$R_{ml} = 450 \ \Omega$$

2.5.6 Current-to-Voltage Converter

At the receiving end of the process-control signal transmission system, we often need to convert the current back into a voltage. This can be done most easily with the circuit shown in Figure 2.25. This circuit provides an output voltage

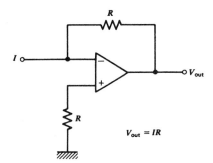

$V_{out} = IR$

Figure 2.25 Current-to-voltage converter.

given by

$$V_{\text{out}} = IR \tag{2.38}$$

provided the op amp saturation voltage has not been reached. The resistor R in the noninverting terminal is employed to provide temperature stability to the configuration.

2.5.7 Integrator

The last regular op amp circuit to be considered is the *integrator*. This configuration, shown in Figure 2.26, consists of an input resistor and feedback capacitor. Using the ideal analysis, we can sum the currents at the summing point as

$$\frac{V_{\text{in}}}{R} + C \frac{dV_{\text{out}}}{dt} = 0 \tag{2.39}$$

which can be solved by integrating both terms so that the circuit response is

$$V_{\text{out}} = -\frac{1}{RC} \int V_{\text{in}}\, dt \tag{2.40}$$

This result shows that the output voltage varies as an integral of the input voltage with a scale factor of $-1/RC$. This circuit is employed in many cases where an integration of a transducer output is desired.

    Other functions also can be implemented, such as a highly linear ramp voltage. If the input voltage is constant, $V_{\text{in}} = K$, and Equation (2.40) reduces to

$$V_{\text{out}} = -\frac{K}{RC} t \tag{2.41}$$

which is a linear ramp, a negative slope of K/RC. Some mechanism of reset through discharge of the capacitor must be provided because otherwise V_{out} will rise to the output saturation value and remain fixed there in time.

Example 2.16

    Use an integrator to produce a linear ramp voltage rising at 10 volts per ms.

    Solution   An integrator circuit, as shown in Figure 2.26, produces a ramp of

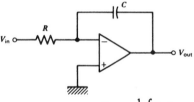

$$V_{\text{out}} = -\frac{1}{RC} \int V_{\text{in}}\, dt$$

Figure 2.26   Integrator circuit.

$$V_{out} = -\frac{V_{in}}{RC}t \qquad (2.41)$$

when the input voltage is constant. If we make $RC = 1$ ms and $V_{in} = -10$ V, then we have

$$V_{out} = (10 \cdot 10^{+3})t$$

which is a ramp rising at 10 volts/ms. A choice of $R = 1$ kΩ and $C = 1$ μF will provide the required RC product.

2.5.8 Linearization

The op amp represents a very effective device to implement linearization. Generally, this is achieved by placing a *nonlinear* element in the feedback loop of the op amp, as shown in Figure 2.27. The summation of currents provides

$$\frac{V_{in}}{R} + F(V_{out}) = 0 \qquad (2.42)$$

where

$$V_{in} = \text{input voltage}$$
$$R = \text{input resistance}$$
$$F(V_{out}) = \text{nonlinear variation of current with voltage}$$

If Equation (2.42) is solved (in principle) for V_{out}, we get

$$V_{out} = G\left(\frac{V_{in}}{R}\right) \qquad (2.43)$$

where

$$V_{out} = \text{output voltage}$$
$$G\left(\frac{V_{in}}{R}\right) = \text{a nonlinear function of the input voltage [actually the inverse function of } F(V_{out})].$$

Thus, as an example, if a diode is placed in the feedback as shown in Figure 2.28,

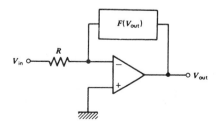

Figure 2.27 A nonlinear amplifier is constructed by placing any nonlinear element in the feedback of the op amp.

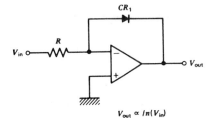

Figure 2.28 When a diode is placed in the feedback leg of an op amp, a nonlinear amplifier is formed whose output is proportional to the natural logarithm of the input.

the function $F(V_{out})$ is an exponential

$$F(V_{out}) = F_0 \exp(\alpha V_{out}) \tag{2.44}$$

where

$F_0 =$ amplitude constant
$\alpha =$ exponential constant

The inverse of this is a logarithm, and thus Equation (2.43) becomes

$$V_{out} = \frac{1}{\alpha} \log_e(V_{in}) - \frac{1}{\alpha} \log_e(F_0 R) \tag{2.45}$$

which thus constitutes a (linear) logarithmic amplifier.

Different feedback devices can produce amplifiers that only smooth out nonlinear variations or provide specified operations such as the logarithmic amplifier.

2.5.9 Special Integrated Circuits (ICs)

A vast line of special integrated circuits (ICs) is available from many manufacturers and is useful to the process-control instrumentation designer. Such special-purpose devices include

1. High-gain differential instrumentation amplifiers
2. Current-to-voltage converters
3. Modulator/demodulators
4. Bridge and null detectors
5. Phase-sensitive detectors

In the following chapters, we often require signal conditioning that can be implemented through the use of these special ICs. In general, we indicate the details of a signal conditioning design, but the reader should always be aware that use of special-purpose ICs may render such a detailed design unnecessary.

2.6 INDUSTRIAL ELECTRONICS

The signal conditioning discussed thus far in this chapter has referred mainly to measurement signal modification. It is often necessary to perform a type of signal conditioning on the controller output to activate the final control element. For example, the 4–20 mA controller output may be required to adjust heat input to a large, heavy-duty oven for baking crackers. Such heat may be provided by a 2-kW electrical heater. Clearly, some sort of conditioning is required to allow such a high-power system to be controlled by a low-energy current signal. In this section we present two devices that are commonly used in process control to provide a mechanism where such energy conversion can occur. The intent here is not to give you all the information needed to construct practical circuits to use these devices, but to familiarize you with them and their specifications.

2.6.1 Silicon-Controlled Rectifier (SCR)

The SCR has become a very important part of high-power electrical signal conditioning and control. In some regards, it is a solid-state replacement for the relay, although there are some problems if that analogy is taken too far. The standard diode is, in the ideal sense, a device that will conduct current in only one direction. The SCR, again in the ideal sense, is like a diode that will not conduct in either direction until it is turned on or "fired." Figure 2.29 shows the schematic symbol of an SCR. Note the similarity to a diode but with the added terminal, called the *gate*. If the SCR is forward biased (that is, positive voltage on the anode with respect to the cathode), it will *not* be conducting. Now suppose a voltage is placed on the gate with respect to the cathode. There will be some positive value of this voltage—the trigger voltage—at which the SCR will start conducting and behave then like a normal diode. Even if the gate voltage is taken away, it will continue to conduct like a diode; that is, once turned on it will stay on, regardless of the gate. The only way to turn the SCR back "off" is to have the forward bias condition taken away. This means the voltage must drop below the forward voltage drop of the SCR so that the current drops below a minimum value, called the holding current, or the polarity from anode to cathode must actually reverse. The fact that the SCR cannot be turned off easily limits its use in dc applications to those cases when some method of reducing the forward current to below the holding values can be provided. In ac circuits, the SCR will automatically turn off in every half cycle when the ac voltage applied to the SCR reverses polarity.

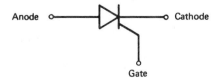

Anode   Cathode

Gate   **Figure 2.29** Symbol for an SCR.

Characteristics and specifications of SCRs are as follows:

1. *Maximum forward current*   There is a maximum current that the SCR can carry in the forward direction without damage. This value varies from a few hundred milliamps to more than a thousand amps for large industrial types.

2. *Peak reverse voltage*   Like a diode, there is a peak reverse bias voltage that can be applied to the SCR without damage. The value varies from a few volts to several thousand volts.

3. *Trigger voltage*   The minimum gate voltage to drive the SCR into conduction varies between types and sizes from a few volts to 40 volts.

4. *Trigger current*   There is a minimum current that the source of trigger voltage must be able to provide before the SCR can be fired. This varies from a few milliamps to several hundred milliamps.

5. *Holding current*   This refers to the minimum anode to cathode current necessary to keep the SCR conducting in the forward conducting state. The value varies from 20 to 100 mA.

Ac operation

Figure 2.30 illustrates the operation of an SCR in varying the rms dc voltage in half-wave operation. The trigger voltage is developed by some circuit that produces a pulse at a certain selected phase of the applied ac signal. Thus, the SCR turns on in a repetitive fashion as shown. The SCR is turned back off, of course, in each half cycle when the ac polarity reverses. By changing the part of the positive half cycle when the trigger is applied, the effective (rms) value of dc voltage applied to the load can be increased. Of course, with this circuit the maximum possible rms dc voltage is that which would be developed by a half-wave rectifier. If more power is required, the SCR can be used with a full-wave bridge circuit. Figure 2.31 shows this type of circuit and the voltages versus time that result. The trigger voltage must now be generated in each half cycle and applied to the SCR trigger (gate) terminal. In a process-control application, the controller output signal would be used to drive a circuit that changed the time at which the pulses were applied to the gates and thus changed the power applied to the load. The voltage applied to the load is pulsating dc. This configuration could not be used with a load that required ac voltage for operation.

Trigger control

To use the SCR in process-control applications, special circuitry to convert control signals into suitable trigger signals to the SCRs is required. These circuits are usually composed of electronic systems that use the control voltage to determine the phase of the ac load voltage at which the SCR should be turned on.

A very elementary example of such a circuit is shown in Figure 2.32. The control signal voltage is used to provide base drive to a transistor via an LED

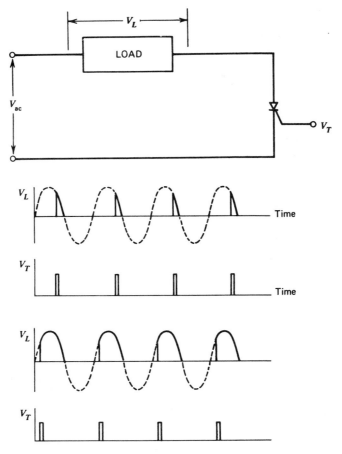

Figure 2.30 Half-wave SCR operation. Changing the time of V_T application changes the dc rms voltage applied to the load V_L.

that assures isolation of the control circuit from the power circuit. At low-base drive the capacitor is charged slowly and will not reach the SCR trigger voltage until late in the cycle (hence low load power).

A large control signal will provide high base drive, and the capacitor will charge much more quickly. Then the SCR will turn on much earlier in the cycle and more power will be delivered to the load.

2.6.2 TRIAC

An extension of the SCR discussed previously is a device that can be triggered to conduct in either direction. The TRIAC can be thought of as two SCRs connected in parallel and reversed, but with the gates connected. A positive trigger will cause it to conduct in one direction, and a negative trigger will cause it to

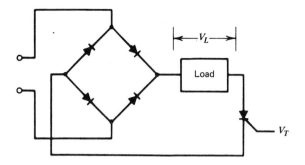

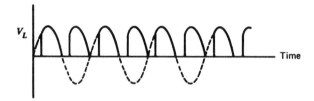

Figure 2.31 Full-wave SCR circuit. The effective rms dc voltage applied to the load is increased because both cycles of the ac are used.

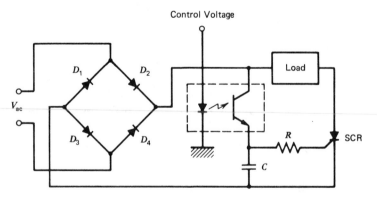

Figure 2.32 This is one example of how an SCR can be triggered from an isolated dc signal voltage.

conduct in the other direction. The TRIAC thus can be used in pure ac applications. Figure 2.33 shows the symbol of a TRIAC and a circuit for a typical application. The voltage across the load, as shown, remains ac. The effective ac rms value of voltage applied to the load can be changed by changing the time in the phase of the cycles when the TRIAC gate is pulsed. The trigger voltage generated must be bipolar, one pulse in one polarity and the next of the opposite polarity.

Specifications of TRIACs are similar to those of SCRs: maximum rms current, peak reverse voltage, trigger voltage, and trigger current.

DIAC

A DIAC is a special kind of two-terminal semiconductor switch that is often used in conjunction with TRIACs for triggering. This device, with a schematic symbol shown in Figure 2.34, is nonconducting (off) in either direction as long as the applied voltage is below a certain value. If that value is exceeded in either polarity, the device will begin to conduct (turn on).

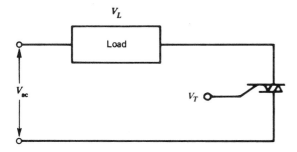

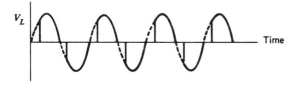

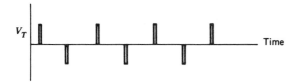

Figure 2.33 The TRIAC can conduct in both directions so that the load voltage remains ac, but the rms value is determined by the time at which the trigger voltages are applied.

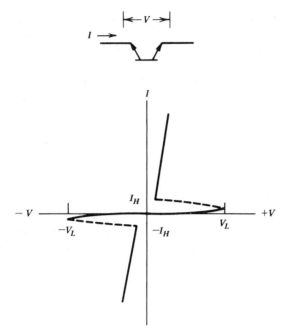

Figure 2.34 The DIAC goes from a nonconducting to conducting state if the voltage exceeds V_T in either polarity. It reverts to nonconducting if the current falls below I_H.

Trigger control

Similar to the SCR, for the TRIAC to be useful in control applications, it is necessary to link the control signal to the application of trigger signals. This involves adjusting the time (or phase) in a cycle of ac voltage when the TRIAC is triggered.

An elementary method of doing this is shown in Figure 2.35. Note the use of a DIAC between the capacitor and the TRIAC trigger terminal. The principle of operation is the same as the SCR system discussed previously, but the load is impressed with an ac voltage and the bridge rectifier is not needed. The capacitor

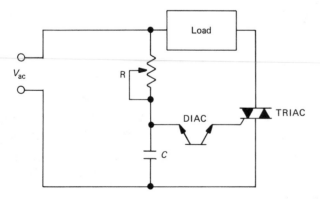

Figure 2.35 The TRIAC can be triggerd from the ac line voltage using the DIAC.

charges in either polarity until the DIAC turns on, which, in turn, triggers the TRIAC.

2.7 DESIGN GUIDELINES

This section discusses typical issues which should be considered when designing an analog signal conditioning system. The examples show how the guidelines can be used to develop a design. The guidelines assure that the problem is clearly understood and that the important issues are included.

Not every guideline will be important in every design, so some will be not applicable. In many cases, not enough information will be available to address an issue properly; then the designer must exercise good technical judgment in accounting for that part of the design.

Figure 2.36 shows the measurement and signal conditioning model. In some cases, the entire system is to be developed, from selecting the sensor to designing the signal conditioning. In other cases, only the signal conditioning will be developed. The guidelines are generalized. Since the sensor is selected from what is available, the actual design is really for the signal conditioning.

GUIDELINES FOR ANALOG SIGNAL CONDITIONING DESIGN

1. *Define the measurement objective.*
   a. *Parameter:*   What is the nature of the measured variable: pressure, temperature, flow, level, voltage, current, resistance, etc.?
   b. *Range:*   What is the range of the measurement: 100 to 200°C, 45 to 85 psi, 2 to 4 V, etc.?
   c. *Accuracy:*   What is the required accuracy: 5% FS, 3% of reading, etc.?
   d. *Linearity:*   Must the measurement output be linear?
   e. *Noise:*   What is the noise level and frequency spectrum of the measurement environment?

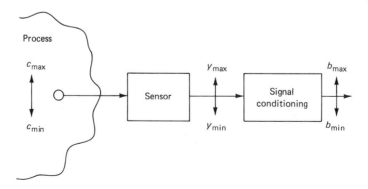

Figure 2.36   Model for measurement and signal conditioning objectives.

2. *Select a sensor* (*if applicable*).
  a. *Parameter:* What is the nature of the sensor output: resistance, voltage, etc.?
  b. *Transfer function:* What is the relationship between the sensor output and the measured variable: linear, graphical, equation, accuracy, etc.?
  c. *Time response:* What is the time response of the sensor: first-order time constant, second-order damping, and frequency?
  d. *Range:* What is the range of sensor parameter output for the given measurement range?
  e. *Power:* What is the power specification of the sensor: resistive dissipation maximum, current draw, etc.?

3. *Design the analog signal conditioning* (*S/C*).
  a. *Parameter:* What is the nature of the desired output? The most common is voltage, but current and frequency are sometimes specified. In the latter cases, conversion to voltage is still often a first step.
  b. *Range:* What is the desired range of the output parameter (e.g., 0 to 5 V, 4 to 20 mA, 5 to 10 kHz)?
  c. *Input impedance:* What input impedance should the S/C present to the input signal source? This is very important in preventing loading of a voltage signal input.
  d. *Output impedance:* What output impedance should the S/C offer to the output load circuit?

4. *Notes on analog signal conditioning design.*
  a. If the input is a resistance change and a bridge or divider must be used, be sure to consider the effect of output voltage nonlinearity with resistance *and* the effect of current through the resistive sensor.
  b. For the op amp portion of the design, the easiest design approach is to develop an equation for output versus input. From this equation, it will be clear what types of circuits may be used. This equation represents the static transfer function of the signal conditioning.
  c. Always consider any possible loading of voltage sources by the signal conditioning. Such loading is a direct error in the measurement system.

The following examples apply these guidelines to measurement signal conditioning problems. In later chapters (4, 5, and 6) on sensors, many other examples will be presented.

Example 2.17

A sensor outputs a voltage from -2.4 to -1.1 V. For interface to an analog-to-digital converter, this needs to be 0 to 2.5 volts. Develop the required signal conditioning.

Solution   For this type of problem, no information is provided about the measured variable, the measurement environment, or the sensor. We are simply asked to provide a voltage-to-voltage conversion. Since the source impedance is not known, it is good design practice to assume it is high and design a high-input-impedance system to avoid loading. Most ADCs have input impedances of at least tens of kilohms, and the output impedance of op amp circuits is quite low, so there is no real concern for the output impedance of the S/C system.

For this type of problem, it is easiest to develop an equation for the output in terms of the input. From this, circuits can be envisioned.

$$V_{out} = mV_{in} + V_0$$

Using the specified information, we form two equations for the unknown slope (gain) m, and offset (bias) V_0.

$$-2.4 = 0m + V_0$$

$$-1.1 = 5m + V_0$$

Clearly, from the first equation we have $V_0 = -2.4$ V, and when this is substituted into the second equation, we get

$$-1.1 = 5m - 2.4$$

Then, solving for m,

$$m = (2.4 - 1.1)/5 = 0.26$$

The transfer function equation is thus

$$V_{out} = 0.26V_{in} - 2.4$$

There are many ways to satisfy this equation. A summing amplifier could be used, but it does not have high input impedance, so a voltage follower would be needed at the input. Also, the summing amplifier inverts, so an inverter would be required to get the correct sign. This circuit is shown in Figure 2.37. Note that the bias has been provided by a divider. A 15-volt supply has been assumed for the divider resistance calculations. The 100-Ω resistor was selected to keep loading by the op amp circuit small. A trimmer (variable) resistor has been used so both loading of the divider by the op amp circuit and variation of the supply from exactly 15 volts can be compensated for by adjusting until the bias is exactly 2.4 volts.

The design could also be accomplished by a differential amplifier. If the 0.26 in the transfer equation is factored, we get

$$V_{out} = 0.26(V_{in} - 9.23)$$

So this is the equation of a differential amplifier with a gain of 0.26 and one input fixed at 9.23 volts. A voltage follower would still be required on the input. (The reader should complete this design.)

Why is it not possible to use a noninverting amplifier? Since it has high input impedance, the voltage follower would not be necessary.

Example 2.18

Temperature is to be measured in the range of 250°C to 450°C with an accuracy of ± 2°C. The sensor is a resistance which varies linearly from 280 Ω to 1060 Ω for this

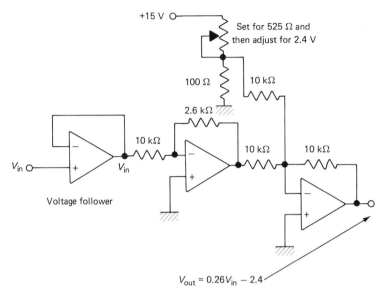

Figure 2.37   One possible solution to Example 2.17.

temperature range. Power dissipated in the sensor must be kept below 5 mW. Develop analog signal conditioning which provides a voltage varying linearly from -5 to $+5$ volts for this temperature range. The load is a high-impedance recorder.

Solution   Following the guidelines, let us first identify all the elements of the problem.

MEASURED VARIABLE PARAMETER: TEMPERATURE
Range: 250 to 450°C
Accuracy: ± 2°C
Noise: unspecified

SENSOR SIGNAL
Parameter: resistance
Transfer function: linear
Time response: unspecified
Range: 280 Ω to 1060 Ω, linear
Power: maximum 5 mW dissipated in sensor

SIGNAL CONDITIONING
Parameter: voltage, linear
Range: -5 to $+5$ volts
Input impedance: keep power in sensor below 5 mW
Output impedance: no problem, high-impedance recorder

The accuracy is $\pm 0.8\%$ at the low end and $\pm 0.44\%$ at the high end. Therefore, we will keep three significant figures to provide 0.1% on values selected.

The 5-mW maximum sensor dissipation means the current must be limited. To find the maximum current, we note that

$$P = I^2R$$

$$0.005 = I^2R$$

$$I = 0.005/R$$

The minimum current will thus occur at the maximum resistance,

$$I_{max} = 0.005/1060 = 2.17 \text{ mA}$$

Thus, the design must always keep the sensor current below 2 mA.

Since the system must be linear, we should set up a linear equation between the sensor resistance and the output voltage. Then it is a matter of determining what circuits will implement the equation.

$$V_{out} = mR_s + V_0$$

We solve for m and V_0 by using the given information,

$$-5 = 280m + V_0$$

$$+5 = 1060m + V_0$$

Subtracting the first equation from the second gives

$$10 = 780m \quad \text{or} \quad m = 0.0128$$

Then, using this in the first equation,

$$-5 = 280(0.0128) + V_0$$

$$V_0 = -8.58$$

So the transfer function equation is

$$V_{out} = 0.0128R_s - 8.58$$

This can be provided by an inverting amplifier with the sensor resistor in the feedback, followed by an inverting summer to get the signs correct. Figure 2.38 shows one possible solution. The fixed input voltage and input resistor of the first op amp have been selected to satisfy the 5-mW maximum power dissipation. This has been done by noting that the current through the sensor is just equal to the current through the input circuit. Thus, by using 1.00 kΩ and 1.00 volt, the current will always be 1 mA and thus less than 2 mA, as required.

As in Example 2.17, trimmers are used in dividers so the fixed voltages can be adjusted to 1.00 and 8.58 volts and thus account for supply voltage differences.

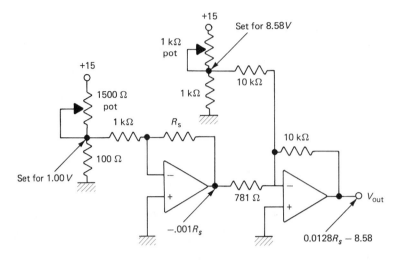

Figure 2.38   One possible solution to Example 2.18.

SUMMARY

The signal conditioning discussed in this chapter relates to the standard techniques employed for providing signal compatibility and measurement in analog systems. The reader was introduced to the basic concepts that form the foundation of such analog conditioning.

To present a complete picture of analog signal conditioning, the following points were considered:

1. The need for analog signal conditioning was reviewed and resolved into the requirements of signal-level changes, linearization, signal conversions, and filtering and impedance matching.

2. Bridge circuits are a common example of a conversion process where a changing resistance is measured either by a current or by a voltage signal. Many modifications of the bridge are used, including *electronic balancing* and techniques of *lead compensation*.

3. The high- and low-pass *RC* filters are passive circuits used to block undesired frequencies from data signals.

4. Operational amplifiers (op amps) are a very special signal conditioning building block around which many special function circuits can be developed. The device was demonstrated in applications involving amplifiers, converters, linearization circuits, integrators, and several other functions.

5. Silicon-controlled rectifiers (SCRs) and TRIACs are semiconductor devices, similar to diodes, that can control large-energy ac or dc signals using low-level imputs.

PROBLEMS

Section 2.3

2.1 A sensor resistance varies from 520 to 2500 Ω. This is used for R_1 in the divider of Figure 2.3, along with $R_2 = 500 \, \Omega$ and $V_s = 10.0$ V. Find (a) the range of the divider voltage V_D and (b) the range of power dissipation by the sensor.

2.2 Prepare graphs of the divider voltage versus transducer resistance for Example 2.1 and Problem 2.1. Does the voltage vary linearly with resistance? Does the voltage increase or decrease with resistance?

2.3 A Wheatstone bridge, as shown in Figure 2.4, nulls with $R_1 = 227 \, \Omega$, $R_2 = 448 \, \Omega$, and $R_3 = 1414 \, \Omega$. Find R_4.

2.4 A sensor with a nominal resistance of 50 Ω is used in a bridge with $R_1 = R_2 = 100$ Ω, $V = 10.0$ V, and $R_3 = 100 \, \Omega$ potentiometer. It is necessary to resolve 0.1 Ω changes of the sensor resistance. (a) At what value of R_3 will the bridge null? (b) What voltage resolution must the null detector possess?

2.5 A bridge circuit is used with a sensor located 100 m away. The bridge is not lead compensated, and the cable to the sensor has a resistance of 0.45 Ω/ft. The bridge nulls with $R_1 = 3400 \, \Omega$, $R_2 = 3445 \, \Omega$, and $R_3 = 1560 \, \Omega$. What is the sensor resistance?

2.6 The bridge of Figure 2.4 has $R_1 = 250 \, \Omega$, $R_3 = 500 \, \Omega$, $R_4 = 340 \, \Omega$, and $V = 1.5$ V. The detector is a galvanometer with $R_G = 150 \, \Omega$. (a) Find the value of R_2 that will null the bridge. (b) Find the offset current that will result if $R_2 = 190 \, \Omega$.

2.7 A current balance bridge, shown in Figure 2.7, has resistances of $R_1 = R_2 = 1 \, k\Omega$, $R_4 = 590 \, \Omega$, $R_5 = 10 \, \Omega$, and $V = 10.0$ V. (a) Find the value of R_3 that nulls the bridge with no current. (b) Find the value of R_3 that balances the bridge with a current of 0.25 mA.

2.8 A potential measurement bridge, such as in Figure 2.8, has $V = 10.0$ V, $R_1 = R_2 = R_3 = 10 \, k\Omega$. Find the unknown potential if the bridge nulls with $R_4 = 9.73 \, k\Omega$.

2.9 An ac Wheatstone bridge with all arms as capacitors nulls when $C_1 = 0.4 \, \mu F$, $C_2 = 0.31 \, \mu F$, and $C_3 = 0.27 \, \mu F$. Find C_4.

2.10 The ac bridge of Figure 2.39 nulls with $R_1 = 1 \, k\Omega$, $R_2 = 2 \, k\Omega$, $R_3 = 100 \, \Omega$, and

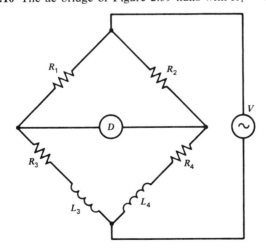

Figure 2.39 Bridge for Problem 2.10.

$L_3 = 250$ mH. (a) Find the values of R_4 and L_4. (b) If the circuit is excited by a 5-V, 1-kHz oscillator, find the offset voltage for $L_4 = 510$ mH. (c) What are the amplitudes of the in-phase and quadrature (90°) components of the offset voltage?

2.11 Develop a low-pass RC filter to attenuate 0.5 MHz noise by 97%. Specify the critical frequency, values of R and C, and the attenuation of a 400-Hz input signal.

2.12 A low-pass RC filter has $f_c = 3.5$ kHz. Find the attenuation of a 1-kHz signal.

2.13 A high-pass RC filter must drive 120 Hz noise down to 1%. Specify the filter critical frequency, values of R and C, and the attenuation of a 30-kHz signal.

2.14 A high-pass filter is found to attenuate a 1-kHz signal by 20 dB. What is the critical frequency?

Section 2.5

2.15 Show how op amps can be used to provide an amplifier with a gain of $+100$ and an input impedance of 1.5 kΩ. Show how this can be done using both inverting and noninverting configurations.

2.16 Specify the components of a differential amplifier with a gain of 22.

2.17 Using an integrator with $RC = 10$ s and any other required amplifiers, develop a voltage ramp generator with 0.5 V/s.

2.18 A control system needs the average of temperature from three locations. Sensors make the temperature information available as voltages, V_1, V_2, and V_3. Develop an op amp circuit that outputs the average of these voltages.

2.19 Use an inverting amplifier, an integrator, and a summing amplifier to develop an output voltage given by

$$V_{out} = 10V_{in} + 4 \int V_{in} \, dt$$

2.20 Develop a voltage-to-current converter that satisfies the requirement $I = 0.0021V_{in}$. If the op amp saturation voltage is ± 12 V and the maximum current delivery is 5 mA, find the maximum load resistance.

Section 2.6

2.21 Figure 2.40 shows an SCR used in a circuit to provide variable electrical power to a resistive load from a high-power square-wave generator. The SCR trigger voltage

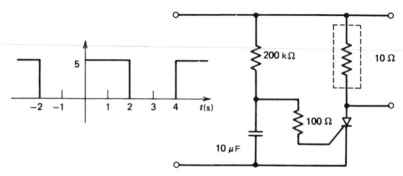

Figure 2.40   Circuit for Problem 2.21.

is $+2.5$ V. Plot the voltage across the capacitor and load as a function of time. The SCR gate-to-cathode resistance goes to zero when triggered. What is the rms power delivery?

2.22 Modify the circuit of Figure 2.40 to use a DIAC with a ± 2-volt ON trigger in place of the 100-Ω resistor. The SCR is replaced with a TRIAC with a 1.5-volt trigger. The input square wave is now bipolar ± 5 volts with the same period. Plot the load voltage versus time. What is the rms power delivery?

Section 2.7

2.23 A bridge circuit has $R_1 = R_2 = R_3 = 120 \ \Omega$ and $V = 10.0$ V. Design a signal conditioning system that provides an output of 0.0 to 5.0 volts as R_4 varies from 120 to 140 Ω. Plot V_{out} versus R_4. Evaluate the linearity.

2.24 Develop signal conditioning for Example 2.1 so an output voltage varies from 0 to 5 volts as the resistance varies from 4 to 12 kΩ.

2.25 Develop signal conditioning for Problem 2.1 so the output voltage varies from 0 to 5 volts as the resistance varies from 520 to 2500 Ω, where 0 V corresponds to 520 Ω.

2.26 A sensor varies from 1 to 5 kΩ. Use this in an op amp circuit to provide a voltage varying from 0 to 8 volts as the resistance changes.

2.27 A process signal varies from 4 to 20 mA. The setpoint is 9.5 mA. Use a current-to-voltage converter and a summing amplifier to get a voltage error signal with a scale factor of 0.5 V/mA.

2.28 Sensor resistance varies from 25 kΩ to 1.5 kΩ as a variable changes from c_{min} to c_{max}. Design a signal conditioning system which provides an output voltage varying from -2 to $+2$ volts as the variable changes from min to max. Power dissipation in the sensor must be kept below 2.5 mW.

2.29 A pressure sensor outputs a voltage varying as 100 mV/psi and has a 2.5-kΩ output impedance. Develop signal conditioning to provide 0 to 2.5 volts as the pressure varies from 50 to 150 psi.

2.30 A system is needed to measure flow, which continuously cycles between 20 to 30 gal/min with a period of 30 seconds. The required output is a voltage varying from -2.5 to $+2.5$ volts for the cycling flow range. The sensor to be used has a transfer function of $1.45 \sqrt{Q}$ volts, where Q is in gal/min, and an output impedance of 2.0 kΩ. Tests show that the output of the sensor has 60 Hz noise of 0.8 volts rms. Design a signal conditioning system, including noise filtering, and evaluate your design as follows: (a) Plot output voltage versus flow and comment on the linearity, and (b) determine the noise on the output as percent FS.

CHAPTER 3

DIGITAL SIGNAL CONDITIONING

INSTRUCTIONAL OBJECTIVES

In this chapter, the basic principles of digital signal processing will be considered with particular emphasis on digital-to-analog (D/A) and analog-to-digital (A/D) conversion techniques and data acquisition systems. After you have read this chapter, you should be able to

1. Develop a Boolean equation for a simple process-control alarm problem.
2. Implement a process-control design of an alarm by digital circuits and comparators.
3. Define the representation of fractional binary and decimal numbers.
4. Diagram a basic DAC and describe its operation.
5. Diagram a successive approximation ADC and describe its operation.
6. Define the conversion resolution of ADCs and DACs.
7. Design a data acquisition system.

3.1 INTRODUCTION

Perhaps the best way to start this chapter is to consider briefly why we are interested in *digital* signal conditioning. An overall survey of electronics applications in industry shows that conversions to digital techniques are occurring rapidly. There are many reasons for this conversion, but two in particular are important. One is the reduction in *uncertainty* when dealing with digitally encoded information compared to analog information. We did not say *accuracy*, we said *uncertainty.* If a system presents analog information, great care must be taken to account for electrical noise influence, drift of amplifier gains, loading effects, and a host of other problems familiar to the analog electronics designer. In a digitally encoded signal, however, a wire carries either a high or low level and is not particularly susceptible to the problems associated with analog processing. Thus, there is an inherent certainty in representing information by digital encoding because of the *isolation* of digital representation from spurious influences. The *accuracy* of this signal in representing the information is discussed later in this chapter (Section 3.3).

A second reason for conversion to digital electronics is the growing desire to use digital computers in the industrial process. The digital computer by nature requires information encoded in digital format before it can be used. The question of the need for digital signal conditioning becomes a question of why computers are so widely used in industry. As discussed in Chapter 11, a few reasons are (1) the ease with which a computer controls a multivariable process-control system, (2) through computer programming, nonlinearities in a sensor output can be linearized, (3) complicated control equations can be solved to determine required control functions, and (4) the ability to microminiaturize rather complex digital processing circuits as integrated circuits (ICs). Finally, the development of microprocessors has brought about quite a transformation of process control to digitally based control systems. With microprocessor-based computers, implementation of computer-based control systems has become practical, and with it, the need for knowledge of digital signal conditioning. This technology not only reduces physical size but both power consumption and failure rate.

With the growing use of computers in process-control technology, it is now clear that any individual trained to work in this field also must be versed in the technology of digital electronics. The basic question is how far such preparation should extend into this related complex field of study. The answer is that a process-control technologist must understand the elements and characteristics of process-control loops. In this context, digital electronics is used as a tool to implement necessary features of process control and therefore should be understood to the extent of knowing how such devices affect the characteristics of the loop. Consider that one does not need to know detailed physics of stretched wires to understand the application of strain gages and to use these devices successfully in process control. Similarly, one does not need to know the internal design of logic gates and microcomputers to use such devices in process control. In this light, the objectives of this chapter have been carefully chosen to provide the reader with sufficient background in digital technology to understand its application to process control.

3.2 REVIEW OF DIGITAL FUNDAMENTALS

A working understanding of the application of digital techniques to process control requires a foundation of basic digital electronics. The design and implementation of control logic systems and microcomputer control systems require a depth of understanding that can only be obtained after several courses devoted to the subject. In this text, we assume a sufficient background so the reader can appreciate the essential features of digital electronic design and its application to process control. A summary of basic digital electronics concepts is presented in Appendix 2.

3.2.1 Digital Information

The use of digital techniques in process control requires that process variable measurements and control information be encoded into a digital form. Digital signals themselves are simply two-state (binary) levels of voltage on a wire, as discussed in Section 1.5.2. We speak, then, of the digital information as a high state (H or **1**) or a low state (L or **0**) on a wire that carries the digital signal.

Digital word

Given the simple binary information that is carried by a digital signal, it is clear that a more complicated arrangement must be used to describe analog information. Generally, this is done by using an assemblage of digital levels to construct a *word*. The individual digital levels are referred to as *bits* of the word. Thus, for example, a 6-bit word consists of six independent digital levels as **101011**, which can be thought of as a six-digit base 2 number. An important consideration, then, is how the process-control information is encoded into this digital word.

Decimal whole numbers

One of the most common schemes for encoding analog data into a digital word is to use the straight counting of decimal (or base 10) and binary representations. The principles of this are reviewed in Appendix 2, together with octal and hexadecimal representations.

Example 3.1

Find the base 10 equivalent of the binary whole number **00100111**.

Solution As in the base 10 system, zeros preceding the first significant digit do not contribute. Thus, the binary number is actually **100111** and so $n = 5$. To find the decimal equivalent, we use Appendix 2 and

$$N_{10} = a_5 2^5 + a_4 2^4 + \cdots a_1 2^1 + a_0 2^0$$

$$N_{10} = (1)2^5 + (0)2^4 + (0)2^3 + (1)2^2 + (1)2^1 + (1)2^0$$

$$N_{10} = 32 + 4 + 2 + 1$$ (A.2.1)

$$N_{10} = 39$$

Example 3.2

Find the binary equivalent of the base 10 number 47.

Solution   Starting the successive division, we get

$$\frac{47}{2} = 23 \text{ with a remainder of } \frac{1}{2} \text{ so that } a_0 = 1$$

then

$$\frac{23}{2} = 11 \text{ with a remainder of } \frac{1}{2} \text{ so that } a_1 = 1$$

then

$$\frac{11}{2} = 5 + \frac{1}{2} \therefore a_2 = 1$$

$$\frac{5}{2} = 2 + \frac{1}{2} \therefore a_3 = 1$$

$$\frac{2}{2} = 1 + 0 \therefore a_4 = 0$$

$$\frac{1}{2} = 0 + \frac{1}{2} \therefore a_5 = 1$$

We find that base 10 number 47 becomes a binary number 101111_2.

The representation of negative numbers in binary format takes on several forms, as discussed in Appendix 2.

Octal and hex numbers

It is quite cumbersome for humans to work with digital words expressed as numbers in the binary representation. For this reason, it has become common to use either the octal (base 8) or hexadecimal (base 16, called hex) representations, which are reviewed in Appendix 2. Octal numbers are conveniently formed from groupings of three binary digits; that is, 000_2 is 0_8 and 111_2 is 7_8. Thus, a binary number like 101011_2 is equivalent to 53_8. Hex numbers are formed easily from groupings of four binary digits; that is, 0000_2 is 0H and 1111_2 is FH. The capital H is used to designate a hex number instead of a subscript of 16. Also recall that the hex counting sequence is 0, 1, 2, 3, 4, 5, 6, 7, 8, 9, A, B, C, D, E, and F to cover the possible states. Because microcomputers must frequently use either 4-bit, 8-bit, or 16-bit words, the hex notation is very commonly used with these machines. In hex, a binary number like 10110110_2 would be written B6H.

3.2.2 Fractional Binary Numbers

Although not as commonly used, it is possible to define a fractional binary number in the same manner as whole numbers using only the **1** and **0** of this counting system. Such numbers, just as in the decimal framework, represent divisions of the counting system to values less than unity. A correlation can be made to decimal numbers in a similar fashion to Equation (A.2.1), as

$$N_{10} = b_1 2^{-1} + b_2 2^{-2} + \cdots b_m 2^{-m} \tag{3.1}$$

where

$$N_{10} = \text{base 10 number less than one}$$
$$b_1 b_2 \cdots b_{m-1} b_m = \text{base 2 number less than one}$$
$$m = \text{number of digits in base 2 number}$$

Example 3.3

Find the base 10 equivalent of the binary number **0.11010₂**.

Solution   This can be found most easily by using

$$N_{10} = b_1 2^{-1} + b_2 2^{-2} + \cdots b_m 2^{-m} \tag{3.1}$$

with

$$m = 5$$

$$N_{10} = (1)2^{-1} + (1)2^{-2} + (0)2^{-3} + (1)2^{-4} + (0)2^{-5}$$

$$N_{10} = \frac{1}{2} + \frac{1}{4} + \frac{1}{16}$$

$$N_{10} = 0.8125_{10}$$

Conversion of a base 10 number that is less than 1 to a binary equivalent employs a procedure where repeated multiplication by 2 is performed. The result of each multiplication will be a fractional part and either a 0 or 1 whole number part, which determines whether that digit is a **0** or **1**. The first multiplication gives the most significant bit b_1, and the last gives either a 0 or 1 for the least significant bit b_m.

Example 3.4

Find the binary, octal, and hex equivalent of 0.3125_{10}.

Solution   Using successive multiplication, we find

$$2(0.3125) = 0.6250 \quad \text{so} \quad b_1 = 0$$

$$2(0.625) = 1.250 \quad \text{so} \quad b_2 = 1$$

$$2(0.25) = 0.5 \quad \text{so} \quad b_3 = 0$$

$$2(0.5) = 1.0 \quad \text{so} \quad b_4 = 1$$

Thus, we find that 0.3125_{10} is equivalent to **0.0101₂**. It can be represented as **0.010100₂** because trailing zeros are not significant in a number less than one and thus as 0.24 octal, because **010₂** $= 2_8$ and **100₂** $= 4_8$. Similarly, this is 0.50H.

3.2.3 Boolean Algebra

In process control, as well as in many other technical disciplines, action is taken on the basis of an evaluation of observations made in the environment. In driving an autombile, for example, we are constantly observing such external factors as traffic, lights, speed limits, pedestrians, street conditions, low-flying aircraft, and such internal factors as how fast we wish to go, where we are going, and many others. We evaluate these factors and take actions predicated on the evaluations. We may see that the light is green, streets good, speed low, no pedestrians or aircraft, we are late, and thus conclude that an action of pressing on the accelerator is required. Then we may observe a parked police radar unit with all other factors the same, negate the aforementioned conclusion, and apply the brake. Many of these parameters can be represented by a *true* or *not true* observation and, in fact, with enough definition, all the observations could be reduced to simple *true* or *false* conditions. When we learn to drive, we are actually setting up internal response to a set of such true/false observations in the environment. In the industrial world, an analogous condition exists relative to the external and internal influences on a manufacturing process, and when we control a process we are in effect teaching a control system response to a set of true/false observations. This teaching may consist of designing electronic circuits that can *logically* evaluate the set of true/false conditions and initiate some appropriate action. To design such an electronic system, we must first be able to mathematically express the inputs, the logical evaluation, and the corresponding outputs. *Boolean algebra* is a mathematical procedure that allows the combinations of true/false conditions in various logical operations so that conclusions can be drawn. For purposes of this text, we do not require expertise in Boolean technique, but an operational familiarity with it that can be applied to a process-control environment.

Before a particular problem in industry can be implemented using digital electronics, it must be analyzed in terms that are amenable to the binary nature of digital techniques. Generally, this is accomplished by stating the problem in the form of a set of true/false-type conditions that must be applied to derive some desired result. These sets of conditions then are stated in the form of one or more Boolean equations. We will see in the next section that a Boolean equation is in a form that is readily implemented with existing digital circuits. The mathematical approach of Boolean algebra allows us to write an analytical expression to represent these stipulations. The fundamentals of Boolean algebra are summarized in Appendix 2.

Let us consider a simple example of how a Boolean equation may result from a practical problem. Consider a mixing tank for which there are three variables of interest: liquid level, pressure, and temperature. The problem is that we must signal an alarm when certain combinations of conditions occur between these variables. Referring to Figure 3.1, we denote level by A, pressure by B, and temperature by C, and assume that setpoint values have been assigned for each variable so that the Boolean variables are either **1** or **0** as the physical quantities are above or below the setpoint values. The alarm will be triggered when the

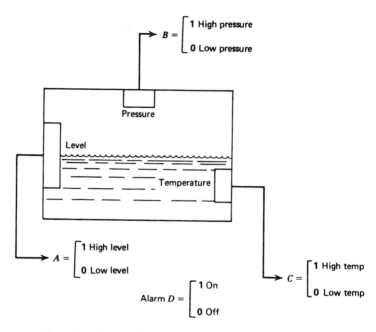

Figure 3.1 System for the application example of Section 3.2.3.

Boolean variable D goes to the logic true state. The alarm conditions are

1. Low level with high pressure
2. High level with high temperature
3. High level with low temperature and high pressure

We now define a Boolean expression with AND operations that will give a $D = 1$ for each condition

1. $D = \overline{A} \cdot B$ will give $D = 1$ for condition (1).
2. $D = A \cdot C$ will give $D = 1$ for condition (2).
3. $D = A \cdot \overline{C} \cdot B$ will give $D = 1$ for condition (3).

The final logic equation results by combining all three conditions so that if any are true, the alarm will sound ($D = 1$). This is accomplished with the OR operation

$$D = \overline{A} \cdot B + A \cdot C + A \cdot \overline{C} \cdot B \tag{3.2}$$

This equation would now form the starting point for a design of electronic digital circuitry that would perform the indicated operations.

3.2.4 Digital Electronics

The electronic building blocks of digital electronics are designed to operate on the binary levels present on digital signal lines. These building blocks are based on families of types of electronic circuits, as discussed in Appendix 2, that have their specific stipulations of power supplies and voltage levels of the **1** and **0** states. The basic structure involves the use of AND/OR logic and NAND/NOR logic to implement Boolean equations.

Example 3.5

Develop a digital circuit using AND/OR gates that implements the equation developed in Section 3.2.3.

Solution   The problem posed in Section 3.2.3 (with Figure 3.1) has a Boolean equation solution of

$$D = \overline{A} \cdot B + A \cdot C + A \cdot \overline{C} \cdot B \tag{3.2}$$

The implementation of this equation using AND/OR gates is shown in Figure 3.2. The AND, OR, and inverter are used in a straightforward implementation of the equation.

Example 3.6

Repeat Example 3.5 using NAND/NOR gates.

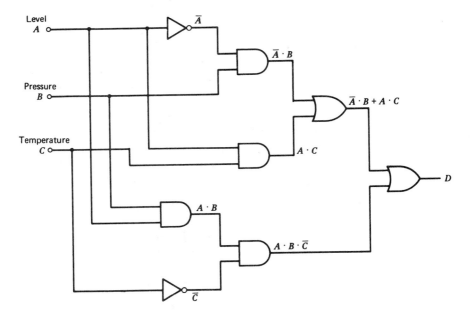

Figure 3.2   Solution for Example 3.5.

Solution   One way to implement the equation in NAND/NOR would be to provide inverters after every gate to, in effect, convert the devices back to AND/OR gates. In this case, the circuit developed would look like Figure 3.2 but with an inverter after every gate. A second approach is to use the Boolean theorems to *reformulate* the equation for better implementation using NAND/NOR logic. For example, if we are to get the desired equation for D as output from a NAND gate, then the inputs must have been

$$\overline{\overline{A} \cdot B + A \cdot C}$$

and

$$\overline{A \cdot B \cdot \overline{\overline{C}}}$$

because NAND between these produces

$$\overline{(\overline{\overline{A} \cdot B + A \cdot C}) \cdot (\overline{A \cdot B \cdot \overline{\overline{C}}})}$$

that, by DeMorgan's theorem, becomes

$$\overline{A} \cdot B + A \cdot C + A \cdot B \cdot \overline{C}$$

that is, the desired output. Working backward from this result allows the circuit to be realized, as shown in Figure 3.3. Many other correct configurations are possible.

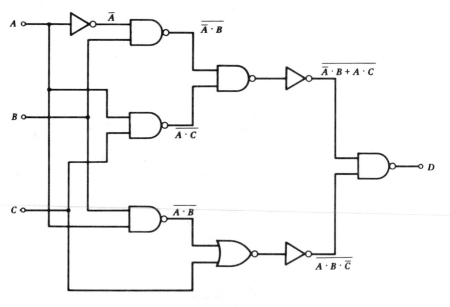

Figure 3.3   Solution for Example 3.6.

3.2.5 Programmable Logic Controllers

The move toward digital logic techniques and computers in industrial control paralleled the development of special controllers called *programmable logic controllers* (PLCs) or simply *programmable controllers* (PCs). These devices are particularly suited to the solution of control problems associated with Boolean equations and binary logic problems in general. They are a computer-based outgrowth of relay sequence controllers. Detailed treatment of this type of control system is given in Chapter 8.

3.2.6 Busses and Tri-State Buffers

Two concepts that deserve special comment are important features of microprocessor-based computers in control applications. These concepts are the use of common busses for data input and output, and the use of tri-state buffers for connections to these busses.

Bus

A data bus is a parallel arrangement of lines connected to a computer, where each line is dedicated to one bit of the data word. Thus, an 8-bit computer may have a data bus with eight lines in parallel. All data input and output to the computer are carried over these lines. Clearly, an important issue is that the data bus lines be free for use by many different devices, including memory, input devices, output devices, and the computer itself. If one of these lines is simply connected to the output of a gate, that line would always represent the state of that gate and be unavailable for other devices.

Tri-state buffers

The problem of allowing many devices to use the bus is solved by using the tri-state buffer. This is a device that acts like a switch. When the switch is closed, its input is placed on the bus line. When the switch is open, the bus line is free for use by other devices. Of course, the switch is actually electronic in nature and very fast in switching between the open and closed state.

The tri-state buffer is described as having three states *on its output*: a logic 1, a logic 0, or a high-impedance state (open circuit). The high-impedance state lasts until an ENABLE command is issued from some other source, usually the computer. The symbol for the tri-state buffer is shown in Figure 3.4. The input state is *not* passed to the output unless the enable line is driven active. Thus, the output of several devices could be connected to the same bus line through tri-state buffers. Only the one whose tri-state buffer is enabled (if any) would have its state impressed on the bus line.

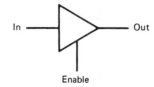

Figure 3.4 Diagram symbol for the tri-state buffer with an active HIGH enable. Active LOW would have the invert symbol on the enable line.

3.3 CONVERTERS

The most important digital tool for the process-control technologist is one that translates digital information to analog and vice versa. Most measurements of process variables are performed by devices that translate information about the variable to an analog electrical signal. To interface this signal with a computer or digital logic circuit, it is necessary first to perform an analog-to-digital (*A/D*) conversion. The specifics of this conversion must be well known so that a unique, known relationship exists between the analog and digital signals. Often, the reverse situation occurs where a digital signal is required to drive an analog device. In this case, a digital-to-analog (*D/A*) converter is required.

3.3.1 Comparators

The most elementary form of communication between the analog and digital is a device (usually an IC) called a *comparator*. This device, which is shown schematically in Figure 3.5, simply compares two analog voltages on its input terminals. Depending on which voltage is larger, the output will be a **1** (high) or **0** (low) digital signal. The comparator is extensively used for alarm signals to computers or digital processing systems. This element also is an integral part of the analog-to-digital and digital-to-analog converter, to be discussed in Section 3.3.2.

A comparator can be constructed from an op amp provided the output is properly clamped to provide the required levels for the logic states (as $+5$ and 0 for transistor–transistor logic [TTL] **1** and **0**). Commercial comparators are designed to have the necessary logic levels on the output.

One of the voltages on the comparator inputs, V_a or V_b in Figure 3.5, will be the variable input and the other a fixed value called a trip, trigger, or reference voltage. The reference value is computed from the specifications of the problem and then applied to the appropriate comparator input terminal, as illustrated in Example 3.7. The reference voltage is provided from a divider from available power supplies.

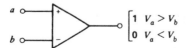

$$\begin{bmatrix} 1 & V_a > V_b \\ 0 & V_a < V_b \end{bmatrix}$$

Figure 3.5 A comparator changes output logic state as a function of the analog input voltages.

Example 3.7

A process-control system specifies that temperature should never exceed 160°C if the pressure also exceeds 10 Pa. Design an alarm system to detect this condition, using temperature and pressure transducers with transfer functions of 2.2 mV/°C and 0.2 V/Pa, respectively.

Solution   The alarm conditions will be a temperature signal of (2.2 mV/°C)(160°C) = 0.352 V coincident with a pressure signal of (0.2 V/Pa)(10 Pa) = 2 volts. The circuit of Figure 3.6 shows how this alarm can be implemented with comparators and one AND gate. The reference voltages could be provided from dividers.

Hysteresis comparator

When using comparators, there is often a problem if the signal voltage has noise or approaches the reference value very slowly. The comparator output may "jiggle" back and forth between high and low as the reference level is reached. This effect is shown in Figure 3.7. Such fluctuation of output may cause problems with the equipment designed to interpret the comparator output signal.

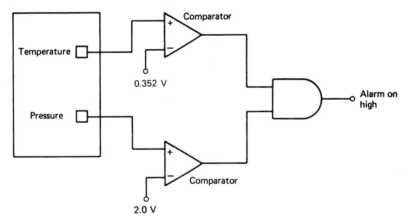

Figure 3.6   Diagram of circuit for Example 3.7.

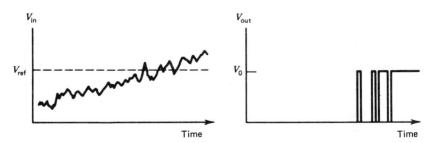

Figure 3.7   Notice how the comparator output "jiggles" as a noisy input passes through the reference voltage value.

   This problem often can be solved by providing a *deadband* or *hysteresis* to the reference level about which output changes occur. Once the comparator has been triggered high, the reference level is automatically reduced so that the signal must fall to some value below the old reference before the comparator goes to the low state.

   There are many ways this hysteresis can be provided, but Figure 3.8 shows one very common technique. Feedback resistor R_f is provided between the output and one of the inputs of the comparator, and that input is separated from the signal by another resistor R. Under the *condition* that $R_f \gg R$, the response of the comparator is shown in Figure 3.8.

   The condition for which the output will go high (V_0) is defined by the condition

$$V_{\text{in}} \geq V_{\text{ref}} \tag{3.3}$$

   Once having been driven high, the condition for the output to drop back to the low ($0V$) state is given by the relation

$$V_{\text{in}} \leq V_{\text{ref}} - (R/R_f)V_0 \tag{3.4}$$

The deadband or hysteresis is given by $(R/R_f)V_0$ and is thus selectable by choice of the resistors, as long as this relation is satisfied. The response of this comparator is shown by the graph of Figure 3.8. The arrows indicate increasing or decreasing input voltage.

Example 3.8

   A transducer converts the liquid level in a tank to voltage according to the transfer function (20 mV/cm). A comparator is supposed to go high (5 V) whenever the level becomes 50 cm. Splashing causes the level to fluctuate by ± 3 cm. Develop a hysteresis comparator to protect against the effects of splashing.

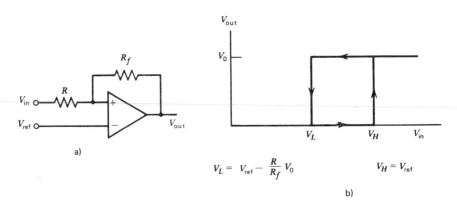

$$V_L = V_{\text{ref}} - \frac{R}{R_f} V_0 \qquad\qquad V_H = V_{\text{ref}}$$

b)

Figure 3.8   The hysteresis comparator has a window within which output changes do not occur.

Solution   The nominal reference for the comparator occurs at 50 cm, which is V_{ref} = (20 mV/cm)(50 cm) = 1 volt. The splashing, however, causes a "noise" of (20 mV/cm)(± 3 cm) = ± 60 mV. This is a total range of 120 mV. We need a deadband of at least 120 mV, but let us make it 150 mV for security. Thus, we have

$$(R/R_f)(5 \text{ V}) = 150 \text{ mV}$$

$$(R/R_f) = 0.03$$

If we make R_f = 100 kΩ, then R = 3 kΩ. Thus, use of these resistors, as shown in Figure 3.8, with a reference of 1 volt will meet the requirement.

3.3.2 Digital-to-Analog Converters (DACs)

A DAC accepts digital information and transforms it into an analog voltage. The digital information is in the form of a binary number with some fixed number of digits. Especially when used in connection with a computer, this binary number is called a binary *word* or *computer* word. The *digits* are called *bits* of the word. Thus, an 8-bit word would be a binary number having eight digits, such as **10110110$_2$**. The D/A converter converts a digital word into an analog voltage by scaling the analog output to be zero when all bits are zero and some maximum value when all bits are one. This can be mathematically represented by treating the binary number that the word represents as a *fractional* number. In this context, the output of the D/A converter can be defined using Equation (3.1) as a *scaling* of some reference voltage.

$$V_{\text{out}} = V_R[b_1 2^{-1} + b_2 2^{-2} + \cdots b_n 2^{-n}] \tag{3.5}$$

where

$$V_{\text{out}} = \text{analog voltage output}$$
$$V_R = \text{reference voltage}$$
$$b_1 b_2 \cdots b_n = n\text{-bit binary word}$$

The minimum V_{out} is zero, and the maximum is determined by the size of the binary word because, with all bits set to one, the decimal equivalent *approaches* V_R as the number of bits increases. Thus, a 4-bit word has a maximum of

$$V_{\text{max}} = V_R[2^{-1} + 2^{-2} + 2^{-3} + 2^{-4}] = 0.9375 V_R$$

and an 8-bit word has a maximum of

$$V_{\text{max}} = V_R[2^{-1} + 2^{-2} + 2^{-3} + 2^{-4} + 2^{-5} + 2^{-6} + 2^{-7} + 2^{-8}] = 0.9961 V_R$$

An alternative equation to Equation (3.5) is often easier to use. This is based on noting that the expression in brackets in Equation (3.5) is really just the fraction of total counting states possible with the n-bits being used. With this recognition, we can write

$$V_{\text{out}} = \frac{N}{2^n} V_R \tag{3.6}$$

where

N = base 10 equivalent of DAC input

Suppose an 8-bit converter with a 5.0-volt reference has an input of $\mathbf{10100111_2}$, or A7H. If this input is converted to base 10, we get $N = 167_{10}$ and $2^8 = 256$. From Equation (3.6), the output of the ADC will be

$$V_{out} = \frac{167}{256} \, 5.0 = 3.2617 \text{ volts}$$

Example 3.9

What is the output voltage of a 10-bit ADC with a 10.0-volt reference if the input is (a) $\mathbf{0010110101_2}$ = 0B5H, (b) 20FH? What input is needed to get a 6.5-volt output?

Solution   Let's use Equation (3.5) for the part (a) and Equation (3.6) for the part (b). Thus, for the 0B5H input we have

$$V_{out} = 10.0[2^{-3} + 2^{-5} + 2^{-6} + 2^{-8} + 2^{-10}]$$

$$= 10.0[0.1767578]$$

$$= 1.767578 \text{ volts}$$

For the (b) case we have 20FH = 527_{10} and $2^{10} = 1024$, so

$$V_{out} = (527/1024) \, 10.0$$

$$= (.514648) \, 10.0$$

$$= 5.14648 \text{ volts}$$

We can use Equation (3.6) to determine what input is needed to get a 6.5-volt output by solving for N,

$$N = 2^n(V_{out}/V_R)$$

$$N = 1024(6.5/10)$$

$$N = 665.6$$

The fact that there is a fractional remainder tells us that we cannot get *exactly* 6.5 volts from the converter. The best we can do is get an output for $N = 665 = 299$H or $666 = 29$AH. The outputs for these two inputs are 6.494 V and 6.504 V, respectively. The only way to get exactly 6.5 volts of output would be to change the value of the reference slightly.

Conversion resolution

The conversion resolution is also a function of the *number* of bits in the word. The *more* bits, the smaller change in analog output for a 1-bit change in binary word and hence the *greater* the resolution. The smallest possible change is simply given by

$$\Delta V_{out} = V_R 2^{-n} \qquad (3.7)$$

where

ΔV_{out} = smallest output change
V_R = reference voltage
n = number of bits in the word

Thus, a 5-bit word D/A converter with a 10-volt reference will provide changes of $\Delta V_{out} = (10)(2^{-5}) = 0.3125$ volts per bit.

Example 3.10

Determine how many bits a D/A converter must have to provide output increments of 0.04 volts or less. The reference is 10 volts.

Solution   One way to find the solution is to continually try word sizes until the resolution falls below 0.04 volts per bit. A more analytical procedure is to use Equation (3.7).

$$\Delta V = 0.04 = (10)(2^{-y})$$

Any *n* larger than the integer part of the exponent of two in this equation will satisfy the requirement. Taking logarithms

$$\log(0.04) = \log[(10)(2^{-y})]$$

$$\log(0.04) = \log(10) - y \log 2$$

$$y = \frac{\log(10) - \log(0.04)}{\log 2}$$

$$y = 7.966$$

Thus, an $n = 8$ will be satisfactory. This can be proved by Equation (3.7).

$$\Delta V_{out} = (10)(2^{-8})$$

$$\Delta V_{out} = \mathbf{0.0390625} \text{ volts}$$

DAC characteristics

For modern applications, most DACs are integrated circuit (IC) assemblies, viewed as a black box having certain input and output characteristics. In Figure 3.9, we see the essential elements of the DAC in terms of required input and output. The associated characteristics can be summarized by reference to this figure.

1. *Digital input*   Typically, a parallel binary word of a number of bits spec-

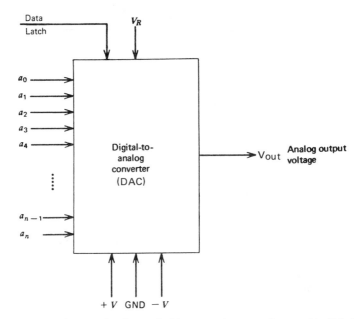

Figure 3.9 Diagram showing typical inputs and outputs for an n-bit digital-to-analog converter (DAC).

ified by the device specification sheet. Typically, TTL logic levels are required unless otherwise noted.

2. *Power supply*   This is bipolar at a level of ± 12 to ± 18 V as required for internal amplifiers. Some DACs operate from a single supply.

3. *Reference supply*   Required to establish the range of output voltages and resolution of the converter. This must be a stable, low-ripple source. In some units, an internal reference is provided.

4. *Output*   A voltage representing the digital input. This voltage changes in steps as the digital input changes by bits with the step determined by Equation (3.7). The actual output may be bipolar if the converter is designed to interpret negative digital inputs.

5. *Offset*   Because the DAC is usually implemented with op amps, there may be the typical output offset voltage with a zero input (see Section 2.4.2). Typically, connections will be provided to facilitate a zeroing of the DAC output with a zero word input.

6. *Data latch*   Many DACs have a data latch built into their inputs. When a logic command is given to latch data, whatever data are on the input bus will be latched into the DAC and the analog output will be updated for that input data. The output will stay at that value until new digital data are latched into the input. In this way the input of the DAC can be connected directly onto the data bus of a computer, but it will only be updated when a latch command is given by the computer.

DAC structure

Generally speaking, a DAC is used as a black box, and no knowledge of the internal workings is required. There is some value, however, to briefly show how such conversion can be implemented. The simplest conversion uses a series of op amps for input for which the gains have been selected to provide an output as given by Equation (3.5). The most common variety, however, uses a *resistive ladder network* to provide the transfer function. This is shown in Figure 3.10 for the case of a 4-bit converter. With the R-$2R$ choice of resistors, it can be shown through network analysis that the output voltage is given by Equations (3.5) or (3.6). The switches are analog electronic switches.

Example 3.11

A control valve has a linear variation of opening as the input voltage varies from 0–10 volts. A microcomputer outputs an 8-bit word to control valve opening using an 8-bit DAC to generate the valve voltage. (a) Find the reference voltage required to

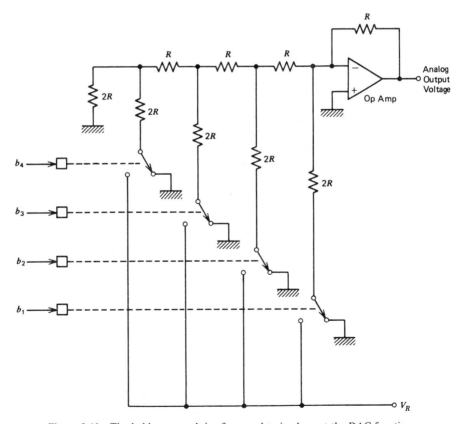

Figure 3.10   The ladder network is often used to implement the DAC function.

obtain a full open valve (10 volts): (b) find the percentage of valve opening for a 1-bit change in the input word.

Solution

(a) The full open valve condition occurs with a 10-volt input. If a 10-volt reference is used, a full digital word **11111111** will not quite give 10 volts, so we use a larger reference. Thus, we have

$$V_{out} = V_R(b_1 2^{-1} + b_2 2^{-2} + \cdots + b_8 2^{-8})$$

$$10 = V_R \left(\frac{1}{2} + \frac{1}{4} + \cdots + \frac{1}{256} \right) \qquad (3.5)$$

$$V_R = \frac{10}{0.9961} = 10.039 \text{ V}$$

(b) The percentage of valve change per step is found first from

$$\Delta V_{out} = V_R 2^{-8}$$

$$\Delta V_{out} = (10.039) \frac{1}{256} \qquad (3.7)$$

$$\Delta V_{out} = 0.0392 \text{ V}$$

Thus,

$$\text{Percent} = \frac{(0.0392)(100)}{10} = \mathbf{0.392\%}$$

Microprocessor compatible DACs

The extensive deployment of microprocessor-based computers in business and industry has prompted the development of special DACs that are designed to be used easily with these systems. The DAC is designed to be connected directly to the data bus, and control lines of the microprocessor or computer.

Data output boards

It is now common and convenient to obtain a printed circuit board which plugs into a common personal computer expansion slot and is a complete data output system. The board has all necessary DACs, address decoding, and bus interface. In most cases, the supplier of the board will also provide elementary software, often written in C, BASIC, or assembly language, as necessary to use the board for data output.

3.3.3 Analog-to-Digital Converters (ADCs)

Although many sensors that provide a direct digital signal output exist and are being developed, most still convert the measured variable into an analog electrical signal. With the growing use of digital logic and computers in process control, it is necessary to employ an ADC to provide a digitally encoded signal for the computer. The transfer function of the ADC can be expressed in the same manner as Equation (3.5) in that some analog voltage is provided as input, and the converter finds a binary number that, when substituted into Equation (3.5), gives the analog input. Thus,

$$V_{\text{in}} \simeq V_R[b_1 2^{-1} + b_2 2^{-2} + \cdots + b_n 2^{-n}] \tag{3.8}$$

where

$$V_{\text{in}} = \text{analog voltage input}$$
$$V_R = \text{reference voltage}$$
$$b_1 b_2 \cdots b_n = n\text{-bit digital outputs}$$

We use an *approximate equality* in this equation because the voltage on the right can change by only a finite step size given by Equation (3.7),

$$\Delta V = V_R 2^{-n} \tag{3.7}$$

Therefore there is an inherent uncertainty of ΔV in any conversion of analog voltage to digital signal. This uncertainty must be taken into account in design applications. If the problem under consideration specifies a certain resolution in analog voltage, then the word size and reference must be selected to provide this in the converted digital number.

Example 3.12

Temperature is measured by a sensor with an output of 0.02 volts/°C. Determine the required ADC reference and word size to measure 0–100°C with 0.1°C resolution.

Solution   At the maximum temperature of 100°C, the voltage output is

$$(0.02 \text{ V/°C})(100°C) = 2 \text{ V}$$

so a 2-V reference is used.
    A change of 0.1°C results in a voltage change of

$$(0.1°C)(0.02 \text{ V/°C}) = 2 \text{ mV}$$

so we need a word size where

$$0.002 \text{ V} = (2)(2^{-y})$$

Choose a size n that is the integer part of y plus one. Thus, solving with logarithms, we find

$$y = \frac{\log(2) - \log(0.002)}{\log 2}$$

$$y = 9.996 \approx 10$$

so a **10**-bit word is required for this resolution. A 10-bit word has a resolution of

$$V = (2)(2^{-10})$$

$$V = 0.00195 \text{ volts}$$

which is better than the minimum required resolution of 2 mV.

Example 3.13

Find the digital word that results from a 3.127-volt input to a 5-bit ADC with a 5-volt reference.

Solution   The relationship between input and output is

$$V_{in} \simeq V_R[a_1 2^{-1} + a_2 2^{-2} + a_3 2^{-3} + a_4 2^{-4} + a_5 2^{-5}] \tag{3.8}$$

Thus, we are to encode a fractional number of V_{in}/V_R or

$$a_1 2^{-1} + a_2 2^{-2} \cdots + a_5 2^{-5} = \frac{3.127}{5} = 0.6254$$

Using the method of successive multiplication defined in Section 3.2.2, we find

$$0.6254\ (2) = 1.2508 \therefore a_1 = 1$$

$$0.2508\ (2) = 0.5016 \therefore a_4 = 0$$

$$0.5016\ (2) = 1.0032 \therefore a_3 = 1$$

$$0.0032\ (2) = 0.0064 \therefore a_4 = 0$$

$$0.0064\ (2) = 0.0128 \therefore a_5 = 0$$

so that the output is $\mathbf{10100_2}$.

It is possible to write Equation (3.8) in a simpler fashion by using the fraction of counting states, as was done for the DAC. In this case, we can write an expression for the actual output N expressed as a base-10 integer.

$$\text{INT}(N) = \frac{V_{in}}{V_R} 2^n \tag{3.9}$$

where

INT(N) = the integer part of N

The integer part of N is then converted to hex and/or binary to determine the actual output of the ADC. In the previous example, we would have

$$\text{INT}(N) = (3.127/5.0)2^5$$

$$= \text{INT}(20.0128) = 20_{10}$$

or an output of 14H = $\mathbf{10100_2}$ as already found.

Example 3.14

The input to a 10-bit ADC with a 2.500-volt reference is 1.45 volts. What is the hex output? Suppose the output was found to be 1B4H. What is the voltage input?

Solution   We will use Equation (3.9) to find the solution to these questions. For the first part, we can form the expression

$$\text{INT}(N) = (1.45/2.5)2^{10}$$

$$= 593.92$$

$$= 593$$

$$= 251\text{H}$$

So the output of the ADC is 251H for a 1.45-volt input. To get the voltage input for a 1B4H output, we solve Equation (3.9) for the voltage

$$V_{\text{in}} = (\text{INT}(N)/2^n)V_R$$

A conversion yields 1B4H = 436_{10},

$$V_{\text{in}} = (436/1024)2.50$$

$$= 1.06445 \text{ volts}$$

However, it is important to realize that any voltage from this to 1.06445 + 2.5/1024 = 1.06689 will give an output of 1B4H. So the correct answer to the question is that the input voltage lies in the range 1.06445 to 1.06689 volts.

Bipolar operation

A bipolar ADC is one which accepts bipolar input voltage for conversion into an appropriate digital output. The most common bipolar ADCs provide an output which is called *offset binary*. This simply means that the normal output is shifted by half the scale so that all zeros corresponds to the negative maximum input voltage instead of 0. In equation form, the relation would be written from Equation (3.9) as

$$\text{INT}(N) = \frac{1}{V_R}[V_{\text{in}} + V_R/2]2^n \tag{3.10}$$

From this equation, you can see that if $V_{\text{in}} = -V_R/2$, then the output is zero, INT(N) = 0. If $V_{\text{in}} = 0$, then the output would be half of 2^n. The output will be the maximum count when the input is $V_R/2 - V_R 2^n$. For example, for 8 bits with a 10.0-volt reference, the step size is $\Delta V_{\text{in}} = (10)2^8 \approx 0.039$ volts. Looking at the possible states, we would have

$$V_{\text{in}} = -5.000 \qquad N = \mathbf{00000000_2}$$

$$V_{\text{in}} = -4.961 \qquad N = \mathbf{00000001_2}$$

$$\text{etc.}$$

$$V_{\text{in}} = -0.039 \qquad N = \mathbf{01111111_2}$$

$$V_{\text{in}} = 0.000 \qquad N = \mathbf{10000000_2}$$

$$V_{in} = +0.039 \qquad N = 10000001_2$$

etc.

$$V_{in} = +4.961 \qquad N = 11111111_2$$

There is an asymmetry to the result so that the converter cannot represent the full range from minus to plus $V_R/2$.

Example 3.15

What is the hex and binary output of a bipolar 12-bit ADC with a 5.00-volt reference for inputs of -0.85 volts and $+1.5$ volts? What input voltage would cause an output of 72H?

Solution   Using Equation (3.10), we get

$$INT(N) = (1/5.00)[-0.85 + 2.50]2^8$$

$$= (1.65/5)256 = 84.48$$

$$= 84_{10}$$

$$= 54H = 01010100_2$$

and

$$INT(N) = (1/5.00)[1.5 + 2.50]256$$

$$= (4/5)256 = 204.8$$

$$= 204_{10}$$

$$= CCH = 11001100_2$$

To get an output of 72H, we solve Equation (3.10) for V_{in}

$$V_{in} = (INT(N)/2^n)V_R - V_R/2$$

$$= (114/256)5.00 - 2.50$$

$$= -0.2734 \text{ V}$$

But of course the actual answer is any voltage between -0.2734 volts and $(-0.2734 + 5/256) = -0.2539$ volts.

A/D structure

Most ADCs are available in the form of integrated circuit (IC) assemblies that can be used as a black box in applications. To fully appreciate the characteristics of these devices, however, it is valuable to examine the standard techniques employed to perform the conversions. There are two methods in use that represent very different approaches to the conversion problem.

Parallel-feedback ADC

The parallel-feedback A/D converter employs a feedback system to perform the conversion, as shown in Figure 3.11. Essentially, a *comparator* is used to compare the input voltage V_x to a feedback voltage V_F that comes from a DAC as shown. The comparator output signal drives a logic network that steps the digital output (and hence DAC input) until the comparator indicates the two signals are the *same* within the resolution of the converter. The most popular parallel-feedback converter is the *successive approximation* device. The logic circuitry is such that it successively sets and tests each bit, starting with the most significant bit of the word. We start with all bits zero. Thus, the first operation will be to set $b_1 = 1$ and test $V_F = V_R 2^{-1}$ against V_x through the comparator.

If V_x is greater, then b_1 will be **1**; b_2 is set to **1** and a test is made of V_x versus $V_V = V_R(2^{-1} + 2^{-2})$, and so on.

If V_x is less than $V_R 2^{-1}$, then b_1 is reset to zero; b_2 is set to **1**, and a test is made for V_x versus $V_R 2^{-2}$. This process is repeated to the least significant bit of the word. The operation can be illustrated best through an example.

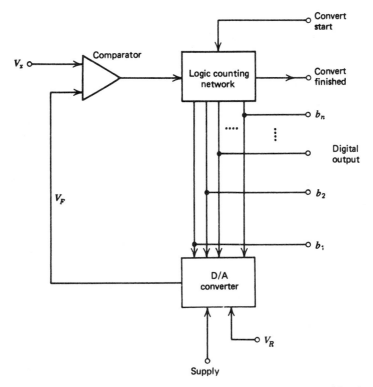

Figure 3.11   The successive approximation ADC is very common and involves the use of an internal DAC.

Example 3.16

Find the successive approximation ADC output for a 4-bit converter to a 3.217-volt input if the reference is 5 volts.

Solution   Following the procedure outlined, we have the following operations: Let $V_x = 3.217$, then

(1) Set $b_1 = 1$ $V_F = 5(2^{-1}) = 2.5$ volts
   $V_x > 2.5$ leave $b_1 = 1$
(2) Set $b_2 = 1$ $V_F = 2.5 + 5(2^{-2}) = 3.75$
   $V_x < 3.75$ reset $b_2 = 0$
(3) Set $b_3 = 1$ $V_F = 2.5 + 5(2^{-3}) = 3.125$
   $V_x > 3.125$ leave $b_3 = 1$
(4) Set $b_4 = 1$ $V_F = 3.125 + 5(2^{-4})$
   $V_x < 3.4375$ reset $b_4 = 0$

By this procedure, we find the output is a binary word of 1010_2.

In addition to the analog input, digital output, power supply, and reference inputs, most A/D converters have a *convert start logic input* and a *conversion complete logic output*, as shown in Figure 3.11.

Ramp A/D

The ramp-type A/D converters essentially compare the input voltage against a linearly increasing ramp voltage. A binary counter is activated that counts ramp steps until the ramp voltage equals the input. The output of the counter is then the digital word representing conversion of the analog input. The ramp itself is typically generated by an op amp integrator circuit, discussed in Section 2.5.8.

Dual slope ramp A/D

This ADC is the most common type of ramp converter. A simplified diagram of this device is shown in Figure 3.12. The principle of operation is based on allowing the input signal to drive the integrator for a fixed time T_1, thus generating an output of

$$V_1 = \frac{1}{RC} \int V_x \, dt \qquad (3.11)$$

or, because V_x is constant,

$$V_1 = \frac{1}{RC} T_1 V_x \qquad (3.12)$$

After time T_1, the input of the integrator is electronically switched to the reference supply. The comparator then sees an input voltage that decreases from V_1 as

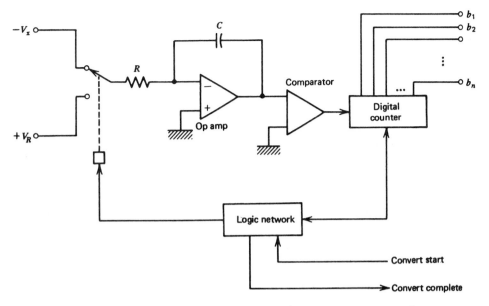

Figure 3.12   A dual slope ADC uses an op amp integrator, comparator, and associated digital circuits.

$$V_2 = V_1 - \frac{1}{RC} \int V_R \, dt \qquad (3.13)$$

or, because V_R is constant and V_1 is given from Equation (3.12),

$$V_2 = \frac{1}{RC} T_1 V_x - \frac{1}{RC} t V_R \qquad (3.14)$$

A counter is activated at time T_1 and counts until the comparator indicates $V_2 = 0$, at which time t_x [Equation (3.14)] indicates that V_x will be

$$V_x = \frac{t_x}{T_1} V_R \qquad (3.15)$$

Thus, the counter time t_x is linearly related to V_x and is independent of the integrator characteristics, that is, R and C. This procedure is shown in the timing diagram of Figure 3.13. Conversion *start* and *complete* digital signals are also used in these devices, and (in many cases) internal or external references may be used.

Example 3.17

A dual slope ADC as shown in Figure 3.12 has $R = 100$ kΩ and $C = 0.01$ μF. The reference is 10 volts, and the fixed integration time is 10 ms. Find the conversion time for a 6.8-volt input.

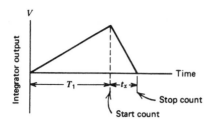

Figure 3.13 The dual slope ADC counts the time required for a zero crossing of the integrator output from a known, fixed input.

Solution We find the voltage after an integration time of 10 ms as

$$V_1 = \frac{1}{RC} T_1 V_x$$

$$V_1 = \frac{(10 \text{ ms})(6.8 \text{ V})}{(100 \text{ k}\Omega)(0.01 \text{ } \mu\text{F})} \tag{3.12}$$

$$V_1 = 6.8 \text{ volts}$$

Then we find the time required to integrate this to zero as $V_2 = 0$ in

$$V_2 = \frac{T_1 V_x}{RC} - \frac{t_x}{RC} V_R \tag{3.14}$$

thus,

$$t_x = \frac{T_1 V_x}{V_R}$$

$$t_x = \frac{(10 \text{ ms})(6.8 \text{ V})}{10 \text{ V}}$$

$$t_x = 6.8 \text{ ms}$$

The total conversion time is then 10 ms + 6.8 ms = **16.8 ms**.

General characteristics

Numerous general features may be indicated for successive approximation A/D converters, important in applications:

1. *Input* Generally an analog voltage level. The most common levels are 0–10 and 0–5 volts or −10 to +10 if bipolar conversion is possible. In some cases, the level is determined by an externally supplied reference.

2. *Output* A parallel or serial binary word that is an encoding of the analog input.

3. *Reference* A stable, low ripple source against which the conversion is performed.

4. *Power supplies* Generally, a bipolar ±12- to ±18-V supply is required for the analog amplifiers and comparators and a +5-V supply for the digital circuitry.

5. *Digital signals*   Most ADCs require an input logic high on a given line to initiate the conversion process. When conversion is complete, the ADC will usually provide a high level on another line as an indicator to following equipment of that status.

6. *Conversion time*   The ADC must sequence through a set of operations before it can find the digital output as described. For this reason, an important part of the specification is the time required for the conversion. The time is typically 10–100 μs, depending on the number of bits and the design of the converter.

Example 3.18

A measurement of temperature using a sensor that outputs 6.5 mV/°C must measure to 100°C. A 6-bit ADC with a 10-volt reference is used. (a) Develop a circuit to interface the sensor and the ADC; (b) find the temperature resolution.

Solution   To measure to 100°C means the sensor output at 100°C will be

$$(6.5 \text{ mV/°C})(100°C) = 0.65 \text{ volts}$$

(a) The interface circuit must provide a gain so that at 100°C the ADC output is **111111**. The input voltage that will provide this output is found from

$$V_x = V_R(a_1 2^{-1} + a_2 2^{-2} + \cdots + a_6 2^{-6})$$

$$V_x = 10 \left(\frac{1}{2} + \frac{1}{4} + \cdots + \frac{1}{64} \right)$$

$$V_x = \mathbf{9.84375 \text{ V}}$$

Thus, the required gain must provide this voltage when the temperature is 100°C.

$$\text{Gain} = \frac{9.84375}{0.65}$$

$$= \mathbf{15.14}$$

The op amp circuit of Figure 3.14 will provide this gain.

(b) The temperature resolution can be found by working backward from the least significant bit (LSB) voltage change of the ADC.

$$\Delta V = V_R 2^{-n} \tag{3.7}$$

$$\Delta V = (10)(2^{-6}) = 0.15625 \text{ V}$$

Working back through the amplifier, this corresponds to a sensor change of

$$\Delta V_T = \frac{0.15625}{15.14} = 0.01032 \text{ V}$$

or a temperature of

$$\Delta T = \frac{0.01032 \text{ V}}{0.0065 \text{ V/°C}} = \mathbf{1.59°C}$$

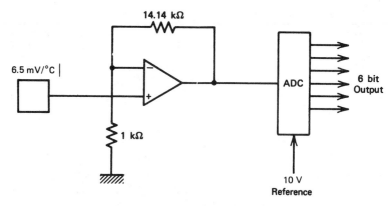

Figure 3.14 Circuit for Example 3.18.

Microprocessor compatible ADCs

Just as with DACs, a whole line of ADCs have been developed that interface easily with microprocessor-based computers. The ADCs have built-in tri-state outputs so that they can be connected directly to the data bus of the computer. Data from the ADC is only placed on the data bus lines when the computer issues an appropriate enable command (often called a READ). Figure 3.15 shows how the ADC appears when connected to the environment of the microprocessor-based computer. The ADC appears much the same as memory. In some cases an ADC input is actually taken by the computer using a memory-read instruction.

The decoding circuitry is necessary to provide the start-convert command, to input the convert-complete response from the ADC, and to issue the tri-state enable back to the ADC.

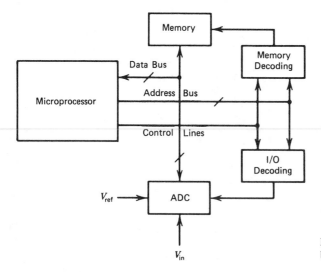

Figure 3.15 A typical microprocessor-based computer with ADC interface.

Data input boards

Just as in the case of DACs, there are now many types of data input boards for personal computers. These boards, which plug into expansion slots, contain the ADC(s), addressing and decoding, bus interface, and the reference for the converter. Often, many channels of input can be taken from a single board. In most cases, the manufacturer will provide user manuals and software in a common language for use in data acquisition systems.

Sample and hold

The ADC requires a certain length of time to determine the appropriate digital output from an analog input. For different ADCs, this time varies from a few microseconds to several milliseconds. It is clear that the input voltage should not change during the conversion.

For slowly varying signals and converters performing the conversion in a few microseconds, this is not a problem. In some cases, however, the signal change rate and conversion rate are of the same order. In these cases it is necessary to "freeze" the input signal during conversion. This can be done using a special op amp circuit called a *sample and hold*.

The circuit in Figure 3.16 shows the essential elements of a sample and hold. The field effect transistor (FET) on the input operates as a simple switch. When it is in the "closed" state, the voltage on the capacitor follows the input voltage. The capacitor voltage is coupled to the output through the very high input impedance voltage follower.

When the FET is driven to the "open" state, the voltage on the capacitor will hold the last value before the open state of the FET occurred. It is possible now to measure this value using the voltage follower, because its high input impedance will not appreciably discharge the capacitor.

This circuit is placed between the signal source and the ADC. It is normally in the closed or sample state. When a conversion is to be made, the FET is first driven to the open or hold state. Then the converter is started and conversion proceeds. When the data acquisition is complete, the FET is driven back to the sample state.

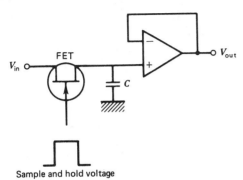

Sample and hold voltage

Figure 3.16 The principal elements of a sample and hold circuit.

3.4 DATA ACQUISITION SYSTEMS

A digital computer can make a tremendous number of calculations in a second, because the typical time required to execute one instruction may be only a few microseconds. For example, a typical microprocessor can add two 8-bit binary numbers in 2 μs. By contrast, most process-control installations involve process variable variations with a time scale on the order of minutes. For this reason and others discussed in Chapter 11, the efficient use of computers in process control means that a single computer may control many variables. To do this, the computer periodically *samples* the value of a variable, evaluates it according to programmed control operations, and outputs an appropriate controlling signal to the final control element. Under the program control, the computer then selects another of the controlled variables, samples, evaluates, and outputs, and so on for all the loops under its control. Getting a sample of a real-world number into the computer is not easy. It requires a combination of hardware and software (programs) to enable the computer to read in a number that may represent some process variable, such as temperature, pressure, and so on. The overall process of doing this, and its reverse of output, is called by the general term *interface*. One can take an ADC and any necessary amplifiers and write the programs required to put together an interface to some computer for a process application. If the computer is to control many loops, we would need such a system for each variable to be inputted. Instead, for input we can use a data acquisition system (DAS) that allows sampled variables from many sources to be inputted to the computer with appropriate programming.

3.4.1 Data Acquisition System (DAS)

There are many different types of data acquisition systems, but it is possible to generalize the essential elements as shown in Figure 3.17. The following paragraphs present general descriptions of each block of the DAS. Most data acquisition systems are available as small modules containing the circuits shown in Figure 3.17. In general, the module accepts a number of analog inputs, called *channels*, as either differential voltage signals (two-wire) or single-ended voltage signals (referred to ground). Typically, a system may have 8 differential input channels or 16 single-ended input channels. The computer can then select any one of the channels under program control for input of data in that channel.

Address decoder

This part of the DAS accepts an input from the computer via the address lines (16 bits for a typical 8-bit microprocessor) that serve to select a particular analog channel to be sampled. The module is often designed so that the association of a particular channel and a computer address word can be selected by the user. In some cases, this is done by making the module channel addresses appear to

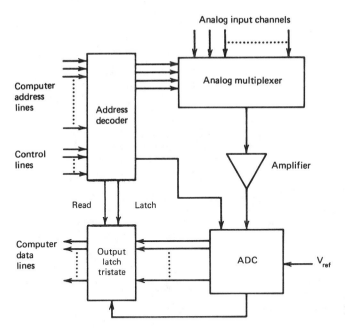

Figure 3.17 Data acquisition system.

the computer as addresses of memory locations. Thus, if the computer executes a command to fetch the contents of some memory location, it may actually be selecting some analog input channel. In other words, the selecting of an input channel is equivalent to reading the contents of a memory location. In other systems, a binary code is sent from the computer through special input/output devices to select an analog channel and input the data on that channel. In such cases, the selection of a channel is done by what may be called a *device select* code.

Analog multiplexer

This element of the DAS is essentially a solid-state switch that takes the decoded address signal and selects the data for the selected channel by closing a switch connected to that analog input line.

As shown in Figure 3.18 for a single-ended system, the multiplexer accepts an input from the address decoder and uses this to close the appropriate switch to allow that channel signal to be passed to the next stage of the DAS. Figure 3.18 shows that channel 2 has been selected, which would probably have been selected by a **10** on the input lines. In this sense, **00** then would select channel 0, **01** channel 1, **10** channel 2, and **11** channel 3. Thus, the address decoder must convert the computer address line to one of these four possibilities when the DAS has been addressed by the computer. The actual switch elements are usually field effect transistors (FETs) that have an "on" resistance of a few hundred ohms and an "off" resistance of hundreds to thousands of megohms.

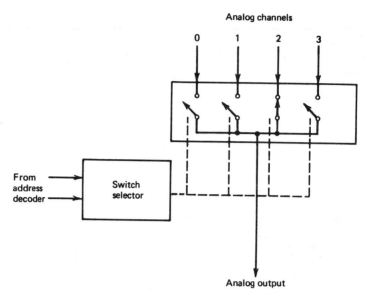

Figure 3.18   Four-channel analog multiplexer.

Amplifier

Most data acquisition systems include a variable gain amplifier that allows the user to compensate for the input level of the signals. The integral ADC usually is designed to work from a definite unipolar or bipolar input range so that input levels must be adjusted to fall within this range. Thus, if the ADC signal input must be in the range of 0 to 5 volts, the amplifier is given a gain to assure that the inputs will be within this range. If there is a great difference between the various input signal levels, some signal conditioning may be required prior to application of the signal to the DAS.

ADC

Of course, an important part of the DAS is an analog-to-digital converter. The converter will accept voltages that span a specific range as provided by the preceding signal conditioning. The converter usually can be configured to accept either unipolar or bipolar inputs. Features such as offset adjustment and full-scale adjustment are common.

3.4.2 Application Notes

There are numerous factors that must be considered when a DAS is to be used. The following paragraphs discuss several of these factors.

Sample and hold

When using the DAS, account must be made for the fact that the signals on the input channels may be changing very rapidly. If the changes are fast enough so that the signal varies during the conversion time, a sample and hold must be used on that channel to hold the value during conversion. This adds to the complexity of the software because account must be taken of commands to the sample and hold module.

Compatibility to computer

In many cases, a data module is designed to work with only one model or type of computer. This is particularly true of microprocessor-based computers where architecture varies greatly between microprocessor models. Thus, it is necessary to select a data module that is compatible with the input/output characteristics of the computer.

Hardware programming

Most data modules give the user many options for use in input/output operations. These options include unipolar/bipolar operation, address selection, amplifier gain, differential/single-ended operation, and others. They are typically selected by wired jumpers between pins of the module or by the attachment of resistors as specified in the specification sheet of the module.

Software programming

Another important aspect of input/output interfacing is the software routines that will use the data modules. These routines must be compatible with the hardware programming and the other characteristics of the module. The programs may include delays waiting for the ADC to complete conversion, for example. This aspect is discussed further in Chapter 11.

Overall response time

A data acquisition system does not provide the digital conversion of data on a selected channel the instant the selection occurs. Rather, there is a delay while the multiplexer acquires the system channel, while the amplifier settles to the value on the channel, and while the ADC performs the conversion operations. This time will be important for a determination of the maximum sampling rate from the DAS. The time may run from tens of microseconds to hundreds of microseconds, depending on the number of bits converted, gains of the amplifiers, and signal-switching speed.

SUMMARY

This chapter provides a digital electronics background to make the reader conversant with the elements of digital signal processing and able to perform simple analysis and design as associated with process control.

1. The use of digital words enables the encoding of analog information into a digital format.

2. It is possible to encode fractional decimal numbers as binary and vice versa using

$$N_{10} = b_1 2^{-1} + b_2 2^{-2} + \cdots + b_m 2^{-m} \qquad (3.1)$$

3. Boolean algebra techniques can be applied to the development of process alarms and elementary control functions.

4. Digital electronic gates and comparators allow the implementation of process Boolean equations.

5. DACs are used to convert digital words into analog numbers using a fractional number representative. The resolution is

$$\Delta V = V_R 2^{-n} \qquad (3.7)$$

6. An ADC of the successive approximations type determines an output digital word for an input analog voltage in as many steps as bits to the word.

7. The dual slope ADC converts analog-to-digital information by a combination of integration and time counting.

8. The data acquisition system (DAS) is a modular device that interfaces many analog signals to a computer. Signal address decoding, multiplexing, and ADC operations are included in the device.

PROBLEMS

Section 3.2

3.1 Convert the following binary numbers into decimal, octal, and hex: (a) 1010_2, (b) 111011_2, (c) 010110_2.

3.2 Convert the following binary numbers into decimal, octal, and hex: (a) 1011010_2, (b) 0.1101_2, (c) 1011.0110_2.

3.3 Convert the following decimal numbers into binary, octal, and hex: (a) 21_{10}, (b) 630_{10}, (c) 427_{10}.

3.4 Convert 27.156_{10} into a binary number with the fractional binary part expressed in 6-bits. What actual decimal does this binary equal?

3.5 Find the 2s complement of (a) 1011_2, (b) 10101100_2.

3.6 Prove by a table of values that $\overline{A \cdot B} = \overline{A} + \overline{B}$ (DeMorgan's theorem).

3.7 Show that the Boolean equation $A \cdot B + A \cdot \overline{A \cdot B})$ reduces to A.

3.8 A process control's moving speed, load weight, and rate of loading in a conveyor system. The variables are provided as high (**1**) and low (**0**) levels for digital control. An alarm should be initiated whenever any of the following occur:

a. Speed is low; both weight and loading rate are high.

b. Speed is high; loading rate is low.

Find a Boolean equation describing the required alarm output. Let the variables be S for speed, W for weight, and R for loading rate.

3.9 Implement Problem 3.8 with (a) AND/OR logic and (b) NAND/NOR logic.

Section 3.3

3.10 A sensor provides temperature data as 360 μV/°C. Develop a comparator circuit that goes high when the temperature reaches 530°C.

3.11 A light level is to trigger a comparator high (5 V) when the intensity reaches 30 W/m^2. Intensity is converted to voltage according to a transfer function of 0.04 V/(W/m^2). Noise is found to contribute ±1.6 W/m^2 of intensity fluctuations. Develop a hysteresis comparator to provide the required output and immunity for noise.

3.12 A 6-bit DAC has an input of **100101$_2$** and uses a 10.0-volt reference. (a) Find the output voltage produced. (b) Specify the conversion resolution.

3.13 A 4-bit DAC must have an 8.00-volt output when all inputs are high. Find the required reference.

3.14 An 8-bit DAC with a 5.00-volt reference connects to a light source with an intensity given by $I_L = 45V^{(3/2)}$ W/m^2.

a. What is the range of intensity which can be produced?

b. What intensities are produced by digital inputs of 1BH, 7AH, 9FH, and E5H?

c. Plot the intensity versus hex input, and comment on linearity.

3.15 An 8-bit ADC with a 10.0-volt reference has an input of 3.797 volts. Find the digital output word. What range of input voltages would produce this same output? Suppose the output of the ADC is **10110111$_2$**. What is the input voltage?

3.16 An ADC that will encode pressure data is required. The input signal is 666.6 mV/psi.

a. If a resolution of 0.5 psi is required, find the number of bits necessary for the ADC. The reference is 10.0 volts.

b. Find the maximum measurable pressure.

3.17 A bipolar ADC has 10 bits and a 10.00-volt reference. What output is produced by inputs of -4.3 V, -0.66 volts, $+2.4$ volts, and $+4.8$ volts? What is the input voltage if the output is 30BH?

3.18 A sample and hold circuit like the one shown in Figure 3.16 has $C = 0.47$ μF, and the ON resistance of the FET is 75 Ω. For what signal frequency is the sampling capacitor voltage down 3 dB from the signal voltage? How does this limit the application of the sample hold?

Section 3.4

3.19 A DAS has an ADC of 8 bits with a 0- to 2.5-volt range of input for 00H to FFH output (i.e., FEH to FFH occurs at 2.50 volts). Inputs are temperature from 20 to 100°C scaled at 40 mV/°C, pressure from 1 to 100 psi scaled at 100 mV/psi, and flow from 30 to 90 gal/min scaled at 150 mV/(gal/min). Develop a block diagram of the required signal conditioning so that each of these can be connected to the DAS and so that the indicated variable range corresponds to 00H to FFH. Specify the resolution of each in terms of the change of each variable that corresponds to an LSB change of the ADC output.

3.20 Design op amp circuits that will provide the signal conditioning specified in Problem 3.19.

3.21 A data acquisition system has eight input channels to be sampled continuously and sequentially. The multiplexer can select and settle on a channel 3.1 μs, the ADC converts in 33 μs, and the computer processes a single channel of data in 450 μs. What is the minimum time between samples for a particular channel?

3.22 Flow is to be measured and inputted to a computer. For maximum resolution, we want the minimum flow, 30 m³/hr, to correspond to 00H, and the maximum, 60 m³/hr, to correspond to FFH from an 8-bit converter. The flow is measured as a voltage given by $V = 0.0022Q^2$, where Q is the flow in m³/hr. Develop the signal conditioning and ADC reference that will provide this specification. What is the resolution of flow when the flow is at the lower limit and at the upper limit? Explain why there is a difference.

3.23 An 8-bit ADC has an 8.00-V reference.
   a. Find the output for inputs of 3.4 volts and 6.7 volts.
   b. What range of inputs could have caused the output to become B7H?

3.24 Tests show that the output of a position sensor is 12 mV/mm, but there is 60-Hz noise on the output of a constant 5-mV rms. The sensor output impedance is 2.5 kΩ. This sensor is to measure work-piece motion, which oscillates between −10 mm and +10 mm with a period of 1.5 seconds. The position is to be interfaced to a 12-bit bipolar, offset binary ADC with a 5.000-volt reference. Design the interface system such that −10 mm corresponds to 000H and +10 mm corresponds to FFFH.
   a. With no filter, determine how many bits are being toggled by noise (i.e., are lost to any real data).
   b. Introduce a filter, reevaluate the effect of noise on the ADC output, and determine how many bits represent real data.

3.25 A sensor linearly changes resistance from 2.35 to 3.57 kΩ over a range of some measured variable. The measurement must have a resolution of at least 1.25 Ω and be interfaced to a computer. Design the signal conditioning and specify the characteristics of the required ADC.

CHAPTER 4

THERMAL SENSORS

INSTRUCTIONAL OBJECTIVES

The objectives of this and the following two chapters stress an understanding beyond that required for simple application of process-control sensors. After you have read this chapter, you should be able to

1. Define thermal energy, the relation of temperature scales to thermal energy, and temperature scale calibrations.
2. Transform a temperature reading between the Kelvin, Rankine, Celsius, and Fahrenheit temperature scales.
3. Design the application of an RTD temperature sensor to specific problems in temperature measurement.
4. Design the application of a thermistor to specific temperature measurement problems.
5. Design the application of a thermocouple to specific temperature measurement problems.
6. Explain the operation of a bimetal strip for temperature measurement.
7. Explain the operation of a gas thermometer and a vapor pressure thermometer.

4.1 INTRODUCTION

Process control is a term used to describe any condition, natural or artificial, by which a physical quantity is regulated. There is no more widespread evidence of such control than that associated with temperature and other thermal phenomena.

In our natural surroundings, some of the most remarkable techniques of temperature regulation are found in the bodily functions of living creatures. On the artificial side, humans have been vitally concerned with temperature control since the first fires were struck for warmth. Industrial temperature regulation has always been of paramount importance and becomes even more so with the advance of technology. In this chapter we will be concerned first with developing an understanding of the principles of thermal energy and temperature, and then developing a working knowledge of the various thermal sensors employed for temperature measurement.

4.2 DEFINITION OF TEMPERATURE

The materials that surround us and, indeed, of which we are constructed are composed of assemblages of atoms. Each of the 92 natural elements of nature is represented by a particular type of atom. The materials that surround us typically are not pure elements, but combinations of atoms of several elements that form molecules. Thus, helium is a natural element consisting of a particular type of atom; water, on the other hand, is composed of molecules, each molecule consisting of a combination of two hydrogen atoms and one oxygen atom. In presenting a physical picture of thermal energy, we need to consider the physical relations or interaction of elements and molecules in a particular material as either solid, liquid, or gas. These statements actually refer to how the molecules of the material are interacting and to the thermal energy of the molecules.

4.2.1 Thermal Energy

Solid

In any solid material, the individual atoms or molecules are quite strongly attracted and bonded to each other, so that no atom is able to move far from its particular location or *equilibrium position*. Each atom, however, is capable of vibration about its particular location. We introduce the concept of thermal energy by considering the molecules' vibration.

Consider a particular solid material in which molecules are exhibiting no vibration; that is, the molecules are at rest. Such a material is said to have zero thermal energy, $W_{TH} = 0$. If we now add energy to this material by placing it on a heater, for example, this energy starts the molecules vibrating about their equilibrium positions. We may say that the material now has some finite thermal energy, $W_{TH} > 0$.

Liquid

If more and more energy is added to the material, the vibrations become more and more violent as the thermal energy increases. Finally, a condition is reached where the bonding attractions that hold the molecules in their equilibrium positions are overcome and the molecules "break away" and move about in the material. When this occurs, we say the material has *melted* and become a liquid. Now, even though the molecules are still attracted to one another, the thermal energy is sufficient to cause the molecules to move about and no longer to maintain the rigid structure of the solid. Instead of vibrating, one considers the molecules as randomly sliding about each other, and the average speed with which they move is a measure of the thermal energy imparted to the material.

Gas

Further increases in thermal energy of the material intensify the velocity of the molecules until finally the molecules gain sufficient energy to completely escape from the attraction of other molecules. Such a condition is manifested by *boiling* of the liquid. When the material consists of such unattached molecules moving randomly throughout a containing volume, we say the material has become a gas. The molecules still collide with each other and the walls of the container, but otherwise move freely throughout the container. The average speed of the molecules is again a measure of the thermal energy imparted to the molecules of the material.

All materials do not undergo these transitions at the same thermal energy, and indeed some not at all. Thus, nitrogen can be solid, liquid, and gas, but paper will experience a breakdown of its molecules before a liquid or gaseous state can occur. The whole subject of thermal transducers is associated with the measurement of the thermal energy of a material or an environment containing many different materials.

4.2.2 Temperature

If we are to measure thermal energy, we must have some sort of units by which to classify the measurement. The original units used were "hot" and "cold." These were quite satisfactory for their time but are inadequate for modern use. The proper unit for energy measurement is the *joules* of the sample in the SI system, but this would depend on the size of the material because it would indicate the total thermal energy contained. Thus, a measurement of the average thermal energy per molecule, expressed in joules, could be used to define thermal energy. We say "could be" because it is not traditionally used. Instead, special sets of units, whose origins are contained in the history of thermal energy measurements, are employed to define the *average energy per molecule* of a material. We will consider the four most common units. In each case, the name used to describe the thermal energy per molecule of a material is related by the statement that the material has a certain *degree of temperature*; the different sets of units are referred to as *temperature scales*.

Calibration

To define the temperature scales, a set of *calibration points* is used; for each, the average thermal energy per molecule is well defined through equilibrium conditions existing between solid, liquid, or gaseous states of various pure materials. Thus, for example, a state of equilibrium exists between the solid and liquid phase of a pure substance when the *rate of phase change* is the same in *either* direction: liquid to solid, and solid to liquid. Some of the standard calibration points are

1. Oxygen: liquid/gas equilibrium
2. Water: solid/liquid equilibrium
3. Water: liquid/gas equilibrium
4. Gold: solid/liquid equilibrium

The various temperature scales are defined by the assignment of numerical values of temperatures to the list and additional calibration points. Essentially, the scales differ in two respects: (1) the location of the 0 of temperature, and (2) the size of one unit of measure; that is, the average thermal energy per molecules represented by one unit of the scale.

The SI definition of the kelvin unit of temperature is in terms of the triple point of water. This is the state at which an equilibrium exists between the liquid, solid, and gaseous state of water maintained in a closed vessel. This system has a temperature of 273.16 K.

Absolute temperature scales

An absolute temperature scale is one that assigns a 0-unit temperature to a material that has no thermal energy, that is, no molecular vibration. There are two such scales in common use: the kelvin scale in kelvin (K) and Rankine scale in degrees Rankine (°R). These temperature scales differ only by the quantity of energy represented by one unit of measure; hence, a simple proportionality relates the temperature in °R to the temperature in K. Table 4.1 shows the values of temperature in kelvin and degrees Rankine at the calibration points introduced earlier. From this table we can determine the transformation of temperature between the water liquid/solid point and water liquid/gas point is 100 K and 180°R, respectively. Because these two numbers represent the same difference of thermal energy, it is clear that 1 K must be larger than 1°R by the ratio of the two numbers.

$$(1 \text{ K}) = \frac{180}{100}(1°\text{R}) = \frac{9}{5}(1°\text{R})$$

Thus, the transformation between scales is given by

$$T(\text{K}) = \frac{5}{9} T(°\text{R}) \qquad\qquad (4.1)$$

TABLE 4.1 Temperature Scale Calibration Points

Calibration Point	Temperature			
	K	°R	°F	°C
Zero thermal energy	0	0	−459.6	−273.15
Oxygen: liquid/gas	90.18	162.3	−297.3	−182.97
Water: solid/liquid	273.15	491.6	32	0
Water: liquid/gas	373.15	671.6	212	100
Gold: solid/liquid	1336.15	2405	1945.5	1063

where

$$T(\text{K}) = \text{temperature in K}$$
$$T(\text{°R}) = \text{temperature in °R}$$

Example 4.1

A material has a temperature of 335 K. Find the temperature in °R.

Solution

$$T(\text{°R}) = \frac{9}{5} T(\text{K})$$

$$\tag{4.1}$$

$$T(\text{°R}) = \frac{9}{5} (335 \text{ K}) = \textbf{603°R}$$

Relative temperature scales

The *relative* temperature scales differ from the *absolute* scales only in a shift of the zero axis. Thus, when these scales indicate a zero of temperature, the thermal energy of the sample is not zero. These two scales are the Celsius (related to the kelvin) and the Fahrenheit (related to the Rankine) with temperature indicated by °C and °F, respectively. Table 4.1 shows various calibration points of these scales. The quantity of energy represented by 1°C is the same as indicated by 1 K, but the zero has been shifted in the Celsius scale, so that

$$T(\text{°C}) = T(\text{K}) - 273.15 \tag{4.2}$$

Similarly, the size of 1°F is the same as the size of 1°R but with a scale shift, so that

$$T(\text{°F}) = T(\text{°R}) - 459.6 \tag{4.3}$$

To transform from Celsius to Fahrenheit, we simply note that the two scales differ by the size of the degree just as in the K and °R, and a scale shift of 32 separates the two; thus

$$T(\text{°F}) = \frac{9}{5} T(\text{°C}) + 32 \tag{4.4}$$

Relation to thermal energy

It is possible to relate temperature to actual thermal energy in joules by using a constant called *Boltzmann's constant*. Although not true in all cases, it is a good approximation to state that the thermal energy, W_{TH}, of a molecule can be found from the absolute temperature in K from

$$W_{TH} = \frac{3}{2} kT \qquad (4.5)$$

where $k = 1.38 \times 10^{-23}$ J/K is Boltzmann's constant. Thus, it is possible to determine the average thermal speed, v_{TH}, of a gas molecule by equating the kinetic energy of the molecule to its thermal energy

$$\frac{1}{2} mv_{TH}^2 = W_{TH} = \frac{3}{2} kT$$

and

$$v_{TH} = \sqrt{\frac{3kT}{m}} \qquad (4.6)$$

where m is the molecule mass in kilograms.

Example 4.2

Given temperature of 144.5°C, express this temperature in (a) K and (b) °F.

Solution

$$\text{(a)} \quad T(\text{K}) = T(°\text{C}) + 273.15$$

$$T(\text{K}) = 144.5 + 273.15 \qquad (4.2)$$

$$T(\text{K}) = \textbf{417.65 K}$$

$$\text{(b)} \quad T(°\text{F}) = \frac{9}{5} T(°\text{C}) + 32$$

$$T(°\text{F}) = \frac{9}{5} (144.5°\text{C}) + 32 \qquad (4.4)$$

$$T(°\text{F}) = \textbf{292.1°F}$$

Example 4.3

A sample of oxygen gas has a temperature of 90°F. If its molecular mass is 5.3×10^{-26} kg, find the average thermal speed of a molecule.

Solution   We first convert the 90°F into K, and then use Equation (4.6) to find the speed.

$$T(°\text{R}) = T(°\text{F}) + 459.6$$

$$T(°\text{R}) = 90°\text{F} + 459.6 \qquad (4.3)$$

$$T(°R) = 549°R$$

$$T(K) = \frac{5}{9} T(°R) = \frac{5}{9} (549.6 \text{ K})$$

$$T(K) = 305.33 \text{ K}$$

(4.1)

Then the velocity is

$$v_{TH} = \sqrt{\frac{3kT}{m}}$$

$$v_{TH} = \left[\frac{(3)(1.38 \times 10^{-23} \text{ J/K})(305.33°K)}{5.3 \times 10^{-26} \text{ kg}} \left(1 \frac{\text{kgm}^2}{\text{s}^2\text{J}} \right) \right]^{1/2}$$

(4.6)

$$v_{TH} = \textbf{488.37 m/s}$$

4.3 METAL RESISTANCE VERSUS TEMPERATURE DEVICES

One of the primary methods for electrical measurement of temperature involves changes in the electrical resistance of certain materials. In this, as well as other cases, the principle measurement technique is to place the temperature-sensing device in contact with the environment whose temperature is to be measured. The sensing device then takes on the temperature of the environment. Thus, a measure of its resistance indicates the temperature of the device and the environment. Time response becomes very important in these cases because the measurement must wait until the device comes into thermal equilibrium with the environment. The two basic devices used are the *resistance-temperature detector* (RTD), based on the variation of metal resistance with temperature, and the *thermistor*, based on the variation of semiconductor resistance with temperature.

4.3.1 Metal Resistance Versus Temperature

A metal is an assemblage of atoms in the solid state in which the individual atoms are in an equilibrium position with superimposed vibration induced by the thermal energy. The chief characteristic of a metal is the fact that each atom gives up one electron, called its valence electron, that can move freely throughout the material; that is, it becomes a conduction electron. We say then, for the whole material, that the valence band of electrons and conduction band of electrons in the material overlap in energy, as shown in Figure 4.1a. Contrast this with a semiconductor where a small gap exists between the top electron energy of the valence band and the bottom electron energy of the conduction band, as shown in Figure 4.1b. In this same scheme, Figure 4.1c shows that an insulator has a very large gap between valence and conduction electrons. When a current is to be passed through a material, it is the conduction band electrons which carry the current.

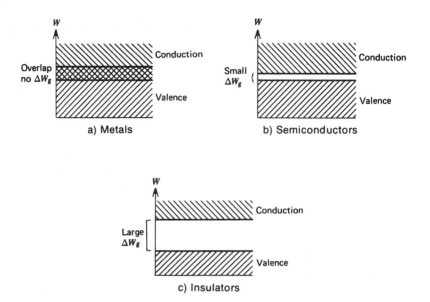

Figure 4.1   Energy bands for solids. Only conduction band electrons are free to carry current.

When we stated earlier that the electrons of a metal "move freely throughout the material," there should have been a condition imposed—that is, that this can be true at absolute zero temperature. In fact, when a certain thermal energy is present in the material and the atoms vibrate, the conduction electrons tend to collide with the vibrating atoms. This impedes the free movement of electrons and absorbs some of their energy; that is, the material exhibits a *resistance* to electrical current flow. Thus, metallic resistance is a function of the vibration of the atoms and thus of the temperature. As the temperature is raised, the atoms vibrate with greater amplitude and frequency, which causes even more collision with electrons, further impeding their flow and absorbing more energy. From this argument, we can see that metallic resistance should increase with temperature, and it does.

The graph in Figure 4.2 shows the effect of increasing resistance with temperature for several metals. To compare the different materials, the graph shows the relative resistance versus temperature. For a specific metal of high purity, the curve of relative resistance versus temperature is highly repeatable, and thus either tables or graphs can be used to determine the temperature from a resistance measurement using that material. It is possible to express the resistance of a particular metal sample at a constant temperature (T) analytically using the equation

$$R = \rho \frac{l}{A} \qquad (T = \text{constant}) \tag{4.7}$$

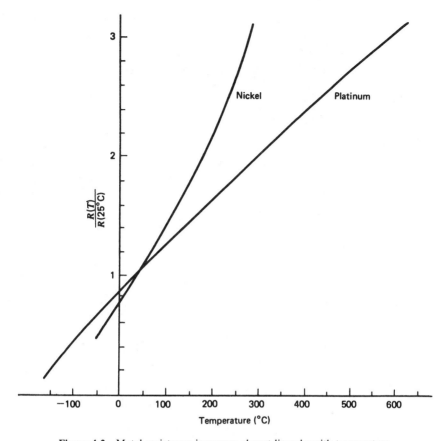

Figure 4.2 Metal resistance increases almost linearly with temperature.

where

R = sample resistance (Ω)
l = length (m)
A = cross-sectional area (m^2)
ρ = resistivity (Ω − m)

In Equation (4.7) the principal increase in resistance with temperature is due to changes in the resistivity (ρ) of the metal with temperature. If the resistivity of some metal is known as a function of temperature, then Equation (4.7) can be used to determine the resistance of any particular sample of that material at the same temperature. In fact, curves such as those in Figure 4.2 are curves of resistivity versus temperature because, for example,

$$\frac{R(T)}{R(75°)} = \frac{\rho(T)l/A}{\rho(75°)l/A} = \frac{\rho(T)}{\rho(75°)} \tag{4.8}$$

The use of either Equation (4.7), resistance versus temperature graphs, or resistance versus temperature tables is only practical when high accuracy is desired. For many applications we can use an analytical approximation of the curves, for which we simply insert the temperature and quickly calculate the resistance, as described in Section 4.3.2.

4.3.2 Resistance Versus Temperature Approximations

An examination of the resistance versus temperature curves of Figure 4.2 shows that the curves are very nearly linear, that is, a straight line. In fact, when only short temperature spans are considered, the linearity is even more evident. This fact is employed to develop approximate analytical equations for resistance versus temperature of a particular metal.

Linear approximation

A linear approximation means that we may develop an equation for a straight line that approximates the resistance versus temperature ($R - T$) curve over some specified span. In Figure 4.3, we see a typical $R - T$ curve of some material. A straight line has been drawn between the points of the curve that represent temperature T_1 and T_2 as shown, and T_0 represents the midpoint temperature. The equation of this straight line is the linear approximation to the curve over the span T_1 to T_2. The equation for this line is typically written as

$$R(T) = R(T_0)[1 + \alpha_0 \, \Delta T] \qquad T_1 < T < T_2 \qquad (4.9)$$

where

$$R(T) = \text{approximation of resistance at temperature } T$$
$$R(T_0) = \text{resistance at temperature } T_0$$
$$\Delta T = T - T_0$$
$$\alpha_0 = \text{fractional change in resistance per degree of temperature at } T_0$$

The reason for using α_0 as the fractional slope of the $R - T$ curve is that this same constant can be used for cases of other physical dimensions (length and cross-sectional area) of the same kind of wire. Note that α_0 depends on the midpoint temperature T_0, which simply says that a straight-line approximation over some other span of the curve would have a different slope.

The value of α_0 can be found from values of resistance and temperature taken either from a graph, as given in Figure 4.2, or a table of resistance versus temperature, as given in Problem 4.8. In general, then

$$\alpha_0 = \frac{1}{R(T_0)} \cdot (\text{slope at } T_0) \qquad (4.10)$$

or, for example, from Figure 4.3

$$\alpha_0 = \frac{1}{R(T_0)} \cdot \left(\frac{R_2 - R_1}{T_2 - T_1} \right) \qquad (4.11)$$

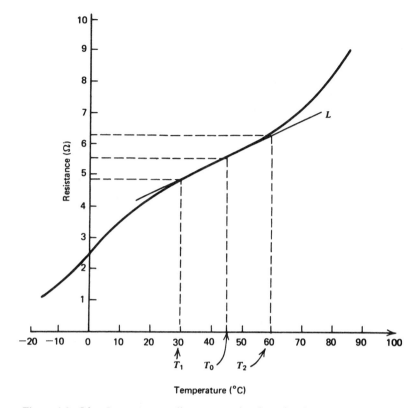

Figure 4.3   Line L represents a linear approximation of resistance versus temperature between T_1 and T_2.

where

    R_2 = resistance at T_2
    R_1 = resistance at T_1

α_0 has units of *inverse* temperature degrees and therefore depends on the temperature scale being used. Thus, the units of α_0 are typically 1/°C or 1/°F.

Quadratic approximation

A quadratic approximation to the $R - T$ curve is a more accurate representation of the $R - T$ curve over some span of temperatures. It includes both a linear term, as before, and a term that varies as the square of the temperature. Such an analytical approximation is usually written as

$$R(T) = R(T_0)[1 + \alpha_1 \Delta T + \alpha_2(\Delta T)^2] \qquad (4.12)$$

where

    $R(T)$ = quadratic approximation of the resistance at T

$R(T_0)$ = resistance of T_0

α_1 = linear fractional change in resistance with temperature

$\Delta T = T - T_0$

α_2 = quadratic fractional change in resistance with temperature

Values of α_1 and α_2 are found from tables or graphs, as indicated in the following examples, using values of resistance and temperature at three points. As before, both α_1 and α_2 depend on the temperature scale being used and have units of 1/°C and $(1/°C)^2$ if Celsius temperature is used, and 1/°F and $(1/°F)^2$ if the Fahrenheit scale is used.

The following examples show how these linear approximations are formed.

Example 4.4

Find a linear approximation for resistance between 30°C and 60°C using the $R - T$ curve of Figure 4.3.

Solution   Using Equations (4.9) and (4.11), we first identify $T_1 = 30°C$, $T_0 = 45°C$, and $T_2 = 60°C$. Then from Figure 4.3, we find $R(30°C) = 4.8\ \Omega$, $R(45°C) = 5.5\ \Omega$, and $R(60°C) = 6.2\ \Omega$.

$$\alpha_0 = \frac{1}{5.5\ \Omega} \frac{6.2\ \Omega - 4.8\ \Omega}{60°C - 30°C}$$

$$\alpha_0 = 0.0085/°C \tag{4.11}$$

Thus

$$R(T) = 5.5[1 + .0085(T - 45)]\Omega$$

with T in °C.

The following typical values of resistance versus temperature are to be used for Examples 4.5 and 4.6.

T(°F)	R(Ω)
60	106.06
65	107.14
70	108.22
75	109.30
80	110.38
85	111.46
90	112.53

Example 4.5

The following equation is a linear approximation for the resistance between 65°F and 75°F from the previous data.

$$R(T) = 108.22[1 + 0.002(T - 70°F)]\Omega$$

Find the resistance at 20°C.

Solution We must first convert the 20°C to its equivalent °F to keep the units consistent.

$$T(°F) = \frac{9}{5}(20°C) + 32$$

$$T(°F) = 68°F$$

(4.4)

Then we use the stated linear equation to find

$$R(68°F) = 108.22[1 + 0.002(68 - 70)]$$

$$R(68°F) = 107.79 \ \Omega$$

Example 4.6

Find a quadratic approximation for resistance versus temperature about 75°F between 60°F and 90°F, using the table of values given between Example 4.4 and Example 4.5.

Solution We can find the quadratic terms in Equation (4.12) by forming two equations using two points about 75°F. Thus, at 75°F the resistance is 109.3 Ω; therefore

$$R(T_0) = 109.3 \ \Omega, \ T_0 = 75°F$$

Now, using 60°F and 90°F as the two points, we get two equations, using

$$R(T) = R(T_0)[1 + \alpha_1 \Delta T + \alpha_2(\Delta T)^2]$$

$$106.06 = 109.3[1 + \alpha_1(60 - 75) + \alpha_2(60 - 75)^2]$$

(4.12)

$$112.53 = 109.3[1 + \alpha_1(90 - 75) + \alpha_2(90 - 75)^2]$$

Solving the equations for α_1 and α_2, we find

$$\alpha_1 = \mathbf{1.973 \times 10^{-3}/°F}$$

$$\alpha_2 = \mathbf{-2.033 \times 10^{-7}/(°F)^2}$$

4.3.3 Resistance-Temperature Detectors

A *resistance-temperature detector* (RTD) is a temperature sensor that is based on the principles discussed in the preceding sections; that is, metal resistance increasing with temperature. Metals used in these devices vary from *platinum*, which is very repeatable, quite sensitive, and very expensive, to *nickel*, which is not quite as repeatable, more sensitive, and less expensive.

Sensitivity

An estimate of RTD *sensitivity* can be noted from typical values of α_0, the linear fractional change in resistance with temperature. For platinum, this number is typically on the order of 0.004/°C, and for nickel a typical value is 0.005/°C. Thus, with platinum, for example, a change of only 0.4 Ω would be expected for a 100-Ω RTD if the temperature is changed by 1°C. Usually, a specification will provide calibration information either as a graph of resistance versus temperature or as a table of values from which the sensitivity can be determined. For the same materials, however, this number is relatively constant because it is a function of resistivity.

Response time

In general, RTD has a response time of 0.5 to 5 seconds or more. The slowness of response is due principally to the slowness of thermal conductivity in bringing the device into thermal equilibrium with its environment. Generally, time constants are specified either for a "free air" condition (or its equivalent) or an "oil bath" condition (or its equivalent). In the former case, there is poor thermal contact and hence slow response, and in the latter, good thermal contact and fast response. These numbers yield a range of response times depending on the application.

Construction

An RTD, of course, is simply a length of wire whose resistance is to be monitored as a function of temperature. The construction is typically such that the wire is wound on a form (in a coil) to achieve small size and improve thermal conductivity to decrease response time. In many cases, the coil is protected from the environment by a *sheath* or protecting tube that inevitably increases response time but may be necessary in hostile environments. A loosely applied standard sets the resistance at multiples of 100 Ω for a temperature of 0°C.

Signal conditioning

In view of the very small fractional changes of resistance with temperature (0.4%), the RTD is generally used in a *bridge* circuit in which a null condition is accurately detected. For process-control applications, the bridge then requires a *self-nulling* property. The output of the nulling circuit then provides the controller output of 4–20 mA or 10–50 mA. Figure 4.4 illustrates the essential features of such a system. The *compensation line* in the R_3 leg of the bridge is required when the lead lengths are so long that thermal gradients along the RTD leg may cause changes in line resistance. These changes show up as false information, suggesting changes in RTD resistance. By using the *compensation* line, the same resistance changes also appear on the R_3 side of the bridge and cause no net shift in the bridge null.

Dissipation constant

Because the RTD is a resistance, there is an I^2R power dissipated by the device itself that causes a slight heating effect, a *self-heating*. This may also cause an erroneous reading or even upset the environment in delicate measurement conditions. Thus, the current through the RTD must be kept sufficiently low and

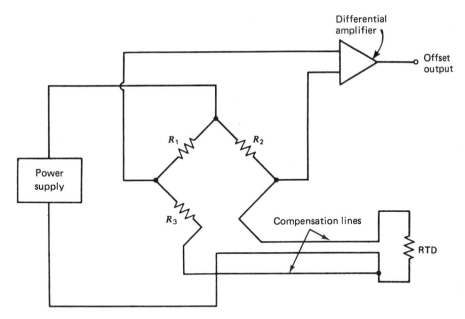

Figure 4.4 Note the compensation line in this typical RTD signal conditioning circuit. Any resistance in the bridge could be used to renull the bridge.

constant to avoid self-heating. Typically, a *dissipation constant* is provided in RTD specifications. This number relates the power required to raise the RTD temperature by one degree of temperature, usually in still air. Thus, a 25-mW/°C dissipation constant shows that if I^2R power losses in the RTD equal 25 mW, then the RTD will be heated by 1°C.

The dissipation constant is usually specified under two conditions: free air and a well-stirred oil bath. This is because of the difference in capacity of the medium to carry heat away from the device. The self-heating temperature rise can be found from the power dissipated by the RTD and the dissipation constant from

$$\Delta T = \frac{P}{P_D} \qquad (4.13)$$

where

ΔT = temperature rise because of self-heating in °C
P = power dissipated in the RTD from the circuit in W
P_D = dissipation constant of the RTD in W/°C

Example 4.7

An RTD has $\alpha_0 = 0.005/°C$, $R = 500 \ \Omega$, and a dissipation constant of $P_D = 30$ mW/°C at 20°C. The RTD is used in a bridge circuit such as in Figure 4.4 with $R_1 = R_2$

= 500 Ω and R_3 a variable resistor used to null the bridge. If the supply is 10 volts and the RTD is placed in a bath at 0°C, find the value of R_3 to null the bridge.

Solution   First we find the value of the RTD resistance at 0°C without including the effects of dissipation. From Equation (4.9), we get

$$R = 500[1 + 0.005(0 - 20)]\Omega$$

$$R = 450 \ \Omega$$

Except for the effects of self-heating, we would expect the bridge to null with R_3 equal to 450 Ω also. Let's see what self-heating does to this problem. First, we find the power dissipated in the RTD from the circuit assuming the resistance is still 450 Ω. The power is

$$P = I^2R$$

and the current I to three significant figures is found from

$$I = \frac{10}{500 + 450} = 0.011 \text{ amps}$$

so that the power is

$$P = (0.011)^2(450) = 0.054 \text{ W}$$

We get the temperature rise from Equation (4.13)

$$\Delta T = \frac{0.054}{0.030} = 1.8°C$$

Thus, the RTD is not actually at the bath temperature of 0°C but at a temperature of 1.8°C. We must find the RTD resistance from Equation (4.9) as

$$R = 500[1 + 0.005(1.8 - 20)]\Omega$$

$$R = 454.5 \ \Omega$$

Thus, the bridge will null with $R_3 = $ **454.5 Ω**.

Range

The effective range of RTDs depends principally on the type of wire used as the active element. Thus, a typical platinum RTD will have a range of −100°C to 650°C, and an RTD constructed from nickel might typically have a specified range of −180°C to 300°C.

4.4 THERMISTORS

The thermistor represents another class of temperature sensor that measures temperature through changes of material resistance. The characteristics of these devices are very different from RTDs and depend on the peculiar behavior of semiconductor resistance versus temperature.

4.4.1 Semiconductor Resistance Versus Temperature

In contrast to metals, electrons in semiconductor materials are bound to each molecule with sufficient strength so that no conduction electrons are contributed from the valence band to the conduction band. We say that a gap of energy ΔW_g exists between valence and conduction electrons, as shown in Figure 4.1b. Such a material behaves as an insulator because there are no conduction electrons to carry current through the material. This is true only when there is no thermal energy present in the sample, that is, at a temperature of 0 K. When the temperature of the material is increased, the molecules begin to vibrate. In the case of a semiconductor, such vibration provides additional energy to the valence electrons. When such energy equals or exceeds the gap energy ΔW_g, these electrons become free of the molecules. Thus, the electron is now in the conduction band and is free to carry current through the bulk of the material. As the temperature is further increased, more and more electrons gain sufficient energy to enter the conduction band. It is then clear that the semiconductor becomes a *better* conductor of current as its temperature is *increased*, that is, as its resistance decreases. From this discussion we form a picture of the resistance of a semiconductor material decreasing from very large values at low temperature to smaller resistance at high temperature. This is just the *opposite* of a metal. An important distinction, however, is that the change in semiconductor resistance is highly nonlinear, as shown in Figure 4.5. The reason semiconductors but not insulators and other materials behave this way is that the energy gap between conduction and valence bands is small enough to allow thermal excitation of electrons across the gap.

It is important to note that the effect just described requires that the thermal energy provide sufficient energy to overcome the band gap energy ΔW_g. In general, a material is classified as a semiconductor when the gap energy is typically 0.01–4 eV (1 eV = 1.6×10^{-19} J). That this is true is exemplified by a consideration of silicon, a semiconductor that has a band gap of $\Delta W_g = 1.107$ eV. When heated, this material passes from insulator to conductor. The corresponding thermal energies that bring this about can be found using Equation (4.5) and the joules to eV conversion, thus:

$$\text{for } T = 0 \text{ K} \qquad W_{TH} = 0.0 \text{ eV}$$

$$\text{for } T = 100 \text{ K} \qquad W_{TH} = 0.013 \text{ eV}$$

$$\text{for } T = 300 \text{ K} \qquad W_{TH} = 0.039 \text{ eV}$$

With average thermal energies as high as 0.039 eV, sufficient numbers of electrons are raised to the conduction level for the material to become a conductor. In *true* insulators, the gap energy is so large that temperatures less than destructive to the material cannot provide sufficient energy to overcome the gap energy.

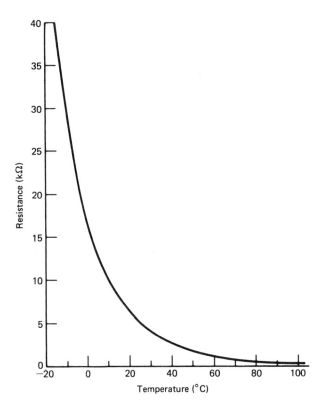

Figure 4.5 This plot of a thermistor resistance versus temperature shows the highly nonlinear characteristic of semiconductor material.

4.4.2 Thermistors

A thermistor is a temperature sensor that has been developed from the principles just discussed regarding semiconductor resistance change with temperature. The particular semiconductor material used varies widely to accommodate temperature ranges, sensitivity, resistance ranges, and other factors. The devices are usually mass produced for a particular configuration, and tables or graphs of resistance versus temperature are provided for calibration. Variation of individual units from these nominal values is indicated as a net percentage deviation or a percentage deviation as a function of temperature.

Sensitivity

The sensitivity of the thermistors is a significant factor in their application. Changes in resistance of 10% per °C are not uncommon. Thus, a thermistor with a nominal resistance of 10 kΩ at some temperature may change by 1 kΩ for a 1°C change in temperature. When used in null-detecting bridge circuits, sensitivity this large can provide for control, in principle, to less than 1°C in temperature.

Construction

Because the thermistor is a bulk semiconductor, it can be fabricated in many forms. Thus, common forms include discs, beads, and rods varying in size from a bead 1 mm in diameter to a disc several centimeters in diameter and several centimeters thick. By variation of doping and use of different semiconducting materials, a manufacturer can provide a wide range of resistance values at any particular temperature.

Range

The temperature range of thermistors depends on the materials used to construct the transducer. In general, there are three range limitation effects: (1) melting or deterioration of the semiconductor, (2) deterioration of encapsulation material, and (3) insensitivity at higher temperatures.

The semiconductor material may melt or otherwise deteriorate as the temperature is raised. This condition generally limits the upper temperature to less than 300°C. At the low end, the principal limitation is that the thermistor resistance becomes very high, into the MΩs, making practical applications difficult. For the thermistor shown in Figure 4.5, if extended, the lower limit is about −80°C, where its resistance has risen to over 3 MΩ! Generally, the lower limit is −50°C to −100°C.

In most cases the thermistor is encapsulated in plastic, epoxy, Teflon®, or some other inert material. This protects the thermistor itself from the environment. This material may place an upper limit on the temperature at which the transducer can be used.

At higher temperatures, the slope of the RT curve of the thermistor goes to zero. The device then is unable to effectively measure temperature because very little change in resistance occurs. You can see this occurring for the thermistor resistance versus temperature curve of Figure 4.5.

Response time

The response time of a thermistor depends principally on the quantity of material present and the environment. Thus, for the smallest bead thermistors in an oil bath (good thermal contact), a response of ½ second is typical. The same thermistor in still air will respond with a typical response time of 10 seconds. When encapsulated, as in Teflon or other materials for protection against a hostile environment, the time response is increased due to the poor thermal contact with the environment. Large disc or rod thermistors may have response times of 10 seconds or more, even with good thermal contact.

Signal conditioning

Because a thermistor exhibits such a large change in resistance with temperature, there are many possible circuit applications. In many cases, however, a bridge circuit with null detection (to maintain a particular temperature) is used because the nonlinear features of the thermistor make its use difficult as an actual measurement device. Because these devices are resistances, care must be taken to ensure that power dissipation in the thermistor does not exceed limits specified or possibly interfere with the environment for which the temperature is being measured. *Dissipation constants* are quoted for thermistors as the power in milliwatts required to raise a thermistor's temperature 1°C above its environment. Typical values vary from 1 mW/°C in free air to 10 mW/°C or more in an oil bath.

Example 4.8

A thermistor is to monitor room temperature. It has a resistance of 3.5 kΩ at 20°C with a slope of − 10%/°C. The dissipation constant is $P_D = 5$ mW/°C. It is proposed to use the thermistor in the divider of Figure 4.6 to provide a voltage of 5.0 volts at 20°C. Evaluate the effects of self-heating.

Solution   It is easy to see that the design seems to work. At 20°C the thermistor resistance will be 3.5 kΩ and the divider voltage will be

$$V_D = \frac{3.5\ k\Omega}{3.5\ k\Omega\ +\ 3.5\ k\Omega}\ 10 = 5\ V$$

Let us now consider the effect of self-heating. The power dissipation in the thermistor will be given by

$$P = \frac{V^2}{R_{TH}} = \frac{(5)^2}{3.5\ k\Omega} = 7.1\ mW$$

The temperature rise of the thermistor now will be found from Equation (4.13)

$$\Delta T = \frac{P}{P_D} = \frac{7.1\ mW}{5\ mW/°C} = 1.42°C$$

But this means the thermistor resistance is really given by

$$R_{TH} = 3.5\ k\Omega\ -\ 1.42°C(0.1/°C)(3.5\ k\Omega)$$

$$= 3.0\ k\Omega$$

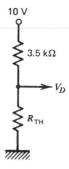

10 V

3.5 kΩ

V_D

R_{TH}

Figure 4.6   Divider circuit for Example 4.8.

and so the divider voltage is actually $V_D = 4.6$ V. The actual temperature of the environment is 20°C, but the measurement indicates that this is not so. Clearly, the system is unsatisfactory.

This example shows the importance of including dissipation effects in resistive-temperature transducers. The real answer to this problem involves a new design that reduces the thermistor current to a value giving perhaps 0.1°C of self-heating.

4.5 THERMOCOUPLES

In previous sections, we considered the change in material resistance as a function of temperature. Such a resistance change is considered a variable parameter property in the sense that the measurement of resistance, and thereby temperature, requires external power sources. There exists another dependence of electrical behavior of materials on temperature that forms the basis of a large percentage of all temperature measurement. This effect is characterized by a voltage-generating sensor in which an electromotive force (emf) is produced that is proportional to temperature. Such an emf is found to be almost linear with temperature and very repeatable for constant materials. Devices that measure temperature on the basis of this thermoelectric principle are called *thermocouples*.

4.5.1 Thermoelectric Effects

The basic theory of the thermocouple effect is found from a consideration of the electrical and thermal transport properties of different metals. In particular, when a temperature differential is maintained across a given metal, the vibration of atoms and motion of electrons is affected so that a difference in potential exists across the material. This potential difference is related to the fact that electrons in the hotter end of the material have more thermal energy than those in the cooler end and thus tend to drift toward the cooler end. This drift varies for different metals at the same temperature because of differences in their thermal conductivities. If a circuit is closed by connecting the ends through another conductor, a current is found to flow in the closed loop.

The proper description of such an effect is to say that an emf has been established in the circuit that is causing the current to flow. In Figure 4.7a, we see a pictorial representation of this effect where two different metals A and B are used to close the loop with the connecting junctions at temperature T_1 and T_2. We could not close the loop with the same metal because the potential differences across each leg would be the same, and thus no net emf would be present. The emf produced is proportional to the difference in temperature between the two junctions. Theoretical treatments of this problem involve the thermal activities of the two metals.

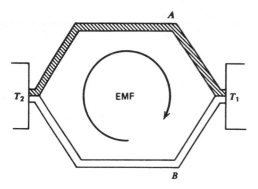

a) Seebeck effect

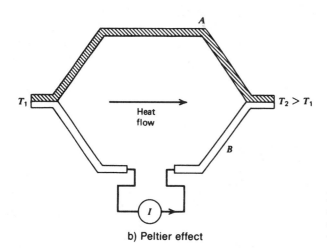

b) Peltier effect

Figure 4.7 The Seebeck and Peltier effects refer to the relation between emf and temperature differences in a two-wire system.

Seebeck effect

Using solid-state theory, the aforementioned situation may be analyzed to show that its emf can be given by an integral over temperature

$$\epsilon = \int_{T_1}^{T_2} (Q_A - Q_B) dT$$

where

$$\epsilon = \text{emf produced in volts}$$
$$T_1, T_2 = \text{junction temperatures in K}$$
$$Q_A, Q_B = \text{thermal transport constants of the two metals}$$

This equation, which describes the *Seebeck effect*, shows that the emf produced is proportional to the *difference* in temperature and, further, to the difference in the metallic thermal transport constants. Thus, if the metals are the same, the emf is zero, and if the temperatures are the same, the emf is also zero.

In practice, it is found that the two constants Q_A and Q_B are *nearly* independent of temperature and that an approximate linear relationship exists as

$$\epsilon = \alpha(T_2 - T_1)$$

where

$$\alpha = \text{constant in volts/K}$$
$$T_1, T_2 = \text{junction temperatures in K}$$

However, the small but finite temperature dependence of Q_A and Q_B is necessary for accurate considerations.

Example 4.9

Find the Seebeck emf for a material with $\alpha = 50 \ \mu V/°C$ if the junction temperatures are 20°C and 100°C.

Solution   The emf can be found from

$$\epsilon = \alpha(T_2 - T_1)$$

$$\epsilon = (50 \ \mu V/°C)(100°C - 20°C)$$

$$\epsilon = \textbf{4 mV}$$

Peltier effect

An interesting and sometimes useful extension of the same thermoelectric properties occurs when the reverse of the Seebeck effect is considered. In this case, we construct a closed loop of two different metals, A and B, as before. Now, however, an external voltage is applied to the system to cause a current to flow in the circuit, as shown in Figure 4.7b. Because of the different electrothermal transport properties of the metals, it is found that one of the junctions will be *heated* and the other *cooled*; that is, the device is a refrigerator! This process is referred to as the Peltier effect. Some practical applications of such a device, such as cooling small electronic parts, have been employed.

4.5.2 Thermocouples

To use the Seebeck effect as the basis of a temperature sensor, we need to establish a definite relationship between the measured emf of the thermocouple and the unknown temperature. We see first that one temperature must already be known because the Seebeck voltage is proportional to the *difference* between junction temperatures. Furthermore, every connection of different metals made in the thermocouple loop for measuring devices, extension leads, and so on will contribute an emf depending on the difference in metals and various junction temperatures. To provide an output that is definite with respect to the temperature to be measured, an arrangement such as that shown in Figure 4.8a is used. This shows that the measurement junction T_M is exposed to the environment whose temperature is to be measured. This junction is formed of metals A and B as shown. Two other junctions then are formed to a common metal C, which then connects to the measurement apparatus. The "reference" junctions are held at a common, known temperature T_R, the reference junction temperature. When an emf is measured, such problems as voltage drops across resistive elements in the loop must be considered. In this arrangement, an open circuit voltage is measured (at high impedance) that is then a function of only the temperature difference ($T_M - T_R$) and the type of metals A and B. The voltage produced has a magnitude dependent on the absolute magnitude of the temperature *difference* and a *polarity* dependent on *which* temperature is larger, reference or measurement junction. Thus, it is *not* necessary that the measurement junction have a higher temperature than the reference junctions, but both magnitude and sign of the measured voltage must be noted.

To use the thermocouple to measure a temperature, the reference temperature must be known and the reference junctions must be held at the same temperature. The temperature should be constant or at least not vary much. In most industrial environments this would be very difficult if the measurement junction and reference junction were close. It is possible to move the reference junctions to a remote location without upsetting the measurement process by the use of *extension wires*, as shown in Figure 4.8b. A junction is formed with the measurement system, but to wires of the same type as the thermocouple. These wires may be stranded and of different gauges, but they must be of the same type of metal as the thermocouple. The extension wires now can be run a significant distance to the actual reference junctions.

Thermocouple types

Certain standard configurations of thermocouples using specific metals (or alloys of metals) have been adopted and given letter designations; examples are shown

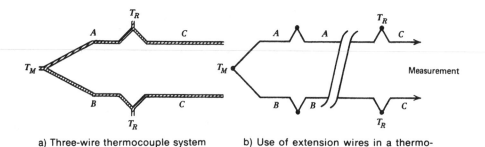

a) Three-wire thermocouple system   b) Use of extension wires in a thermo-
                                                couple system

Figure 4.8 Practical measurements with a thermocouple system employ a third wire type and extension wires to carry the emf to the measurement device.

in Table 4.2. Each type has its particular features, such as range, linearity, inertness to hostile environments, sensitivity, and so on, and is chosen for specific applications accordingly. In each type, various sizes of conductors may be employed for specific cases, such as oven measurements, highly localized measurements, and so on. The curves of voltage versus temperature in Figure 4.9 are shown for a reference temperature of 0°C and for several types of thermocouples. We wish to note several important features from these curves.

First, we see that the type J and E thermocouples are noted for their rather

TABLE 4.2   Standard Thermocouples

Type	Materials[a]	Normal Range
J	Iron-constantan[b]	−190°C to 760°C
T	Copper-constantan	−200°C to 371°C
K	Chromel-alumel	−190°C to 1260°C
E	Chromel-constantan	−100°C to 1260°C
S	90% Platinum + 10% rhodium − platinum	0°C to 1482°C
R	87% Platinum + 13% rhodium − platinum	0°C to 1482°C

[a] First material is more positive when the measurement temperature is more than the reference temperature.
[b] Constantan, chromel, and alumel are registered trade names of alloys.

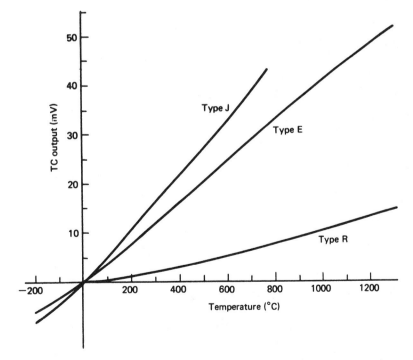

Figure 4.9   These curves of thermocouple (TC) voltage versus temperature for a 0°C reference show the different sensitivities and nonlinearities of three types.

large slope, that is, high sensitivity—making measurements easier for a given change in temperature. We note that the type R thermocouple has much less slope and is appropriately less sensitive. It has the significant advantages of a much *larger possible range* of measurement, including very high temperatures, and is very inert material. Another important feature is the observation that these curves are not exactly linear. To take advantage of the inherent accuracy possible with these devices, comprehensive tables of voltage versus temperature have been determined for many types of thermocouples. Such tables are found in Appendix 3.

Thermocouple tables

The thermocouple tables simply give the voltage that results for a particular type of thermocouple when the reference junctions are at a particular reference temperature, and the measurement junction is at a temperature of interest. Referring to the tables, for example, we see that for a type J thermocouple at 210°C with a 0°C reference, the voltage is

$$V(210°C) = 11.34 \text{ mV} \quad \text{(type J, 0°C ref.)}$$

Conversely, if we measured a voltage of 4.768 mV with a type S and a 0°C reference, we find from the table

$$T(4.768 \text{ mV}) = 555°C \quad \text{(type S, 0°C ref.)}$$

In most cases, the measured voltage does not fall exactly on a table value as in this case. When this happens, it is necessary to *interpolate* between table values that bracket the desired value. In general, the value of temperature can be found using the following interpolation equation:

$$T_M = T_L + \left[\frac{T_H - T_L}{V_H - V_L} \right] (V_M - V_L) \tag{4.14}$$

The measured voltage V_M lies between a higher voltage V_H and a lower voltage V_L, which are in the tables. The temperatures corresponding to these voltages are T_H and T_L, respectively, as shown in Example 4.10.

Example 4.10

A voltage of 23.72 mV is measured with a type K thermocouple (TC) at a 0°C reference. Find the temperature of the measurement junction.

Solution From the table we find that $V_M = 23.72$ lies between $V_L = 23.63$ mV and $V_H = 23.84$ mV with corresponding temperatures of $T_L = 570°C$ and $T_H = 575°C$, respectively. The junction temperature is found from Equation (4.14).

$$T_M = 570°C + \frac{(575°C - 570°C)}{(23.84 - 23.63 \text{ mV})} (23.72 \text{ mV} - 23.63 \text{ mV})$$

$$T_M = 570°C + \frac{5°C}{0.21}(0.09 \text{ mV}) \tag{4.14}$$

$$T_M = \mathbf{572.1°C}$$

The reverse situation, although not as common in practice, may occur when the voltage for a particular temperature T_M, which is not in the table, is desired. Again, an interpolation equation can be used, such as

$$V_M = V_L + \left[\frac{V_H - V_L}{T_H - T_L}\right](T_M - T_L) \tag{4.15}$$

where all terms are as defined for Equation (4.14).

Example 4.11

Find the voltage of a type J TC with a 0°C reference if the junction temperature is $-172°C$.

Solution   We do not let the signs bother us but merely apply the interpolation relation directly. From the tables, we see that the junction temperature lies between a high (algebraically) $T_H = -170°C$ and a low $T_L = -175°C$. The corresponding voltages are $V_H = -7.12$ mV, $V_L = -7.27$ mV. The TC voltage will be

$$V_M = -7.27 \text{ mV} + \frac{-7.12 + 7.27}{-170 + 180}(-172°C + 175°C)$$

$$V_M = -7.27 \text{ mV} + \frac{0.15 \text{ mV}}{5°C}(3°C) \tag{4.15}$$

$$V_M = \mathbf{-7.18 \text{ mV}}$$

Change of table reference

It has already been pointed out that thermocouple tables are prepared for a particular junction temperature. It is possible to use these tables with a thermocouple (TC) that has a different reference temperature by an appropriate shift in the table scale. The key point to remember is that the voltage is proportional to the difference between the reference and measurement junction temperature. Thus, if a new reference is greater than the table reference, all voltages of the table will be less for this TC. The amount less will be just the voltage of the new reference as found on the table. Perhaps a few examples are in order here. Suppose we have a type J TC with a 30°C reference. On the 0°C reference table, a type J at 30°C will produce 1.54 mV. This means that any temperature with this TC will generate a voltage 1.54 mV less than those in the table. Thus, referring to the table

400°C results in $V = 21.85 - 1.54 = 20.31$ mV (type J, 30°C)

150°C results in $V = 8.00 - 1.54 = 6.46$ mV (type J, 30°C)

-90°C results in $V = -4.21 - 1.54 = -5.75$ mV (type J, 30°C)

In a similar fashion, if the new reference is lower than the reference, all of the table voltages will be larger. For example, consider a type K thermocouple with a reference at -26°C. First, by interpolation, we find the voltage that this corresponds to on the 0°C reference tables

$$V(-26°C) = -1.14 + \frac{-0.95 + 1.14}{-25 + 30}(-26 + 30)$$

$$V(-26°C) = -0.98 \text{ mV (type K, 0°C ref)}$$

Thus, every voltage on the table must be increased by 0.98 mV, so

400°C results in $V = 16.40 + 0.98 = 17.38$ mV

150°C results in $V = 6.13 + 0.98 = 7.11$ mV

-90°C results in $V = -3.19 + 0.98 = -2.21$ mV

In effect, we are sliding the curves of TC voltage versus temperature along the temperature axis to give a zero voltage at the reference being used. This is shown in Figure 4.10. The shifts made, as in the previous examples, are not exact because

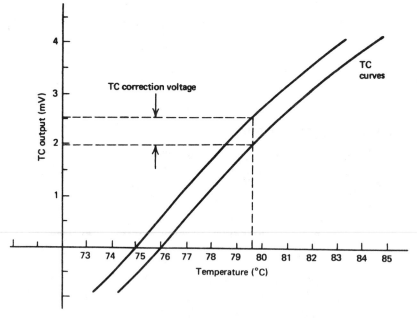

Figure 4.10 When a TC reference is changed, a correction voltage must be applied. If the tables are for 75°C and the actual reference is 76°C, a correction of 0.5 mV is needed.

of the dependence on temperature of the metallic thermoelectric constants. If a very large difference in temperature exists between the table reference temperature and the reference being used, inaccuracies will probably exist.

4.5.3 Thermocouple Sensors

The use of a thermocouple for a temperature sensor has evolved from an elementary process with crudely prepared thermocouple constituents into a very precise and exacting technique.

Sensitivity

A review of the tables shows that the range of thermocouple voltages is typically less than 100 mV. The actual sensitivity strongly depends on the type of signal conditioning employed and on the TC itself. We see from Figure 4.9 the following worst and best case of sensitivity:

- Type J: 0.05 mV/°C (typical)
- Type R: 0.006 mV/°C (typical)

Construction

A thermocouple by itself is, of course, simply a welded or even twisted junction between two metals and, in many cases, that is the construction. There are cases, however, where the TC is sheathed in a protective covering or even sealed in glass to protect the unit from a hostile environment. The size of the TC wire is determined by the application and can range from #10 wire in rugged environments to fine #30 AWG wires or 0.02-mm microwire in refined biological measurements of temperature.

Range

The thermocouple temperature sensor has the greatest range of all the types considered. The tables in Appendix 3 show that the general-purpose, Type J thermocouple is usable from 150°C to 745°C. The Type S is usable up to 1765°C. Other special types have ranges above and below these.

Signal conditioning

The key element in the use of thermocouples is that the output voltage is very small, typically less than 50 mV. This means that considerable amplification will be necessary for practical application. In addition, the small signal levels make the devices susceptible to electrical noise. In most cases the thermocouple is used with a high-gain differential amplifier. In addition, the input impedance should be high, particularly if extension wires are used, to reduce errors from voltage drops because of current drawn from the thermocouple.

Another problem with the practical use of thermocouples is the necessity of knowing the reference temperature. Because the TC voltage is proportional to the difference between the measurement and reference junction temperatures, variations of the reference temperature show up as direct errors in the measurement temperature determination. The following techniques are employed for the reference junction:

1. *Controlled temperature reference block* In some cases, particularly when many thermocouples are in use, extension wires bring all reference junctions to a temperature-controlled box in the control room.

2. *Reference compensation circuits* Circuits have been designed, using precision thermistors, that output a voltage that properly corrects the thermocouple voltage for variations of reference junction temperature. The temperature-sensing element is connected to the reference junction block (see Section 4.6.5).

3. *Software reference correction* In computer-based measurement systems, the reference junction temperature can be measured by a precision thermistor and provided as an input to the computer. Software routines then can provide necessary corrections to the thermocouple temperature signal that is also an input to the computer.

A thermistor can be used in the latter two cases because the variation of reference temperature is small so that the nonlinearity of the thermistor resistance versus temperature is an insignificant factor.

Noise

Perhaps the biggest obstacle to the use of thermocouples for temperature measurement in industry is their susceptibility to electrical noise. First, the voltages generated generally are less than 50 mV and often are only 2 or 3 millivolts, and in the industrial environment it is common to have hundreds of millivolts of electrical noise generated by large electrical machines in any electrical system. Second, a thermocouple constitutes an excellent antenna for pickup of noise from electromagnetic radiation in the radio, TV, and microwave bands. In short, a bare thermocouple may have many times more noise than temperature signal at a given time.

To use thermocouples effectively in industry, a number of noise reduction techniques are employed. The following three are the most popular:

1. The extension or lead wires from the thermcouple to the reference junction or measurement system are twisted and then wrapped with a grounded foil sheath.

2. The measurement junction itself is grounded at the point of measurement. The grounding is typically to the inside of the stainless steel sheath that

covers the actual thermocouple. This means that that voltage generated must be measured by a differential system that cancels noise that appears in equal magnitude on each wire of the thermocouple.

3. An instrumentation amplifier that has excellent common mode rejection is employed for measurement. This goes hand-in-hand with the use of a grounded junction.

4.6 OTHER THERMAL SENSORS

The transducers discussed in the previous sections cover a large fraction of the temperature measurement techniques used in process control. There remain, however, numerous other devices or methods of temperature measurement that may be encountered. Pyrometric methods involve measurement of temperature by the electromagnetic radiation that is emitted in proportion to temperature. This technique is discussed in detail in Chapter 6. Several other techniques are discussed in this section.

4.6.1 Bimetal Strips

This type of temperature transducer has the characteristics of being relatively inaccurate, having hysteresis, having relatively slow time response, and being low cost. Such devices are used in numerous applications, particularly where an on/off cycle rather than smooth or continuous control is desired.

Thermal expansion

We have seen that greater thermal energy causes the molecules of a solid to execute greater-amplitude and higher-frequency vibrations about their average positions. It is natural to expect that an expansion of the volume of a solid would accompany this effect, as the molecules tend to occupy more volume on the average with their vibrations. This effect varies in degree from material to material because of many factors, including molecular size and weight, lattice structure, and others. Although one can speak of a volume expansion, as described, it is more common to consider a length expansion when dealing with solids, particularly in the configuration of a rod or beam. Thus, if we have a rod of length l_0 at temperature T_0 as shown in Figure 4.11, and the temperature is raised to a new value T, then the rod will be found to have a new length l given by

$$l = l_0[1 + \gamma \Delta T] \tag{4.16}$$

where $\Delta T = T - T_0$ and γ is the linear thermal expansion coefficient appropriate to the material of which the rod is produced. Several different expansion coefficients are given in Table 4.3.

T_0

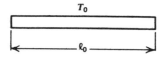

ℓ_0

$T > T_0$

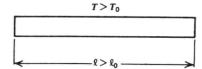

$\ell > \ell_0$

Figure 4.11 A solid object experiences a physical expansion in proportion to temperature. The effect is highly exaggerated in this drawing.

TABLE 4.3 Thermal Expansion Coefficients

Material	Expansion Coefficient
Aluminum	$25 \times 10^{-6}/°C$
Copper	$16.6 \times 10^{-6}/°C$
Steel	$6.7 \times 10^{-6}/°C$
Beryllium/copper	$9.3 \times 10^{-6}/°C$

Bimetallic sensor

The thermal sensor exploiting the effect discussed previously occurs when two materials with grossly different thermal expansion coefficients are bonded together. Thus, when heated, the different expansion of rates cause the assembly curve shown in Figure 4.12. This effect can be used to close switch contacts or to actuate an on/off mechanism when the temperature increases to some appropriate setpoint. The effect also is used for temperature indicators, by means of assemblages, to convert the curvature into dial rotation.

T_0

γ_1

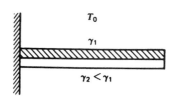

$\gamma_2 < \gamma_1$

$T > T_0$

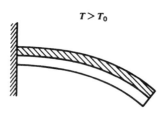

Figure 4.12 A bimetal strip will curve when exposed to a temperature change because of the different thermal expansion coefficients. The thickness of the metal strips has been exaggerated.

Example 4.12

How much will an aluminum rod of 10-m length at 20°C expand when the temperature is changed from 0°C to 100°C?

Solution   First find the length of 0°C and at 100°C, then find the difference. Using Equation (4.16) at 0°C, we get

$$l_1 = (10 \text{ m})[1 + (2.5 \times 10^{-5}/°C)(0°C - 20°C)] \tag{4.16}$$

$$l_1 = 9.995 \text{ m}$$

and at 100°C

$$l_2 = 10 \text{ m } 1 + (2.5 \times 10^{-5}/°C)(100°C - 20°C)$$

$$l_2 = 10.02 \text{ m}$$

Thus, the expansion is

$$l_2 - l_1 = 0.025 \text{ m} = \textbf{25 mm}$$

4.6.2 Gas Thermometers

The operational principle of the gas thermometer is based on a basic law of gases. In particular, if a gas is kept in a container at constant volume and the pressure and temperature vary, then the ratio of gas pressure and temperature is a constant

$$\frac{p_1}{T_1} = \frac{p_2}{T_2} \tag{4.17}$$

where

p_1, T_1 = absolute pressure and temperature (in K) in state 1
p_2, T_2 = absolute pressure and temperature in state 2

Example 4.13

A gas in a closed volume has a pressure of 120 psi at a temperature of 20°C. What will the pressure be at 100°C?

Solution   First, we convert the temperature to the absolute scale of Kelvin using Equation (4.2). We get 293.15 K and 373.15 K. Then, we use Equation (4.17) to find the pressure in state 2.

$$p_2 = \frac{T_2}{T_1} p_1$$

$$p_2 = \frac{373.15}{293.15}(120 \text{ psi})$$

$$p_2 = \textbf{153 psi}$$

Because the gas thermometer converts temperature information directly into pressure signals, it is particularly useful in pneumatic systems. Such transducers are also advantageous because there are no moving parts and no electrical stimulation is necessary. For electronic analog or digital process-control applications, however, it is necessary to devise systems for converting the pressure to electrical signals. This type of transducer is often used with Bourdon tubes (see Chapter 5) to produce direct indicating temperature meters and recorders. The gas most commonly employed is nitrogen. Time response is quite slow in relation to electrical devices because of the greater mass that must be heated.

4.6.3 Vapor Pressure Thermometers

A vapor pressure thermometer converts temperature information into pressure as does the gas thermometer, but it operates by a different process. If a closed vessel is partially filled with liquid, then the space above the liquid will consist of evaporated vapor of the liquid at a pressure that depends on the temperature. If the temperature is raised, more liquid will vaporize and the pressure will increase. A decrease in temperature also will result in condensation of some of the vapor, and the pressure will decrease. Thus, vapor pressure depends on temperature. Different materials will have different curves of pressure versus temperature, and there is no simple equation like that for a gas thermometer. Figure 4.13 shows a curve of vapor pressure versus temperature for methyl chloride, which is often employed in these transducers. The pressure available is substantial as the temperature rises. As in the case of gas thermometers, the range is not great, and response time is slow (20 seconds and more) because the liquid and the vessel must be heated.

Example 4.14

Two methyl chloride vapor pressure temperature transducers will be used to measure the temperature difference between two reaction vessels. The nominal temperature is 85°C. Find the pressure difference per degree Celsius at 85°C from the graph in Figure 4.13.

Solution   We are just estimating the slope of the graph in the vicinity of 85°C. To do this, let us find the slope between 80°C and 90°C,

$$\frac{\Delta p}{\Delta T} = \frac{400 - 320 \text{ psi}}{90 - 80°C}$$

$$\frac{\Delta p}{\Delta T} = 8 \text{ psi/°C}$$

4.6.4 Liquid-Expansion Thermometers

Just as a solid experiences an expansion in dimension with temperature, a liquid also shows an expansion in volume with temperature. This effect forms the basis for the traditional liquid-in-glass thermometers that are so common in temperature

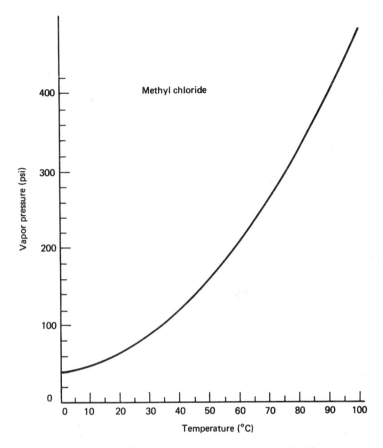

Figure 4.13   Vapor pressure curve for methyl chloride.

measurement. The relationship that governs the operation of this device is

$$V(T) = V(T_0)[1 + \beta \Delta T] \qquad (4.18)$$

where

$$
\begin{aligned}
V(T) &= \text{volume at temperature } T \\
V(T_0) &= \text{volume at temperature } T_0 \\
\Delta T &= T - T_0 \\
\beta &= \text{volume thermal expansion coefficient}
\end{aligned}
$$

In actual practice, the expansion effects of the glass container must be accounted for to obtain high accuracy in temperature indications. This type of temperature transducer is not commonly used in process-control work because further transduction is necessary to convert the indicated temperature into an electrical signal.

4.6.5 Solid-State Temperature Sensors

Many integrated circuit manufacturers now market solid-state temperature sensors for consumer and industrial applications. These devices offer voltages which vary linearly with temperature over a specified range. They function by exploiting the temperature sensitivity of doped semiconductor devices such as diodes and transistors. One common version is essentially a zener diode in which the zener voltage increases linearly with temperature.

The operating temperature of these sensors is typically in the range of $-50°C$ to $150°C$. The time constant in good thermal contact varies in the range of 1 to 5 seconds, while in poor thermal contact it may increase to 60 seconds or more. The dissipation constant is in the range of 2 to 20 mW/°C depending on the case, conditions, and heat sinking.

These sensors are easy to interface to control systems and computers and are becoming popular for measurements within the somewhat limited range they offer. An important application is to provide automatic reference temperature compensation for thermocouples. This is provided by connecting the sensor to the reference junction block of the TC and providing signal conditioning so that the reference corrections are automatically provided to the TC output. The following example illustrates a typical application.

Example 4.15

A type J TC is to be used in a measurement system which must provide an output of 2.00 volts at 200°C. A solid-state temperature sensor system will be used to provide reference temperature correction. The sensor has three terminals: supply voltage V_S, output voltage V_T, and ground. The output voltage varies as 8 mV/°C.

Solution   A type J TC with a 0°C reference will output 10.78 mV at 200°C. Therefore, the overall gain required will be

$$2.00 \text{ V}/0.01078 = 185.5$$

For compensation, the sensor will be physically connected to the reference block of the TC. The tables show that the a type J TC has a slope of approximately 50 μV/°C. Thus the output of the sensor with temperature is (8 mV/°C)/(50 μV/°C) = 160 times larger than the required correction. So we can provide the correction by amplifying the TC output by a gain of 160 and then adding the sensor reference correction. To make up the rest of the required gain of 185.5, we will need an amplifier with a gain of 185.5/160 = 1.159. The circuit is shown in Figure 4.14. The output equation is given by

$$V_{\text{out}} = 1.159[160 V_{TC} + V_C]$$

To see how well this works, suppose we consider three cases: 50°C, 150°C, and 200°C. Suppose the actual reference temperature is 20°C. The following table shows (1) the expected output voltage if the reference was 0°C and a reference correction was applied from the tables and then the output was amplified to get 2.00 at

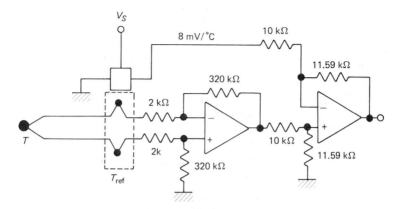

Figure 4.14   One possible solution to Example 4.15.

200°C, (2) the compensation circuit output for a reference of 20°C as determined by the circuit of Figure 4.14 and the preceding equation, and (3) the percent difference.

T (°C)	V_{TC} (0°C)	$V_{TC} \times 185.5$	V_{out} (20°C)	Difference (%)
50	2.58 mV	0.479 V	0.475 V	−0.8
100	5.27 mV	0.976 V	0.974 V	−0.2
150	8.00 mV	1.484 V	1.480 V	−0.3
200	10.78 mV	1.999 V	1.995 V	−0.2

You can see that the compensated output differs from the output expected with an actual 0°C reference by less than 1%. This example illustrates the great success of using solid-state temperature sensors for reference correction when using thermocouples.

4.7 DESIGN CONSIDERATIONS

In the design of overall process-control systems, specific requirements are set up for each element of the system. The design of the elements themselves, which constitutes subsystems, involves careful matching of the elemental characteristics to the overall system design requirements. Even in the design of monitoring systems, where no integration of subsystems is required, it is necessary to match the transducer to the measurement environment and to the required output signal. In keeping with these considerations, we can treat temperature transducer design procedures by the following steps.

   1. *Identify the nature of the measurement*   This includes the nominal value and range of the temperature measurement, the physical conditions of the environment where the measurement is to be made, required speed of measurement, and any other features that must be considered.

2. *Identify the required output signal*   In most applications the output will be either a standard 4–20 mA current or a voltage that is scaled to represent the range of temperature in the measurement. There may be further requirements related to isolation, output impedance, or other factors. In some cases, a specific digital encoding of the output may be specified.

3. *Select an appropriate sensor*   Based primarily on the results of the first step, a sensor that matches the specifications of range, environment, and so forth is selected. To some extent, factors such as cost and availability will be important in the selection of a sensor. The requirements of output signals also enter into this selection, but with lower significance because signal conditioning generally provides the required signal transformations.

4. *Design the required signal conditioning*   Using the signal conditioning techniques treated in Chapters 2 and 3, the direct transduction of temperature is converted into the required output signal. The specific type of signal conditioning depends, of course, on the type of sensor employed as well as the nature of the specified output signal characteristics.

In one form or another, these steps are required of any temperature sensor application, although they are not necessarily performed in the sequence indicated. The primary concern of this text is to enable the reader to do transducer selection and signal conditioning associated with a particular requirement. In a particular situation where a thermal sensor is required, there is no unique design to fit the application. There are, in fact, so many different designs possible that one must adjust his or her thinking from searching for *the* solution to searching for *a* solution to a design problem. A solution is *any* arrangement that satisfies *all* of the specified requirements of the problem. The following examples illustrate several problems in thermal design and typical solutions.

Example 4.16

Develop a system that turns on an alarm LED when the temperature in a chamber reaches 10 ± 0.5°C. When the temperature drops below about 8°C, the LED should be turned off.

Solution   This looks like a natural for a hysteresis comparator, as discussed in Chapter 3. What we need to do is develop a measurement to give a voltage that rises with temperature to trigger the comparator at 10°C and then allow the hysteresis to leave it on until the temperature falls to 8°C. We will work with three significant figures.

Because we are interested in only two specific temperatures and not general measurement, let us use a thermistor. The nonlinearity will not matter, as we are interested in only two specific values of temperature/resistance. The thermistor shown in Figure 4.5 has resistances of 10 kΩ at 10°C and 11 kΩ at 8°C. The ±0.5°C requirement means that the self-heating must be kept below 0.5°C; just to be sure, let us use 0.25°C. Because the dissipation constant for this thermistor was given as

5 mW/°C, we can determine the maximum allowable power dissipated by the transducer from Equation (4.13).

$$P = P_D \Delta T = (5 \text{ mW/°C})(0.25°C)$$

$$P = 1.25 \text{ mW}$$

This represents a maximum current at 10°C of

$$I = [P/R]^{1/2} = [1.25 \text{ mW}/10 \text{ k}\Omega]^{1/2}$$

$$I = 0.354 \text{ mA}$$

Since this is a maximum, we will use $I = 0.35$ mA. We will convert the resistance variation to voltage variation using a divider, such as shown in Figure 2.3, and then feed the divider voltage to a hysteresis comparator. A common supply is $+5$ volts, so using this for V_S, we will use the thermistor for R_1 (so V_D will increase with decreasing resistance which is increasing temperature). R_2 will be determined by requiring the current to be 0.35 mA. Thus, the voltage dropped across the thermistor at 10°C is

$$V_{TH} = IR_1 = (0.35 \text{ mA})(10 \text{ k}\Omega) = 3.5 \text{ V}$$

Therefore, the value of R_2 is

$$R_2 = V/I = (5 - 3.5)/0.35 \text{ mA} = 4.28 \text{ k}\Omega$$

We will simply use 4.3 kΩ, as it is a common fixed resistance value. We now have the two voltages of interest from the divider,

$$\text{For } 10°C, \ V_D = 1.50 \text{ V}$$

$$\text{For } 8°C, \ V_D = 1.41 \text{ V}$$

Because the difference is 0.09 V, this must be the required hysteresis voltage. Assuming the comparator output is 5.0 V in the high state, the ratio of input to feedback resistance can be found from Equation (3.4), where $V_{ref} = 1.50$ V.

$$(R/R_f)(5.0 \text{ V}) = 0.09 \text{ V}$$

$$(R/R_f) = 0.018$$

We will pick $R_f = 500$ kΩ, then $R = 9$ kΩ. The final design is shown in Figure 4.15. The reference voltage has been achieved by a divider with a trimmer resistor. As usual, many other designs are possible.

Example 4.17

Figure 4.16 shows an industrial process. Vapor flows through a chamber containing a liquid at 100°C. A control system will regulate the vapor temperature, so a measurement must be provided to convert 50°C to 80°C into 0 to 2.0 volts. The error should not exceed ± 1°C. If the liquid level rises to the tip of the transducer, its temperature will rise suddenly to 100°C. This event should cause an alarm comparator output to go high.

Solution   This is an example of a midrange temperature measurement. Let us use

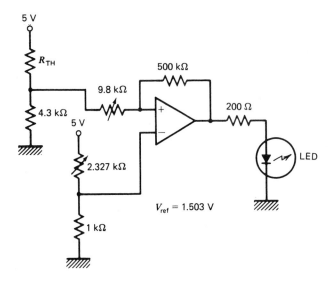

Figure 4.15 Circuit solution for Example 4.16. Trimmer resistors are used to obtain nonstandard values.

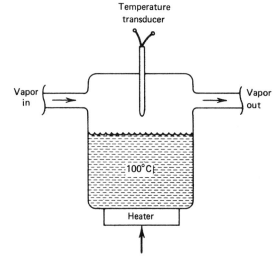

Figure 4.16 Vapor temperature-control process for Example 4.17.

an RTD because the output over the 30° range will be substantially linear. Here are the specifications:

$$R \text{ at } 65°C = 150 \ \Omega$$

$$\alpha \text{ at } 65°C = 0.004/°C$$

$$P_D = 30 \text{ mW/°C}$$

The three resistances of interest are at 50°C, 80°C, and 100°C. From the linear RTD relation for resistance [Equation (4.9)], we find

$$\text{At } 50°C \text{ RTD} = 150[1 + 0.004(50 - 65)] = 141 \ \Omega$$

$$\text{At } 80°C \text{ RTD} = 150[1 + 0.004(80 - 65)] = 159 \ \Omega$$

$$\text{At } 100°C \text{ RTD} = 150[1 + 0.004(100 - 65)] = 171 \ \Omega$$

For a 1°C error because of self-heating, we can now find the maximum current through the RTD. The maximum power is found from Equation (4.13)

$$P + P_D \Delta T = (30 \text{ mW/°C})(1°C) = 30 \text{ mW}$$

and then the maximum current from

$$I = [P/R]^{1/2} = [30 \text{ mW/159 } \Omega]^{1/2}$$

$$I = 13.7 \text{ mA}$$

Although an op amp could be used, let us place the RTD in a bridge circuit and use the offset voltage for measurement. The small range of resistance will not cause any appreciable nonlinear effects, and the bridge can be nulled at 50°C, which will simplify the signal conditioning.

The bridge will be excited from a 5.0-volt source because this value is common. We will use the RTD as R_4 of Figure 2.4. The value of R_2 is determined by the requirement that the current be below 13.7 mA. The voltage across the RTD at 80°C will be

$$V = IR = (13.7 \text{ mA})(159 \ \Omega) = 2.17 \text{ V}$$

Therefore, R_2 is found from

$$R_2 = (5 - 2.17)/13.7 \text{ mA} = 206.5 \ \Omega$$

Let us just use 220 Ω for R_2 because this is a standard value and will assure that the current is low and the error condition is satisfied. To null the bridge at 50°C, we will make R_1 = 220 Ω and use a trimmer to set R_3 to 141. The bridge is shown in Figure 4.17.

The bridge offset voltages will be found from Equation (2.8) for the endpoints and at 100°C

$$\text{At } 50°C, \Delta V = 5 \frac{141}{220 + 141} - 5 \frac{141}{220 + 141}$$

$$\Delta V = 0 \text{ (just as designed)}$$

$$\text{At } 80°C, \Delta V = 5 \frac{159}{220 + 159} - 5 \frac{141}{220 + 141}$$

$$\Delta V = 0.1447 \text{ V}$$

$$\text{At } 100°C, \Delta V = 5 \frac{171}{220 + 171} - 5 \frac{141}{220 + 141}$$

$$\Delta V = 0.2338 \text{ V}$$

All we need now is an amplifier to boost the 80°C voltage to 2.0 volts. The required gain is (2/0.1447) = 13.8. Also, because the 5-volt source used for the bridge is ground

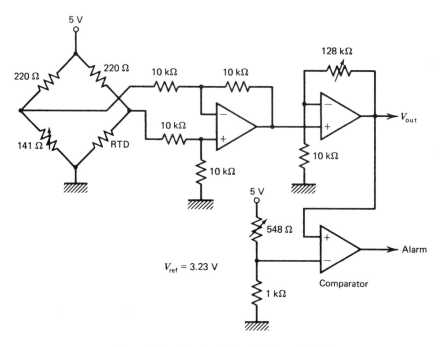

Figure 4.17 Circuit solution for Example 4.17.

referenced, we must use a differential amplifier for the bridge offset voltage. Figure 4.17 shows the required amplifier with gain. The comparator reference voltage is $V_{\text{ref}} = 13.8(0.2338) = 3.23$ V.

Example 4.18

Temperature for a plating operation must be measured for control within a range of 500°F to 600°F. Develop a measuring system that scales this temperature into 0 to 5 volts for input to an 8-bit ADC and a computer control system; measurement must be to within ±1°F.

Solution The given temperature range corresponds to 260°C–315.6°C. For this high a temperature, we will use a Type J thermocouple, although a platinum RTD also could be used.

   The reference is assumed to be 25 ± 0.5°C to satisfy the required measurement accuracy. This can be supplied by a commercial reference, a correction circuit, or by measuring the reference for input to the computer and software adjustment.

   With a 25°C reference, the Type J tables and interpolation can be used to find the voltages of the thermocouple as

$$\text{For } 260°C, \ V_{TC} = 12.84 \text{ mV}$$

$$\text{for } 315.6°C, \ V_{TC} = 15.90 \text{ mV}$$

Signal conditioning can be developed by first finding an equation for the final output

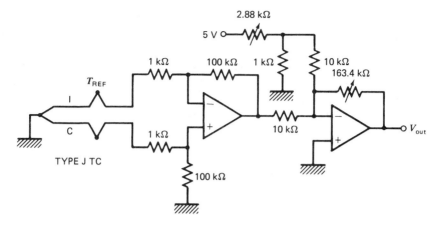

Figure 4.18   Circuit solution for Example 4.18.

from this input. We have

$$V_{ADC} = mV_{TC} + V_0$$

using the two conditions

$$0 = m(0.01284) + V_0$$

$$5 = m(0.01590) + V_0$$

we find the values $m = 1634$ and $V_0 = -21$.

This gain is too high for a single amplifier: anyway, we need a differential amplifier on the front end for the thermocouple. Let us use a differential amplifier with a gain of -100 to produce an intermediate voltage, $V_1 = -100V_{TC}$. Then the equation becomes

$$V_{ADC} = 16.34V_1 - 21$$

or

$$V_{ADC} = 16.34(V_1 - 1.29)$$

This is simply a summing amplifier. The final circuit is shown in Figure 4.18. You can see from this example that the reference requirement for thermocouples presents some problems.

SUMMARY

1. The measurement and control of temperature plays a very important role in the process-control industry. The class of transducers that performs this measurement consists primarily of three types: (1) the resistance temperature detector

(RTD), (2) the thermistor, and (3) the thermocouple. In this chapter, the basic operating principles and application information have been provided for these transducers. Several other transducers have been briefly described.

2. The concept of temperature is contained in the representation of a body's thermal energy as the average thermal energy per molecule expressed in units of degrees of temperature. Four units of temperature are in common use. Two of these scales are called absolute because an indication of zero units corresponds to zero thermal energy. These scales, which are designated by kelvin (K) and degrees Rankine (°R), differ only in the amount of energy represented by each unit. The amount of energy represented by 1 K corresponds to 9/5°R. Thus, we can transform temperatures using

$$T(\text{K}) = \frac{5}{9} T(°\text{R}) \tag{4.1}$$

3. The other two scales are called relative because their zero does not occur at a zero of thermal energy. The Celsius scale (°C) corresponds in degree size to the kelvin but has a shift of the zero so that

$$T(°\text{C}) = T(\text{K}) - 273.15 \tag{4.2}$$

4. In a similar fashion, the Fahrenheit (°F) and Rankine scales are related by

$$T(°\text{F}) = T(°\text{R}) - 459.6 \tag{4.3}$$

5. The RTD is a transducer that depends on the increase in metallic resistance with temperature. This increase is very nearly linear, and analytical approximations are used to express the resistance versus temperature as either a linear equation

$$R(T) = R(T_0)[1 + \alpha_0 \Delta T] \tag{4.9}$$

or a quadratic relationship

$$R(T) = R(T_0)[1 + \alpha_1 \Delta T + \alpha_2 (\Delta T)^2] \tag{4.12}$$

6. When greater accuracy is desired, tables or graphs of resistance versus temperature are used. Because of the small fractional change in resistance with temperature, the RTD is usually used in a bridge circuit with a high-gain null detector.

7. The thermistor is based on the decrease of semiconductor resistance with temperature. This device has a highly nonlinear resistance versus temperature curve and is not typically used with any analytical approximations. Such transducers can exhibit a very large change in resistance with temperature and hence make very sensitive temperature change detectors. Many circuit configurations are used, including bridges and operational amplifiers.

8. A thermocouple is a junction of dissimilar metal wires usually joined to a third metal wire through two reference junctions. A voltage is developed across the common metal wires which is proportional, almost linearly, to the difference in temperature between the measurement and reference junctions. Extensive tables of temperature versus voltage for numerous types of TCs using standard metals and alloys allow an accurate determination of temperature at the reference junctions. The voltage must be measured at high impedance to avoid loading effects on the measured voltage. Thus, potentiometric, operational amplifiers, or other high-impedance techniques are employed in signal conditioning.

9. A bimetal strip converts temperature into a physical motion of metal elements. This flexing can be used to close switches or cause dial indications.

10. Gas and vapor pressure temperature transducers convert temperature into gas pressure, which then is converted to an electrical signal or is used directly in pneumatic systems.

PROBLEMS

Section 4.1

4.1 Convert 453.1°R into K, °F, and °C.

4.2 Convert −222°F into °C, °R, and K.

4.3 Convert 150°C into K and °F.

4.4 A process temperature is found to change by 33.4°F. Calculate the change in °C. (*Hint*: A change in temperature does not involve a scale shift.)

4.5 A sample of hydrogen gas has a temperature of 500°C. Calculate the average molecular speed in m/s. Express this also in ft/s. *Note*: Gaseous hydrogen exists as the molecule H_2 with a mass of 3.3×10^{-27} kg.

4.6 Temperature is to be controlled in the range 350°C to 550°C. What is this expressed in °F?

Section 4.2

4.7 An RTD has α (20°C) = 0.004/°C. If $R = 106 \ \Omega$ at 20°C, find the resistance at 25°C.

4.8 The RTD of Problem 4.7 is used in the bridge circuit of Figure 4.4. If $R_1 = R_2 = R_3 = 100 \ \Omega$ and the supply voltage is 10.0 volts, calculate the voltage the detector must be able to resolve in order to resolve a 1.0°C change in temperature.

4.9 Use the values of RTD resistance versus temperature shown to find the equations for the linear and quadratic approximations of resistance between 100°C and 130°C. Assume $T_0 = 115$°C.

$T(°C)$	$R(\Omega)$
90.0	562.66
95.0	568.03
100.0	573.40
105.0	578.77
110.0	584.13
115.0	589.48
120.0	594.84
125.0	600.18
130.0	605.52

What is the error between table resistance at 105°C and that determined from the two approximations?

4.10 Suppose the RTD of Problem 4.7 has a dissipation constant of 25 mW/°C and is used in a circuit which puts 8 mA through the sensor. If the RTD is placed in a bath at 100°C, what resistance will the RTD have? What then is the indicated temperature?

Section 4.4

4.11 In Problem 4.8, the RTD is replaced by a thermistor with R (20°C) $= 100\ \Omega$ and an R versus T of $-10\%/°C$ near 20°C. Calculate the voltage resolution of the detector needed to resolve a 1.0°C change in temperature.

4.12 Modify the divider of Example 4.8 so that self-heating is reduced to 0.1°C. What is the divider voltage for 20°C? What is the divider voltage for 19°C and 21°C?

4.13 The thermistor of Figure 4.5 is used as shown in Figure 4.19 to convert temperature into voltage. Plot V_{out} versus temperature from 0°C to 80°C. Is the result linear? What is the maximum self-heating if $P_D = 5$ mW/°C?

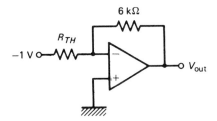

Figure 4.19 Circuit for Problem 4.13.

Section 4.5

4.14 A type J thermocouple measures 22.5 mV with a 0°C reference. What is the junction temperature?

4.15 A type S TC with a 21°C reference measures 12.120 mV. What is the junction temperature?

4.16 If a type J TC is to measure 500°C with a $-10°C$ reference, what voltage will be produced?

4.17 A type K TC with a 0°C reference will monitor an oven temperature at about 300°C.
   a. What voltage would be expected?

b. Extension wires of 1000-ft length and 0.01 Ω/ft will be used to connect to the measurement site. Determine the minimum voltage measurement input impedance if the error is to be within 0.2%.

4.18 We need to get 1.5 volts from a candle flame at about 700°C. How many type K TCs will be required in series if the reference is assumed to be nominal room temperature, 70°F?

Section 4.6

4.19 A thermoelectric switch is composed of a copper rod 10 cm long at 20°C which is to touch an electrical contact at a temperature of 150°C. What distance should there be between the rod end and the contact at 20°C?

4.20 A gas thermometer has a pressure of 125 kPa at 0°C and 215 kPa at some unknown temperature. Determine the temperature in °C.

4.21 A methyl chloride vapor pressure thermometer will be used between 70°F and 200°F. What pressure range corresponds to this temperature range?

4.22 Find a linear approximation of methyl chloride pressure versus temperature from 70°C to 90°C. Find the maximum error between the approximation and actual pressure in this range.

Section 4.7

4.23 Using an RTD with $\alpha = 0.0034$/°C and $R = 100$ Ω at 20°C, design a bridge and op amp system to provide a 0.0- to 10.0-volt output for a 20°C to 100°C temperature variation. The RTD dissipation constant is 28 mW/°C.

4.24 A type K TC with a 0°C reference will be used to measure temperature between 200°C and 350°C. Devise a system that will convert this temperature range into an 8-bit digital word with conversion from 00H to 01H at 200°C and the change from FEH to FFH occuring at 350°C. An ADC is available with a 2.500-volt internal reference.

4.25 Solve Example 4.17 using a type K TC with a reference compensated to 0°C.

4.26 Solve Example 4.18 using an RTD with $R = 500$ Ω and $\alpha = 0.003$/°C at 260°C. The dissipation constant is 25 mW/°C.

4.27 You have been commissioned to design a thermistor-based digital temperature measurement system. The ADC has a 5.00-volt reference and is 8 bits. The thermistor specifications are $R = 5.00$ kΩ at 90°F, $P_D = 5$ mW/°C, and a slope between 90°F and 110°F of -8 Ω/°C. The design should be made so that 90°F gives an ADC output of 5AH (90_{10}) and 110°F gives 6EH (110_{10}).

4.28 Solve Example 4.17 using the RTD as the feedback element of an inverting amplifier instead of using a bridge circuit.

4.29 A type K TC measurement system must provide an output of 0 to 2.5 volts for a temperature variation of 500 to 700°C. A three-terminal solid-state sensor with 12 mV/°C will be used to provide reference compensation. Develop the complete circuit.

4.30 A calibrated RTD with $\alpha = 0.0041$/°C, $R = 306.5$ Ω at 20°C, and $P_D = 30$ mW/°C will be used to measure a critical reaction temperature. Temperature must be measured between 50 and 100°C with a resolution of 0.1°C. Devise a signal conditioning system which will provide an appropriate digital output to a computer. Specify the requirements on the ADC and appropriate analog signal conditioning to interface to your ADC.

CHAPTER 5

MECHANICAL SENSORS

INSTRUCTIONAL OBJECTIVES

The objectives of this chapter are confined to the subject of mechanical transducers. After you have read this chapter, you should be able to

1. **Define the relationship between acceleration, velocity, and position.**
2. **Define the characteristics of vibration and shock.**
3. **Draw and label a typical stress-strain curve.**
4. **Design the application of an LVDT to a displacement measurement problem.**
5. **Describe the types of accelerometers and the characteristics of each.**
6. **Design a system of strain measurement using metal foil strain gauges.**
7. **Define two types of pressure measurement with electrical signal output.**
8. **Diagram a system of flow measurement using differential pressure measurement.**

5.1 INTRODUCTION

The class of sensors used for the measurement of *mechanical* phenomena is of
special significance because of the extensive use of these devices throughout the
process-control industry. In many instances, an interrelation exists by which a
transducer designed to measure some mechanical variable is used to measure
another variable. To learn to use mechanical sensor transducers, it is important

to understand the mechanical phenomena itself, the operating principles of the transducer, and the application details of the transducer. All of these will be discussed for the measurement devices presented in this chapter.

The objectives of this chapter are confined to the subject of mechanical transducers. Each of the mechanical dynamic variables presented is itself a complex field of study. Our purpose here is to give an overview of the essential features associated with each to make the reader conversant with the principal transducers used to measure mechanical variables, the characteristics of each, and appropriate application notes. As in previous chapters, an expert understanding of the phenomenon is not required to effectively employ transducers for its measurement.

5.2 DISPLACEMENT, LOCATION, OR POSITION SENSORS

The measurement of displacement, position, or location is an important topic in the process industries. Examples of industrial requirements to measure these variables are many and varied, and the required transducers are also of greatly varied designs. To give a few examples: (1) location and position of objects on a conveyor system, (2) orientation of steel plates in a rolling mill, (3) liquid/solid level measurements, (4) location and position of work piece in automatic milling operations, and (5) conversion of pressure to a physical displacement that is measured to indicate pressure. In the following sections, the basic principles of several common types of displacement, position, and location sensors are given.

5.2.1 Potentiometric

The simplest type of displacement sensor involves the action of displacement in moving the wiper of a potentiometer. This device then converts linear or angular motion into a changing resistance that may be converted directly to voltage and/ or current signals. Such potentiometric devices often suffer from the obvious problems of mechanical wear, friction in the wiper action, limited resolution in wirewound units, and high electronic noise. (See Figure 5.1.)

Example 5.1

A potentiometric displacement sensor is to be used to measure work-piece motion from 0 to 10 cm. The resistance changes linearly over this range from 0 to 1 kΩ. Develop signal conditioning to provide a linear, 0- to 10-volt output.

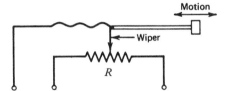

Figure 5.1 Potentiometric displacement sensor.

Solution The key thing here is not to lose the linearity of the resistance versus displacement. We cannot use a voltage divider because the voltage versus resistance variation is *not* linear for that circuit. We can use an op amp circuit, however, because the gain and therefore the output voltage is linearly dependent on the feedback resistor. Therefore, the circuit of Figure 5.2 will satisfy this problem. The output voltage is

$$V_{\text{out}} = -(R_D/1 \text{ k}\Omega)(-10 \text{ V})$$

$$V_{\text{out}} = 0.01R_D$$

As R_D varies from 0 to 1 kΩ, the output will change linearly from 0 to 10 volts.

5.2.2 Capacitive and Inductive

A second class of sensors for displacement measurement involves changes in capacity or inductance.

Capacitive

The basic operation of a capacitive sensor can be seen from the familiar equation for a parallel-plate capacitor

$$C = K\epsilon_0 \frac{A}{d}$$

where

K = the dielectric constant
ϵ_0 = permittivity = 8.85 pF/m
A = plate common area
d = plate separation

There are three ways to change the capacity: variation of the distance between the plates (d), variation of the shared area of the plates (A), and variation of the

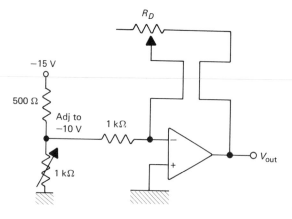

Figure 5.2 Circuit for Example 5.1.

dielectric constant (K). The former two methods are shown in Figure 5.3. The last method is illustrated later in this chapter. An ac bridge circuit or other active electronic circuit is employed to convert the capacity change to a current or voltage signal.

Inductive

If a permeable core is inserted into an inductor as shown in Figure 5.4, the net inductance is increased. Every new position of the core produces a different inductance. In this fashion, the inductor and movable core assembly may be used as a displacement transducer. An ac bridge or other active electronic circuit sensitive to inductance then may be employed for signal conditioning.

5.2.3 Variable Reluctance

The class of variable reluctance displacement sensors differs from the inductive in that a moving core is used to vary the magnetic flux coupling between two or more coils, rather than changing an individual inductance. Such devices find ap-

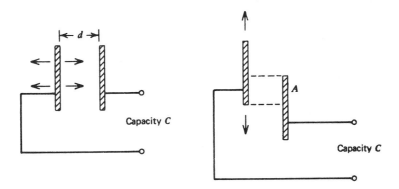

Figure 5.3 Capacity varies with the distance between the plates and the common area. Both effects are used in sensors.

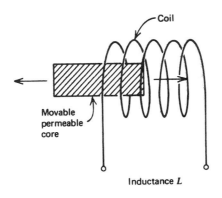

Figure 5.4 Differential core positions in the coil produce changes in inductance which can be employed for displacement measurements.

plication in many circumstances for the measure of both translational and angular displacements. Many configurations of this device exist, but the most common and extensively used is called a *linear variable differential transformer* (LVDT).

LVDT

The basic structure of the LVDT is a movable core of permeable material and three coils, as shown in Figure 5.5. The inner core is the primary that provides magnetic flux through its excitation by some ac source. The two secondary coils have voltages induced because of the flux linkages with the primary. When the core is centrally located, the voltage induced in each secondary is the same. But when the core is displaced, the change in flux linkage causes one secondary voltage to increase and the other to decrease. The secondary windings are generally connected in series opposition so that the voltages induced in each are out of phase with the other. In this case, as shown in Figure 5.6, the output voltage amplitude is zero when the core is centrally located and increases as the core is moved either in or out. It happens that the voltage amplitude is *linear* with core displacement over some range of core travel. Furthermore, there is a phase shift as the core moves from the central location, so that phase measurement relates the direction of core motion.

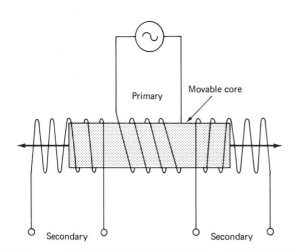

Figure 5.5   A linear variable differential transformer (LVDT) has a primary excitation coil and two series-connected secondary coils with a movable core.

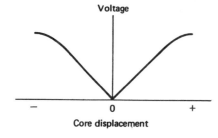

Figure 5.6   Over some range of core motion, the LVDT voltage amplitude versus core displacement is very linear. This is for opposition-connected secondaries.

A signal conditioning circuit providing a dc level proportional to amplitude and a polarity proportional to direction is easily designed. The LVDT is a device that outputs a *bipolar* voltage *linearly proportional* to displacement. A simple circuit is shown in Figure 5.7.

Commercial LVDTs that can resolve displacements as small as 0.002 mm or 2 μm are available. Some have built-in oscillator excitation, a phase-sensitive detector, and demodulation in an IC so that only a power supply is required. The dc voltage output is linear with displacement. Many variations of this device have been designed for special applications, including angular displacement measurements.

5.2.4 Level Sensors

The measurement of solid or liquid level is a special class of displacement sensors. The level measured is most commonly associated with material in a tank or hopper. A great variety of measurement techniques exist, as the representative sample of the following examples show.

Mechanical

One of the most common techniques for level measurement, particularly for liquids, is a float that is allowed to ride up and down with level changes. This float, as shown in Figure 5.8a, is connected by linkages to a secondary displacement measuring system such as a potentiometric device or an LVDT core.

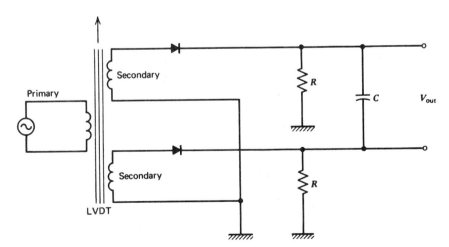

Figure 5.7   This LVDT circuit provides a dc output voltage giving the direction of core motion by the polarity, and the extent of motion by the amplitude.

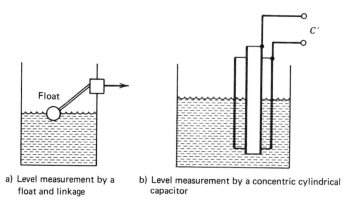

a) Level measurement by a   b) Level measurement by a concentric cylindrical
   float and linkage   capacitor

Figure 5.8   There are many level measurement techniques.

Electrical

There are several purely electrical methods of measuring level. For example, one may use the inherent conductivity of a liquid or solid to vary the resistance seen by probes inserted into the material. Another common technique is illustrated in Figure 5.8b. In this case two concentric cylinders are contained in a liquid tank. The level of the liquid partially occupies the space between the cylinders, with air in the remaining part. This device acts like two capacitors in parallel, one with the dielectric constant of air (≈ 1) and the other with that of the liquid. Thus, variation of liquid level causes variation of the electrical capacity measured between the cylinders.

Ultrasonic

The use of ultrasonic reflection to measure level is favored because it is a "noninvasive" technique, that is, it does not involve placing anything in the material. Figures 5.9a and 5.9b show the external and internal techniques. Obviously, the external technique is better suited to solid-material level measurement. In both cases the measurement depends on the length of time taken for reflections of an ultrasonic pulse from the surface of the material. Ultrasonic techniques based on reflection time also have become popular for ranging measurements.

Pressure

For liquid measurement, it is also possible to make a noncontact measurement of level if the density of the liquid is known. This method is based on the well-known relationship between pressure at the bottom of a tank, and the height and density of the liquid. This is addressed further in Section 5.5.1.

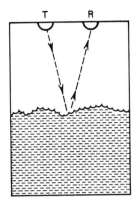

a) Solid or liquid, above surface
   measurement

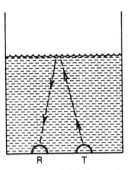

b) Liquid material, below surface
   material

Figure 5.9   Ultrasonic level measurement needs no physical contact with the material, just a transmitter T and receiver R.

Example 5.2

The level of ethyl alcohol is to be measured from 0 to 5 m using a capacitive system such as that shown in Figure 5.8b. The following specifications define the system:

For ethyl alcohol: $K = 26$ (for air, $K = 1$)

Cylinder separation: $d = 0.5$ cm

Plate area: $A = \pi RL$

where

$R = 5.75$ cm = average radius
L = distance along cylinder axis

Find the range of capacity variation as the alcohol level varies from 0 to 5 m.

Solution   We saw earlier that the capacity is given by $C = K\epsilon_0(A/d)$. Therefore, all we need to do is find the capacity for the entire cylinder with *no* alcohol and then multiply that by 26.

$$A = 2\pi RL = 2\pi(0.0575 \text{ m})(5 \text{ m}) = 1.806 \text{ m}^2$$

Thus, for air

$$C = (1)(8.85 \text{ pF/M})(1.806 \text{ m}^2/0.005 \text{ m})$$

$$C = 3196 \text{ pF} \approx 0.0032 \text{ } \mu\text{F}$$

With the ethyl alcohol, the capacity becomes

$$C = 26(0.0032 \text{ } \mu\text{F})$$

$$C = \textbf{0.0832 } \boldsymbol{\mu}\textbf{F}$$

The range is 0.0032 to 0.0832 μF.

5.3 STRAIN SENSORS

Although not obvious at first, the measurement of strain in solid objects is very common in process control. The reason it is not obvious is that strain sensors are used as a secondary step in transducers to measure many other process variables, including flow, pressure, weight, and acceleration. Strain measurements have been used to measure pressures from over a million pounds per square inch to those within living biological systems. We will first review the concept of strain and how it is related to the forces that produce it, and then discuss the sensors used to measure strain.

5.3.1 Strain and Stress

Strain is the result of the application of forces to solid objects. The forces are defined in a special way described by the general term, *stress*. For those readers needing a review of force principles, Appendix 4 discusses elementary mechanical principles including force. In this section we will define stress and the resulting strain.

Definition

A special case exists for the relation between force applied to a solid object and the resulting deformation of that object. Solids are assemblages of atoms in which the atomic spacing has been adjusted to render the solid in equilibrium with all external forces acting on the object. This spacing determines the physical dimensions of the solid. If the applied forces are changed, the object atoms rearrange themselves again to come into equilibrium with the *new* set of forces. This rearrangement results in a change in physical dimensions that is referred to as a *deformation* of the solid.

The study of this phenomena has evolved into a very exact technology. The effect of applied force is referred to as a *stress* and the resulting deformation as a *strain*. To facilitate a proper analytical treatment of the subject, stress and strain are carefully defined to emphasize the physical properties of the material being stressed and the specific type of stress applied. We delineate here the three most common types of *stress-strain* relationships.

Tensile stress-strain

In Figure 5.10a, the nature of a tensile force is shown as a force applied to a sample of material so as to elongate or pull apart the sample. In this case, the stress is defined as

$$\text{Tensile Stress} = \frac{F}{A} \qquad (5.1)$$

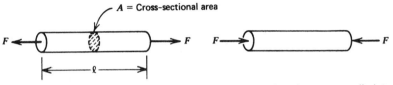

a) Tensile stress applied to a rod

b) Compressional stress applied to a rod

Figure 5.10   Tensile and compressional stress can be defined in terms of forces applied to a rod.

where

F = applied force in N
A = cross-sectional area of the sample in m^2

We see that the units of stress are N/m^2 in the SI units (or lb/in^2 in the English units) and are thus like a pressure.

The *strain* in this case is defined as the *fractional change in length* of the sample

$$\text{Tensile Strain} = \frac{\Delta l}{l} \qquad (5.2)$$

where

Δl = change in length in m (in)
l = original length in m (in)

Strain is thus a unitless quantity.

Compressional stress-strain

The only differences between *compressional* and *tensile* stress are the direction of the applied force and the polarity of the change in length. Thus, in a compressional stress, the force *compresses* the sample, as shown in Figure 5.10b. The compressional stress is defined as in Equation (5.1).

$$\text{Compressional Stress} = \frac{F}{A} \qquad (5.3)$$

The resulting strain is also defined as the fractional change in length but where the sample will now decrease in length.

$$\text{Compressional Strain} = \frac{\Delta}{l/l}$$

Shear stress-strain

Figure 5.11a shows the nature of the shear stress. In this case, the force is applied as a *couple* (that is, *not* along the same line), tending to shear off the solid object that separates the force arms. In this case, the stress is again

$$\text{Shear Stress} = \frac{F}{A} \tag{5.4}$$

where

F = force in N
A = cross-sectional area of sheared member in m^2

The strain in this case is defined as the fractional change in dimension of the sheared member. This is shown in the cross-sectional view of Figure 5.11b.

$$\text{Shear Strain} = \frac{\Delta x}{l} \tag{5.5}$$

where

Δx = deformation in m (as shown in Figure 5.11b)
l = width of a sample in m

Stress-strain curve

If a specific sample is exposed to a range of applied stress and the resulting strain is measured, a graph similar to Figure 5.12 results. This graph shows that the relationship between stress and strain is *linear* over some range of stress. If the stress is kept *within* the *linear* region, the material is essentially *elastic* in that if the stress is removed, the deformation is also gone. But if the elastic limit is exceeded, *permanent* deformation results. The material may begin to "neck" at some location and finally break. Within the *linear* region, it is found that a specific type of material will always follow the same curves despite different physical dimensions. Thus, we can say that the linearity and slope are a constant of the *type of material* only. In tensile and compressional stress, this constant is called the modulus of elasticity or Young's modulus, as given by

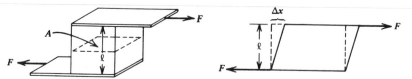

a) Shear stress results from a force couple

b) Shear stress tends to deform an object as shown

Figure 5.11   Shear stress is defined through the elements of this figure.

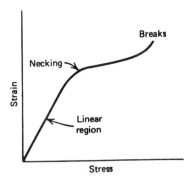

Figure 5.12 A typical stress-strain curve showing the linear region, necking, and eventual breaking.

$$E = \frac{\text{Stress}}{\text{Strain}} = \frac{F/A}{\Delta l/l} \tag{5.6}$$

where

stress $= F/A$ in N/m^2 (or lb/in^2)
strain $= \Delta l/l$ unitless
$E =$ Modulus of elasticity in N/m^2

The modulus of elasticity has units of stress, that is, N/m^2. Table 5.1 gives the modulus of elasticity for several materials. In an exactly similar fashion the shear modulus is defined for shear stress-strain as

$$M = \frac{\text{Stress}}{\text{Strain}} = \frac{F/A}{\Delta x/l} \tag{5.7}$$

where Δx is defined in Figure 5.11b and all other units have been defined in Equation (5.6).

Example 5.3

Find the strain that results from a tensile force of 1000 N applied to a 10-m aluminum beam having a 4×10^{-4} m^2 cross-sectional area.

Solution The modulus of elasticity of aluminum is found from Table 5.1 to be $E = 6.89 \times 10^{10}$ N/m^2. Now we have

$$E = \frac{F/A}{\Delta l/l} \tag{5.6}$$

TABLE 5.1   Modulus of Elasticity

Material	Modulus (N/m^2)
Aluminum	6.89×10^{10}
Copper	11.73×10^{10}
Steel	20.70×10^{10}
Polyethylene (plastic)	3.45×10^{8}

so that

$$\text{Strain} = \frac{F}{EA}$$

$$= \frac{10^3 \text{ N}}{(4 \times 10^{-4} \text{ m}^2)(6.89 \times 10^{10} \text{ N/m}^2)}$$

$$= \mathbf{3.63 \times 10^{-5}} \quad \text{or} \quad \mathbf{36.3 \ \mu m/m} \text{ (see next paragraph)}$$

Strain units

Although strain is a unitless quantity, it is common practice to express the strain as the ratio of two length units, for example, as m/m or in/in; also, because the strain is usually a very small number, a micro (μ) prefix is often included. In this sense, a strain of 0.001 would be expressed as 1000 μin/in, or 1000 μm/m. In the previous example, the solution is stated as 36.3 μm/m. In general, the smallest value of strain encountered in most applications is 1 μm/m. Strain is a unitless quantity so that it is not necessary to do unit conversions. A strain of 153 μm/m also could be written in the form of 153 μin/in or even 153 μfurlongs/furlong.

5.3.2 Strain Gauge Principles

In Section 4.2.1, we saw that the resistance of a metal sample is given by

$$R_0 = \rho \frac{l_0}{A_0} \tag{4.7}$$

where

R_0 = sample resistance Ω
ρ = sample resistivity Ω-m
l_0 = length in m
A_0 = cross-sectional area in m^2

Suppose this sample is now stressed by the application of a force F as shown in Figure 5.10a. Then we know that the material elongates by some amount Δl so that the new length is $l = l + \Delta l$. It is also true that in such a stress-strain condition, although the sample lengthens, its volume will remain nearly constant. Because the volume unstressed is $V = l_0 A_0$ it follows that if the volume remains constant and the length increases, then the area must *decrease* by some amount ΔA

$$V = l_0 A_0 = (l_0 + \Delta l)(A_0 - \Delta A) \tag{5.8}$$

Because *both* length and area have changed, we find that the resistance of the sample will have also changed

$$R = \rho \frac{l_0 + \Delta l}{A_0 - \Delta A} \tag{5.9}$$

Using Equations (5.8) and (5.9), the reader can verify the new resistance is approximately given by

$$R \simeq \rho \frac{l_0}{A_0} \left(1 + 2 \frac{\Delta l}{l_0} \right) \tag{5.10}$$

from which we conclude that the change in resistance is

$$\Delta R \simeq 2 R_0 \frac{\Delta l}{l_0} \tag{5.11}$$

Equation (5.11) is the basic equation that underlies the use of metal strain gauges because it shows that the strain $\Delta l/l$ converts directly into a *resistance change*.

Example 5.4

Find the change in a nominal wire resistance of 120 Ω that results from a strain of 1000 μm/m.

Solution   We can find the change in gauge resistance from

$$\Delta R \simeq 2 R_0 \frac{\Delta l}{l_0}$$

$$\Delta R \simeq (2)(120)(10^{-3}) \tag{5.11}$$

$$\Delta R = \mathbf{0.24 \ \Omega}$$

Example 5.4 shows a very significant factor regarding strain gauges. The change in resistance is very small for typical strain values. For this reason, resistance change measurement methods used with strain gauges must be very sophisticated.

Measurement principles

The basic technique of strain gauge (SG) measurement involves attaching (gluing) a metal wire or foil to the element whose strain is to be measured. As stress is applied and the element deforms, the SG material experiences the *same* deformation if it is securely attached. Because strain is a *fractional* change in length, the change in SG resistance reflects the strain of both the gauge and the element to which it is secured.

Temperature effects

If not for temperature compensation effects, the aforementioned method of SG measurement would be useless. To see this we need only note that the metals used in SG construction have linear temperature coefficients of $\alpha \cong 0.004/°C$, typical for most metals. Temperature changes of 1°C are not uncommon in measurement conditions in the industrial environment. If the temperature change in Example 5.4 had been 1°C, substantial change in resistance would have resulted.

Thus, from Chapter 4,

$$R(T) = R(T_0)[1 + \alpha_0 \Delta T] \qquad (4.9)$$

or

$$\Delta R_T = R_0 \alpha \Delta T$$

where

ΔR_T = resistance change because of temperature change
$\alpha_0 \simeq 0.004/°C$ in this case
$\Delta T \simeq 1°C$ in this case
$R(T_0) = 120\ \Omega$ nominal resistance

Then, we find $\Delta R_T = 0.48\ \Omega$, which is *twice the change* because of strain! Obviously, temperature effects can mask the strain effects we are trying to measure. Fortunately, we are able to compensate for temperature and other effects, as shown in the signal conditioning methods in the next section.

5.3.3 Metal Strain Gauges (SGs)

Metal SGs are devices that operate on the principles discussed earlier. The following items are important to understanding SG applications.

Gauge factor

The relation between strain and resistance change of Equation (5.11) is only approximately true. Impurities in the metal, the type of metal, and other factors lead to slight corrections. An SG specification always indicates the correct relation through statement of a *gauge factor* (GF), which is defined as

$$GF = \frac{\Delta R/R}{\text{Strain}} \qquad (5.12)$$

where

$\Delta R/R$ = fractional change in gauge resistance because of strain
strain = $\Delta l/l$ = fractional change in length

For metal gauges, this number is always close to 2. For some special alloys and carbon gauges, the GF may be as large as 10. A high gauge factor is desirable because it indicates a large change in resistance for a given strain and is easier to measure.

Construction

Strain gauges are used in two forms, wire and foil. The basic characteristics of each type are the same in terms of resistance change for a given strain. The design of the SG itself is such as to make it very long to give a large enough nominal resistance (to be practical) and to make the gauge of sufficiently fine wire or foil so as to not resist strain effects. Finally, it is necessary to make the gauge sensitivity *unidirectional* so that it responds to strain *only in one direction*. In Figure 5.13, we see the common *pattern* of SGs that provides these characteristics. By folding the material back and forth as shown, we achieve a long length to provide high resistance. Further, if a strain is applied transversely to the SG length, the pattern will tend to *unfold* rather than stretch with no change in resistance. These gauges are usually mounted on a paper backing that is bonded (using epoxy) to the element whose strain is to be measured. The normal SG resistances available are typically 60, 120, 240, 350, 500, and 1000 Ω.

Signal conditioning

Two effects are critical in the signal conditioning techniques used for SGs. The first is the small, fractional changes in resistance that require carefully designed resistance *measurement* circuits. A good SG system might require a resolution of 2 μm/m strain. From Equation (5.11), this would result in a ΔR of only 4.8 $\times$ 10^{-4} Ω for a nominal gauge resistance of 120 Ω.

The second effect is the need to provide some compensation for temperature effects to eliminate masking changes in strain.

The bridge circuit provides the answer to both areas. The sensitivity of the bridge circuit for detecting small changes in resistance is well known. Furthermore, by using a dummy gauge as shown in Figure 5.14a, we can provide the required temperature compensation. In particular, the dummy is mounted in an insensitive orientation (Figure 5.14b), but in the same proximity as the active SG.

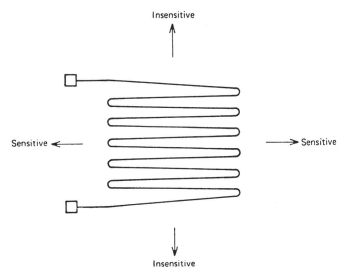

Figure 5.13   The most common foil and wire strain gauge pattern. This device is sensitive to strain in only one direction.

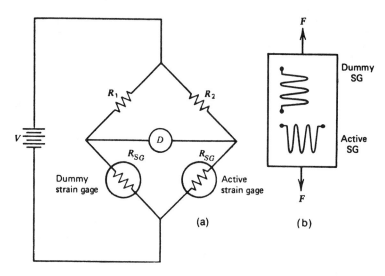

Figure 5.14 It is necessary to employ temperature compensation in strain gauge bridges. This is usually accomplished by the use of a dummy gauge, as shown.

Then, both gauges change in resistance from temperature effects, but the detector does not respond to a change in both strain gauges. Only the active SG responds to strain effects. Other configurations are used where more than one gauge is active, but the basic idea of temperature compensation is retained.

Example 5.5

A strain gauge with a GF = 2.03 and R = 350 Ω is used in the bridge of Figure 5.14a. The bridge resistors are $R_1 = R_2 = 350$ Ω, and the dummy gauge has $R = 350$ Ω. If a strain of 1450 μm/m is applied, find the bridge offset voltage if $V_S = 10.0$ volts.

Solution   With no strain, the bridge is balanced. When the strain is applied, the gauge resistance will change by a value given by

$$\text{GF} = \frac{\Delta R/R}{\text{Strain}} \tag{5.12}$$

Thus,

$$\Delta R = (\text{GF})(\text{strain})(R)$$

$$\Delta R = (2.03)(1.45 \times 10^{-3})(350\ \Omega)$$

$$\Delta R = 1.03\ \Omega$$

Thus, the new resistance $R = 351$ Ω. The bridge offset voltage is

$$\Delta V = \frac{RV}{R_1 + R} - \frac{R_{SG}V}{R_{SG} + R_2}$$

Thus,

$$\Delta V = 5 - \frac{(351)(10)}{701}$$

$$\Delta V = -\mathbf{0.007} \text{ V}$$

so that a 7-mV offset results.

5.3.4 Semiconductor Strain Gauges (SGs)

The use of semiconductor material, notably silicon, for SG application has increased over the past few years. There are presently several disadvantages to these devices compared to the metal variety, but numerous advantages for their use.

Principles

As in the case of the metal SGs, the basic effect is a change of resistance with strain. In the case of a semiconductor, the resistivity also changes with strain along with the physical dimensions. This is due to changes in electron and hole mobility with changes in crystal structure as strain is applied. The net result is a much *larger* gauge factor than is possible with metal gauges.

Gauge factor

The semiconductor device gauge factor (GF) is still given by

$$\text{GF} = \frac{\Delta R/R}{\text{Strain}} \tag{5.12}$$

The value of the semiconductor gauge factor varies between -50 and -200. Thus, resistance changes will be factors of from 25 to 100 times those available with metal SGs. It must also be noted, however, that these devices are highly *nonlinear* in resistance versus strain. In other words, the gauge factor is not a constant as the strain takes place. Thus, the gauge factor may be -150 with no strain but drop (nonlinearly) to -50 at 5000 μm/m. The resistance change will be nonlinear with respect to strain. To use the semiconductor strain gauge to measure strain, we must have a curve or table of values of gauge factor versus resistance.

Construction

The semiconductor strain gauge physically appears as a band or strip of material with electrical connection, as shown in Figure 5.15. The gauge is either bonded directly onto the test element or, if encapsulated, is attached by the encapsulation material. These SGs also appear as IC assemblies in configurations used for other measurements.

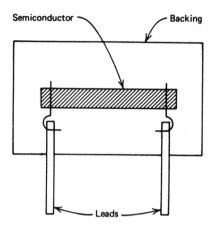

Figure 5.15 Typical semiconductor strain gauge configuration.

Signal conditioning

The signal conditioning is still typically a bridge circuit with temperature compensation. An added problem is the need for linearization of the output because the basic resistance versus strain characteristic is nonlinear. This is usually provided by *active* linearization circuits.

Example 5.6

(a) Contrast the resistance change produced by a 150-μm/m strain in a metal gauge with GF = 2.13 with

(b) A semiconductor SG with GF = −151. Nominal resistances are *both* 120 Ω.

Solution   From the basic equation

$$GF = \frac{\Delta R/R}{\text{Strain}}$$

(a) We find for the *metal* gauge SG

$$\Delta R = (120 \ \Omega)(2.13)(0.15 \times 10^{-3})$$

$$\Delta R = 0.038 \ \Omega$$

(b) For the semiconductor gauge, the change is

$$\Delta R = (120 \ \Omega)(-151)(0.15 \times 10^{-3})$$

$$\Delta R = -2.72 \ \Omega$$

5.3.5 Load Cells

One important direct application of SGs is for the measurement of force or weight. These transducer devices, called *load cells*, measure deformations produced by the force or weight. In general, a beam or yoke assembly is used that has several strain gauges mounted so that the application of a force causes a strain in the assembly that is measured by the gauges. A common application uses one of these devices in support of a hopper or feed of dry or liquid materials. A measure of the weight through a load cell yields a measure of the quantity of material in the hopper. Generally, these devices are calibrated so that the force (weight) is directly related to the resistance change. Forces as high as 5 MN (approximately 10^6 lb) can be measured with an appropriate load cell.

Example 5.7

Figure 5.16 shows a simple load cell consisting of an aluminum post of 2.500-cm radius with a detector and compensation strain gauges. The strain gauges are used in the bridge of Figure 5.14, with $V = 2$ volts, $R_1 = R_2 = R_G = 120.0\ \Omega$, and GF $= 2.13$. Find the variation of bridge offset voltage for a load of 0 to 5000 lb.

Solution   We can find the strain for a 5000-lb load, then the resulting change in resistance, and, from that, the bridge offset voltage. First we change the force to newtons.

$$(5000\ \text{lb}/0.2248\ \text{lb/N}) = 22240\ \text{N}$$

The cross-sectional area of the post is

$$A = \pi r^2 = \pi(0.025\ \text{m})^2 = 1.963 \times 10^{-3}\ \text{m}^2$$

From Table 5.1, the modulus of elasticity of aluminum is $E = 6.89 \times 10^{10}\ \text{N/m}^2$. From Equation (5.6), we find the strain

$$\Delta l/l = F/EA$$

$$\Delta l/l = (22240\ \text{N})/[(6.89 \times 10^{10}\ \text{N/m}^2)(1.963 \times 10^{-3}\ \text{m})]$$

$$\Delta l/l = 1.644 \times 10^{-4} = 164.4\ \mu\text{m/m (or }\mu\text{in/in)}$$

The relation between resistance and strain is given by Equation (5.12) (GF = $(\Delta R/R)/(\Delta l/l)$), so the resistance is given by

$$\Delta R/R = 2.13(1.644 \times 10^{-4})$$

$$= 3.502 \times 10^{-4}$$

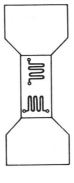

Figure 5.16   Load cell for Example 5.7.

$$= 3.502 \times 10^{-4}$$

Because $R = 120.0\ \Omega$, $\Delta R = 0.04203\ \Omega$. To get the bridge offset voltage, we note that the post is under compression and, therefore, the resistance will decrease. With no strain the bridge is nulled. Under a 5000-lb load, the active gauge has $R = 119.958$ Ω. Thus, the offset voltage of the bridge is

$$\Omega V = 2\,\frac{120}{120 + 120} - 2\,\frac{119.958}{120 + 119.958}$$

$$\Delta V = 1.750 \times 10^{-4}\ \text{V} = 175.0\ \mu\text{V}$$

As the force varies from 0 to 5000 lb, the offset voltage varies from 0 to 175 μV.

5.4 MOTION SENSORS

A special class of transducer is used to measure the *velocity* and *acceleration* of objects in industrial processes and testing. Often, these variables are not under specific control but are used to evaluate the performance, durability, and failure modes of manufactured products and the processes that produce them. The basic physical concepts of motion are discussed in Appendix 4. In the following sections, several special types of motion are discussed, and then the transducers used for measurement of these variables are studied.

5.4.1 Types of Motion

The design of a transducer to measure motion is often tailored to the type of motion that is to be measured. It will help you understand these transducers if you have a clear understanding of the types of motion.

Rectilinear

This type of motion is characterized by velocity and acceleration, which is composed of straight-line segments. Thus, objects may accelerate forward to a certain velocity, deaccelerate to a stop, reverse, and so on. There are many types of transducers designed to handle this type of motion. Typically, maximum accelerations are less than 10 **gs**, and no angular motion (in a curved line) is allowed. It is perhaps too strong to say that no angular motion is allowed. Rather, if there is angular motion, then several rectilinear motion transducers must be used, each sensitive to only one line of motion. Thus, if vehicle motion is to be measured, two transducers may be used, one to measure motion in the forward direction of vehicle motion and the other perpendicular to the forward axis of the vehicle.

Angular

Some transducers are designed to measure *only* rotations about some axis, such as the angular motion of the shaft of a motor. Such devices cannot be used to measure the physical displacement of the whole shaft, but only its rotation.

Vibration

In the normal experiences of daily living, a person rarely experiences accelerations that vary from 1 **g** by more than a few percent. Even the severe environments of a rocket launching involve accelerations of only 1 **g** to 10 **g**. If an object is placed in periodic motion about some equilibrium position as in Figure 5.17, very large *peak* accelerations may result that reach to 100 **g** or more. This motion is called *vibration*. Clearly, the measurement of acceleration of this magnitude is very important to industrial environments, where vibrations are often encountered from machinery operations. In general, vibrations are somewhat random in both the frequency of periodic motion and the magnitude of displacements from equilibrium. For analytical treatments, vibration is defined in terms of a regular periodic motion where the position of an object in time is given by

$$x(t) = x_0 \sin \omega t \qquad (5.13)$$

where

    $x(t)$ = object position in m
    x_0 = peak displacement from equilibrium in m
    ω = angular frequency in rad/s

The definition of ω as angular frequency is consistent with the reference to ω as angular speed because they are the same. If an object rotates, we define the time to complete one rotation as a period T that corresponds to a frequency $f = 1/T$. The frequency represents the number of revolutions per second and is measured in hertz (Hz), where 1 Hz = 1 revolution per second. An angular rate of one revolution per second corresponds to an angular velocity of 2π rad/s because one revolution sweeps out 2π radians. From this argument, we see that f and ω are related by

$$\omega = 2\pi f \qquad (5.14)$$

Because f and ω are related by a constant, we refer to ω as both angular frequency as well as angular velocity.

Now we can find the vibration velocity as a derivative of Equation (5.13)

$$v(t) = -\omega x_0 \cos \omega t \qquad (5.15)$$

and we can get the vibration acceleration from a derivative of (5.15)

$$a(t) = -\omega^2 x_0 \sin \omega t \qquad (5.16)$$

Vibration position, velocity, and acceleration are all periodic functions having the *same* frequency and period. Of particular interest is the *peak* acceleration:

Figure 5.17 An object in periodic motion about an equilibrium at $x = 0$. The peak motion is x_0.

$$a_{peak} = \omega^2 x_0 \qquad (5.17)$$

We see that the peak acceleration is dependent on ω^2, the angular frequency squared. This may result in very large acceleration values, even with *modest* peak displacements, as Example 5.8 shows.

Example 5.8

A water pipe vibrates at a frequency of 10 Hz with a displacement of 0.5 cm. Find (a) the peak acceleration in m/s^2, and (b) **g** acceleration.

Solution   The peak acceleration will be given by

(a) $$a_{peak} = \omega^2 x_0 \qquad (5.17)$$

where

$$\omega = 2\pi f = 20\ \pi \text{ rad/s and } x_0 = 0.5 \text{ cm} = 0.005 \text{ m}$$
$$a_{peak} = (20\ \pi)^2 (0.005)$$
$$a_{peak} = \textbf{19.7 m/s}^2$$

(b) Noting that 1 **g** = 9.8 m/s^2, we get

$$a_{peak} = (19.7 \text{ m/s}^2)\left(\frac{1 \text{ g}}{9.8 \text{ m/s}^2}\right)$$

$$a_{peak} = \textbf{2.0 g}$$

A 2-**g** vibrating excitation of any mechanical element can be very destructive, yet this is generated under the modest conditions of Example 5.5. A special class of transducers has been developed for measuring vibration acceleration.

Shock

A very special type of acceleration occurs when an object that may be in uniform motion or modestly accelerating is suddenly brought to rest, such as from a collision. Such phenomena are the result of very large accelerations or actually *deaccelerations*, such as an object being dropped from some height onto a hard surface. The name *shock* is given to deaccelerations that are characterized by very short times, typically in the order of milliseconds with peak accelerations over 500 **g**. In Figure 5.18 we have a typical acceleration graph as a function of time for a shock experiment. This graph is characterized by a maximum or peak deacceleration a_{peak}, a shock duration T_d, and bouncing. We can find an average shock by knowing the velocity of the object and the shock duration, as considered in Example 5.19.

Example 5.9

A TV set is dropped from a 2-m height. If the shock duration is 5 ms, find the average shock in **g**.

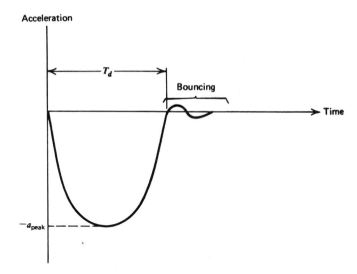

Figure 5.18   A typical shock graph showing acceleration versus time. Shock duration is T_d.

Solution   The TV accelerates at 9.8 m/s^2 for a 2 m. We find the velocity as

$$v^2 = 2 \, \mathbf{g}x$$

$$v^2 = (2)(9.8 \text{ m/s}^2)(2 \text{ m})$$

$$v = 6.2 \text{ m/s}$$

If the duration is 5 ms, we have

$$\bar{a} = \frac{6.2 \text{ m/s}}{5 \times 10^{-3} \text{ s}}$$

$$\bar{a} = 1200 \text{ m/s}^2$$

or 122 **g**! No wonder that the TV breaks apart when it hits ground.

5.4.2 Accelerometer Principles

Many accelerometers operate according to one basic principle. The variations in design involve only how this principle is implemented. The basic principle involves Newton's law ($F = ma$).

Spring-mass system

The principle of acceleration measurement is based upon a marriage of Newton's law (relating force and acceleration) and Hooke's law (relating force and spring action: see Appendix 4). In Figure 5.19a we see the combination of a mass free to move and a spring attached to some base. If the entire assembly is accelerated

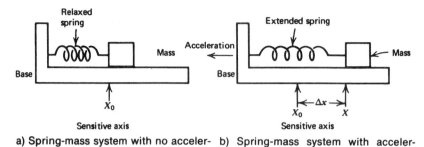

a) Spring-mass system with no acceleration

b) Spring-mass system with acceleration

Figure 5.19   The basic accelerometer is a spring-mass system.

to the left, Newton's law states that the mass must be under the influence of a force, $F = ma$. This force is provided by the spring that extends until the force provided by Hooke's law matches that required by the acceleration. As long as the system accelerates in this manner, the spring-mass system remains in this state, as shown in Figure 5.19b. We can then form an equality

$$k \, \Delta x = ma \qquad (5.18)$$

where

$k =$ spring constant in N/m
$\Delta x =$ spring extension in m
$m =$ mass in kg
$a =$ acceleration in m/s^2

Equation (5.18) allows the measurement of acceleration to be reduced to a measurement of spring extension (linear displacement) because

$$a = \frac{k}{m} \, \Delta x \qquad (5.19)$$

If the acceleration is reversed, the same physical argument would apply, except that the spring is compressed instead of extended. Equation (5.19) describes the relationship between spring displacement and acceleration.

The *spring-mass principle* applies to all common accelerometer designs. The mass that converts the acceleration to spring displacement is referred to as the *test mass* or *seismic mass*. We see then that acceleration measurement reduces to linear displacement measurement; most designs differ in how this displacement measurement is made.

Natural frequency and damping

On closer examination of the simple principle described earlier, we find another characteristic of spring-mass systems that complicates the analysis. In particular, a system consisting of a spring and attached mass always exhibits oscillations at some characteristic *natural frequency*. Experience tells us that if we pull a mass back and then release it (in the absence of acceleration), it will be pulled back by the spring, overshoot the equilibrium, and oscillate back and forth. Only friction associated with the mass and base eventually brings the mass to rest. Any displacement measuring system must respond to this oscillation as if an actual acceleration occurs; in fact, none is applied. This natural frequency is given by

$$f_N = \frac{1}{2\pi} \sqrt{\frac{k}{m}} \qquad (5.20)$$

where

f_N = natural frequency in Hz
k = spring constant in N/m
m = seismic mass in kg

The friction that eventually brings the mass to rest is defined by a *damping coefficient* α, which has the units of s^{-1}. In general, the effect of oscillation is called *transient response*, described by a periodic damped signal, as shown in Figure 5.20, whose equation is

$$X_T(t) = X_0 e^{-\alpha t} \sin(2\pi f_N t) \qquad (5.21)$$

where

$X_T(t)$ = transient mass position
X_0 = peak position, initially
α = damping coefficient
f_N = natural frequency

The parameters, natural frequency, and damping coefficient in Equation (5.21) have a profound effect on the application of accelerometers.

Vibration effects

The effect of natural frequency and damping on the behavior of spring-mass accelerometers is best described in terms of an applied vibration. If the spring-mass system is exposed to a vibration, then the resultant acceleration is given by Equation (5.16)

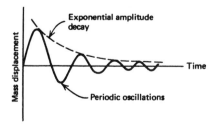

Figure 5.20 A spring-mass system exhibits a natural oscillation with damping as a transient response to an impulse input.

$$a(t) = -\omega^2 x_0 \sin \omega t \qquad (5.16)$$

If this is used in Equation (5.18), we can show that the mass motion is given by

$$\Delta x = -\frac{mx_0}{k} \omega^2 \sin \omega t \qquad (5.22)$$

where all terms were previously defined and $\omega = 2\pi f$, with f the applied frequency.

The derivation of Equation (5.22), which has ignored the natural frequency, indicates that the mass displacement amplitude varies as the *applied* frequency is squared. To see the effect of the natural frequency and damping on this result, we plot the actual displacement amplitude versus applied frequency in Figure 5.21. This figure shows that the behavior predicted by Equation (5.22) is correct for frequencies lower than the natural frequency but deviates as the natural frequency is reached. A peak, or resonance, occurs at the natural frequency. After further increases in applied frequency, we find that the amplitude is *independent* of frequency.

The effect of damping is seen as a reduction of the resonance peak and constant displacement at higher frequency. Thus, we can make the following observations concerning the effects of natural frequency and an applied vibration frequency:

1. $f < f_N$—For an applied frequency less than the natural frequency, the natural frequency has little effect on the basic spring-mass response given by Equations (5.18) and (5.22). A rule of thumb states that a safe maximum applied frequency is $f < 1/2.5 f_N$.

2. $f > f_N$—For an applied frequency much larger than the natural frequency, the accelerometer output is independent of the applied frequency. Although not shown in Figure 5.17, the accelerometer becomes a measure of *vibration*

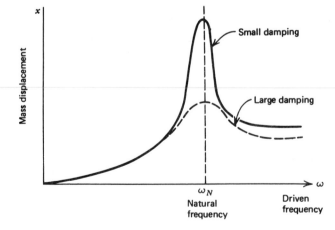

Figure 5.21 A spring-mass accelerometer driven by a periodic acceleration of varying frequency shows a peak mass displacement or resonance at the natural frequency.

displacement x_0 of Equation (5.13) under these circumstances. It is interesting to note that the seismic mass is stationary in space in this case and the housing, which is driven by the vibration, moves about the mass. A general rule sets $f > 2.5f_N$ for this case.

Generally, accelerometers are *not* used near their natural frequency because of high nonlinearities in output.

Example 5.10

An accelerometer has a seismic mass of 0.05 kg and a spring constant of 3.0×10^3 N/m. Maximum mass displacement is ± 0.02 m (before the mass hits the stops). Calculate (a) the maximum measurable acceleration in **g**, and (b) the natural frequency.

Solution   We find the maximum acceleration when the maximum displacement occurs

(a)

$$a = \frac{k}{m} \Delta x$$

$$a = \left(\frac{3.0 \times 10^3 \text{ N/m}}{0.05 \text{ kg}} \right)(0.02 \text{ m}) \qquad (5.19)$$

$$a = 1200 \text{ m/s}^2$$

or because

$$1 \text{ g} = 9.8 \text{ m/s}^2$$

$$a = (1200 \text{ m/s}^2)\left(\frac{1 \text{ g}}{9.8 \text{ m/s}^2} \right)$$

$$a = 122 \text{ g}$$

(b) The natural frequency is

$$f_N = \frac{1}{2\pi} \sqrt{\frac{k}{m}}$$

$$f_N = \frac{1}{2\pi} \sqrt{\frac{3.0 \times 10^3 \text{ N/m}}{0.05 \text{ kg}}}$$

$$f_N = \mathbf{39 \text{ Hz}}$$

5.4.3 Types of Accelerometers

The variety of accelerometers used results from different applications with requirements of range, natural frequency, and damping. In this section, various accelerometers with their special characteristics are reviewed. The basic difference is in the method of mass displacement *measurement*. In general, the specification sheets for an accelerometer will give the natural frequency, damping coefficient, and a scale factor that relates the output to an acceleration input. The values of test mass and spring constant are seldom known or required.

Potentiometric

This simplest accelerometer type measures mass motion by attaching the mass to the wiper arm of a potentiometer. In this manner, the mass position is conveyed as a changing resistance. The natural frequency of these devices is generally less than 30 Hz, limiting its application to *steady-state* acceleration or *low-frequency* vibration measurement. Numerous signal conditioning schemes are employed to convert the resistance variation into a voltage or current signal.

LVDT

A second type of accelerometer takes advantage of the natural linear displacement measurement of the LVDT (Section 5.2.3) to measure mass displacement. In these instruments, the LVDT core itself is the seismic mass. Displacements of the core are converted directly into a linearly proportional ac voltage. These accelerometers generally have a natural frequency less than 80 Hz and are commonly used for steady-state and low-frequency vibration.

Variable reluctance

This accelerometer type falls in the same general category as the LVDT in that an inductive principle is employed. Here, the test mass is usually a permanent magnet. The measurement is made from the voltage induced in a surrounding coil as the magnetic mass moves under the influence of an acceleration. This accelerometer is used in vibration studies only, because it has an output *only* when the mass is in motion. Its natural frequency is typically less than 100 Hz. This type of accelerometer often is used in oil exploration to pick up vibrations reflected from underground rock strata. In this form, it is commonly referred to as a *geophone*.

Piezoelectric

The piezoelectric accelerometer is based on a property exhibited by certain crystals where a voltage is generated across the crystal when stressed. This property is also the basis for such familiar transducers as crystal phonograph cartridges and crystal microphones. For accelerometers, the principle is shown in Figure 5.22. Here, a piezoelectric crystal is spring-loaded with a test mass in contact with the crystal. When exposed to an acceleration, the test mass stresses the crystal by a force ($F = ma$), resulting in a voltage generated across the crystal. A measure of this voltage is then a measure of the acceleration. The crystal per se is a very high-impedance source and thus requires a high-input impedance low-

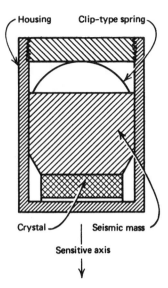

Figure 5.22 A piezoelectric accelerometer.

noise detector. Output levels are typically in the millivolt range. The natural frequency of these devices may exceed 5 kHz, so that they can be used for vibration and shock measurements.

5.4.4 Applications

A few notes about the application of accelerometers are of aid in understanding how the selection of a transducer is made in a particular case.

Steady-state acceleration

In steady-state accelerations, we are interested in a measure of acceleration that may vary in time but that is *nonperiodic*. Thus, the stop–go motion of an automobile is an example of a steady-state acceleration. For these steady-state accelerations, we select a transducer having (1) adequate range to cover expected acceleration *magnitudes* and (2) a natural frequency sufficiently high that its period is shorter than the characteristic time span over which the measured acceleration changes. By using electronic integrators, the basic accelerometer can provide both velocity (first integration) and position (second integration) information.

Example 5.11

An accelerometer outputs 14 mV per **g**. Design a signal conditioning system that provides (a) a velocity signal scaled at 0.25 volt for every m/s, and (b) determine the gain of the system and the feedback resistance ratio.

Solution   First we note that 14 mV/**g** becomes

$$\left(14 \, \frac{\text{mV}}{\text{g}}\right)\left(\frac{1 \, \text{g}}{9.8 \, \text{m/s}^2}\right) = 1.43 \, \frac{\text{mV}}{\text{m/s}^2}$$

Now we need an integrator to get the velocity and amplifier to provide the proper scale. Such a circuit is shown in Figure 5.23. We chose $T = RC = 1$ so that the integrator output is scaled at

$$\left(1.43 \, \frac{\text{mV}}{\text{m/s}^2}\right)\left(\frac{-1}{1 \, \text{s}}\right) = -1.43 \, \frac{\text{mV}}{\text{m/s}}$$

in units of output voltage velocity.

Then, the required gain is found from

$$G = \frac{0.25 \, \text{V/m/s}}{-1.43 \, \text{mV/m/s}} = -175$$

Thus, we can make $R_2 = \mathbf{175 \, k\Omega}$ and $R_1 = \mathbf{1 \, k\Omega}$ so that

$$G = -\frac{R_2}{R_1} = -175$$

Vibration

The application of accelerometers for vibration first requires that the applied frequency is less than the natural frequency of the accelerometer. Second, one must be sure the stated range of acceleration measured will never exceed that of the specification for the device. This assurance must come from a consideration of Equation (5.22) under circumstances of maximum frequency and vibration displacement.

Shock

The primary elements of importance in shock measurements are that the device have a natural frequency that is greater than 1 kHz and a range typically greater

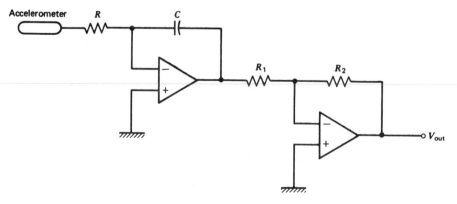

Figure 5.23 An integrator and inverter can be used to obtain velocity information from an accelerometer.

than 500 **g**. The only accelerometer that can usually satisfy these requirements is the piezoelectric type (Section 5.4.3).

5.5 PRESSURE SENSORS

The measurement and control of fluid (liquid and gas) pressure has to be one of the most common of all the process industries. Because of the great variety of conditions, ranges, and materials for which pressure must be measured, there are many different types of pressure sensor designs. In the following paragraphs, the basic concepts of pressure are presented, and a brief description is given of the most common types of pressure sensors. You will see that pressure measurement is often accomplished by conversion of the pressure information to some intermediate form, such as displacement, that is then measured by a sensor to determine the pressure.

5.5.1 Pressure Principles

Pressure is simply the force per unit area that a fluid exerts on its surroundings. If it is a gas, then the pressure of the gas is the force per unit area that the gas exerts on the walls of the container that holds it. If the fluid is a liquid, then the pressure is the force per unit area that the liquid exerts on the container in which it is contained. Obviously, the pressure of a gas will be uniform on all the walls that must enclose the gas completely. In a liquid, the pressure will vary, being greatest on the bottom of the vessel and zero on the top surface, which need not then be enclosed.

Static pressure

The statements made in the previous paragraph are explicitly true for a fluid that is not moving in space, that is not being pumped through pipes or flowing through a channel. The pressure in cases where no motion is occurring is referred to as *static* pressure.

Dynamic pressure

If a fluid is in motion, the pressure that it exerts on its surroundings *depends* on the motion. Thus, if we measure the pressure of water in a hose with the nozzle closed, we may find a pressure of, say, 40 pounds per square inch (note: force per unit area). If the nozzle is opened, the pressure in the hose will drop to a different value, say, 30 pounds per square inch. For this reason, a thorough description of pressure must note the circumstances under which it is measured. Pressure can depend on flow, compressibility of the fluid, external forces, and numerous other factors.

Units

Since pressure is force per unit area, we describe it in the SI system of units by newtons per square meter. This unit has been named the *pascal* (Pa), so that 1 Pa = 1 N/m². As will be seen later, this is not a very convenient unit, and it is often used in conjunction with the SI standard prefixes, as kPa or MPa. You will see the combination N/cm² used, but use of this combination should be avoided in favor of Pa with the appropriate prefix. In the English system of units, the most common designation is the pound per square inch, lb/in². This is usually just written *psi*. The conversion is that 1 psi is approximately 6.895 kPa. For very low pressures, such as may be found in vacuum systems, the unit *Torr* is often used. One Torr is approximately 133.3 Pa. Again, use of the pascal with appropriate prefix is preferred. Other units that you may encounter in the pressure description are the *atmosphere* (at), which is 101.325 kPa or ≈14.7 psi, and the *bar*, which is 100 kPa. The use of inches or feet of water and millimeter of mercury will be discussed later.

Gauge pressure

In many cases the absolute pressure is not the quantity of major interest in describing the pressure. The atmosphere of gas that surrounds the earth exerts a pressure, because of its weight, at the surface of the earth of approximately 14.7 psi, as noted in the atmosphere unit. If a closed vessel at the earth's surface contained a gas at an absolute pressure of 14.7 psi, then it would exert *no effective pressure* on the walls of the container because the atmospheric gas exerts the same pressure from the outside. In cases like this, it is more appropriate to describe pressure in a relative sense, that is, compared to atmospheric pressure. This is called *gauge pressure* and is given by

$$p_g = p_{abs} - p_{at} \tag{5.23}$$

where

$$p_g = \text{gauge pressure}$$
$$p_{abs} = \text{absolute pressure}$$
$$p_{at} = \text{atmospheric pressure}$$

In the English system of units, the abbreviation *psig* is used to represent the gauge pressure.

Head pressure

For liquids, the expression *head pressure* or *pressure head* is often used to describe the pressure of the liquid in a tank or pipe. This refers to the *static* pressure produced by the weight of the liquid above the point at which the pressure is being described. This pressure depends *only* on the height of the liquid above that point and the liquid density (mass per unit volume). In terms of an equation, if a liquid is contained in a tank, then the pressure at the bottom of the tank is given by

$$p = \rho g h \qquad (5.24)$$

where

  p = pressure in Pa
  ρ = density in kg/m^3
  g = acceleration due to gravity (9.8 m/s^2)
  h = depth of liquid in m

This same equation could be used to find the pressure in the English system, but it is common practice to express the density in this system as the weight density ρ_w in lb/ft^3, which includes the gravity term of Equation (5.24). In this case, the relation between pressure and depth becomes

$$p = \rho_w h \qquad (5.25)$$

where

  p = pressure in lb/ft^2
  ρ_w = weight density in lb/ft^3
  h = depth in ft

If the pressure is desired in psi, then the ft^2 would be expressed as 144 in^2. Because of the common occurrence of liquid tanks and the necessity to express the pressure of such systems, it has become common practice to describe the pressure directly in terms of the *equivalent* depth of a particular liquid. Thus, the term *mm of mercury* means that the pressure is equivalent to that produced by so many millimeters of mercury depth, which could be calculated from Equation (5.24) using the density of mercury. In the same sense, the expression "inches of water" or "feet of water" means the pressure that is equivalent to some particular depth of water using its weight density.

Now you can see the basis for level measurement on pressure mentioned in Section 5.2.4. Equation (5.24) shows that the level of liquid of density ρ is directly related to the pressure. From level measurement we pass to pressure measurement, which is usually done by some type of displacement measurement.

Example 5.12

A tank holds water with a depth of 7.0 feet. What is the pressure at the tank bottom in psi and Pa? (density = 10^3 kg/m^3)

Solution   We can find the pressure in Pa directly by converting the 7.0 ft into meters, thus, (7.0 ft)(0.3048 m/ft) = 2.1 m. From Equation (5.24),

$$p = (10^3 \text{ kg/m}^3)(9.8 \text{ m/s}^2)(2.1 \text{ m})$$

$$p = 21 \text{ kPa}   \text{(note significant figures)}$$

To find the pressure in psi, we can convert the pressure in Pa to psi or use Equation

(5.25). Let's use the latter. The weight density is found from

$$\rho_w = (10^3 \text{ kg/m}^3)(9.8 \text{ m/s}^2) = 9.8 \times 10^3 \text{ N/m}^3$$

or

$$= (9.8 \times 10^3 \text{ N/m}^3)(0.3048 \text{ m/ft}^3)(0.2248 \text{ lb/N})$$

$$= 62.4 \text{ lb/ft}^3$$

The pressure is

$$p = (62.4 \text{ lb/ft}^3)(7.0 \text{ ft}) = 440 \text{ lb/ft}^2$$

$$p = \mathbf{3 \ psi}$$

5.5.2 Pressure Sensors ($p >$ one atmosphere)

In general, the design of pressure sensors employed for measurement of pressure higher than one atmosphere differs from those employed for pressure less than one atmosphere. In this section the basic operating principles of many types of pressure sensors used for the higher pressures are considered. You should be aware that this is not a rigid separation, because you will find many of these same principles employed in the lower (vacuum) pressure measurements.

Most pressure sensors used in process control require the transduction of pressure information into a physical displacement. Measurement of pressure requires techniques for producing the displacement and means for converting such displacement into a proportional electrical signal. This is *not* true, however, in the *very low* pressure region ($p < 10^{-3}$ atm) where many purely electronic means of pressure measurement may be used.

Diaphragm

One common element used to convert pressure information into a physical displacement is the diaphragm shown in Figure 5.24. If a pressure p_1 exists on one side of the diaphragm and p_2 on the other, then a net force is exerted given by

$$F = (p_2 - p_1)A \tag{5.26}$$

where

$$A = \text{diaphragm area in m}^2$$
$$p_1, p_2 = \text{pressure in N/m}^2$$

A diaphragm is like a spring and therefore extends or contracts until a Hooke's

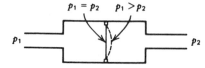

Figure 5.24   A diaphragm is used in many pressure measurement systems. The displacement is proportional to pressure difference.

law force is developed which balances the pressure difference force. This is shown in Figure 5.24 for p_1 greater than p_2. A *bellows* shown in Figure 5.25 is another device much like the diaphragm that converts a pressure differential into a physical displacement, except that here the displacement is much more a straight-line expansion.

Figure 5.25 also shows how an LVDT can be connected to the bellows so that pressure measurement is converted directly from displacement to a voltage. In addition, the displacement and pressure are very nearly linearly related, and because the LVDT voltage is linear with displacement, the voltage and pressure are also linearly related.

Bourdon tube

A very special and common pressure-to-displacement conversion is accomplished by a specially constructed tube, shown in Figure 5.26. If a section of tubing is partially flattened and coiled as shown, then the application of pressure inside the tube causes the tube to uncoil. This then provides a displacement that is proportional to pressure.

Electronic conversions

Many techniques are used to convert the displacements generated in the previous examples into electronic signals. The simplest technique is to use a mechanical linkage connected to a potentiometer. In this fashion, pressure is related to a

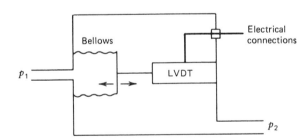

Figure 5.25 A bellows transforms pressure difference into displacement. In this example, the displacement is measured by an LVDT.

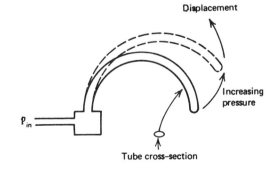

Figure 5.26 The Bourdon tube is a very common device for converting pressure into displacement.

resistance change. Other methods of conversion employ strain gages directly on a diaphragm. LVDTs and other inductive devices are used to convert bellows or Bourdon tube motions into proportional electrical signals.

Often pressure measurement is accomplished using a diaphragm in a special feedback configuration, shown in Figure 5.27. The feedback system keeps the diaphragm from moving, using an induction motor. The error signal in the feedback system provides an electrical measurement of the pressure.

Solid-state pressure sensors

Intergrated circuit manufacturers have developed composite pressure sensors which are particularly easy to use. These devices commonly employ a semiconductor diaphragm on to which a semiconductor strain gauge and temperature compensation sensor have been grown. Appropriate signal conditioning is included in integrated circuit form, providing a dc voltage or current linearly proportional to pressure over a specified range.

These devices are available for absolute, gauge, and differential pressure measurement. The are simple to use, often needing only three connections, a dc supply, ground, and signal output. Of course, the connection to the measurement environment is made through a fitting or welded pipe connection.

5.5.3 Pressure Sensors ($p <$ one atmosphere)

Measurements of pressure less than 1 atm are most conveniently made using purely electronic methods. There are three common methods of electronic pressure measurements.

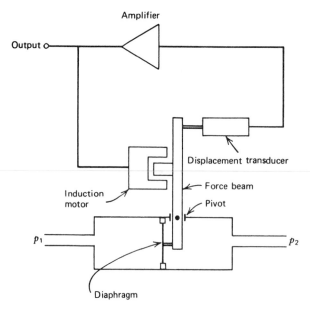

Figure 5.27 A differential pressure (DP) cell measures pressure difference with a diaphragm. In this example, diaphragm displacement is minimized by an electronic feedback system.

The first two devices are useful for pressure less than 1 atm down to about 10^{-3} atm. They are both based on the rate at which heat is conducted and radiated away from a heated filament placed in the low-pressure environment. The heat loss is proportional to the number of gas molecules per unit volume, and thus, under constant filament current, the filament temperature is proportional to gas pressure. We have thus transduced a pressure measurement to a temperature measurement.

Pirani gauge

This gauge determines the filament temperature through a measure of filament *resistance* in accordance with the principles established in Section 4.3. Filament excitation and resistance measurement are both performed with a bridge circuit. The response of resistance versus pressure is highly nonlinear.

Thermocouple

A second pressure transducer or gauge measures filament temperature using a thermocouple directly attached to the heated filament. In this case, ambient room temperature serves as a reference for the thermocouple, and the voltage output, which is proportional to pressure, is highly nonlinear. Calibration of both Pirani and thermocouple gauges depends on the type of gas for which the pressure is being measured.

Ionization gauge

This device is useful for the measurement of very low pressures from about 10^{-3} atm to 10^{-13} atm. This gauge employs electrons, usually from a heated filament, to ionize the gas whose pressure is to be measured, and then measures the current flowing between two electrodes in the ionized environment, as shown in Figure 5.28. The number of ions per unit volume depends on the gas pressure, and hence the current also depends on gas pressure. This current is then monitored as an approximately *linear* indication of pressure.

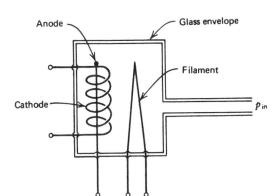

Figure 5.28 The ionization gauge is used to measure very low pressures, down to 10^{-13} atm.

5.6 FLOW SENSORS

The measurement and control of flow can be said to be the very heart of process industries. Continuously operating manufacturing processes involve the movement of raw materials, products, and waste throughout the process. All such functions can be considered flow, whether automobiles through an assembly line or methyl chloride through a pipe. The methods of measurement of flow are at least as varied as the industry. It would be unreasonable to try to present every type of flow sensor, and in this section we will consider flow on three broad fronts—solid, liquid, and gas. Like pressure, we will find that flow information is often translated into an intermediate form that is then measured using techniques developed for that form.

5.6.1 Solid Flow Measurement

The most common solid flow measurement occurs when material in the form of small particles, such as crushed material or powder, is carried by a conveyor belt system or by some other host material. For example, if solid material is suspended in a liquid host, the combination is called a *slurry*, which is then pumped through pipes like a liquid. We will consider the conveyor system and leave the slurry to be treated as liquid flow.

Conveyor flow concepts

For solid objects, the flow usually is described by a specification of the mass or weight per unit time that is being transported by the conveyor system. The units will be in many forms, for example, kg/min or lb/min. To make a measurement of flow, it is only necessary to weight the quantity of material on some fixed length of the conveyor system. Knowing the speed of the conveyor allows calculation of the material flow rate.

In Figure 5.29, a typical conveyor system is shown where material is drawn from a hopper and transported by the conveyor system. Assuming that the material can flow freely from the hopper, the faster the conveyor is moved, the faster material will flow from the hopper, and the greater the material flow rate on the conveyor. In this case, flow rate can be calculated from

$$Q = \frac{WR}{L} \tag{5.27}$$

where

Q = flow in kg/min
W = weight of material on section of length L
R = conveyor speed in m/min
L = length of weighing platform in m

Flow sensor

In the example with which we are working in Figure 5.29, it is evident that the flow sensor is actually the assembly of conveyor, hopper opening, and weighing platform. It is the actual weighing platform that performs the measurement from which flow rate is determined, however. We see that flow measurement becomes weight measurement. In this case, we have suggested that this weight is measured by means of a load cell, which is then a strain gauge measurement. Another popular device for weight measurement of moving systems like this is an LVDT that measures the droop of the conveyor at the point of measurement because of the material that it carries.

Example 5.13

A coal conveyor system moves at 100 ft/min. A weighing platform is 5.0 ft in length, and a particular weighing shows that 75 lb of coal are on the platform. Find the coal delivery in lb/hr.

Solution   We can use Equation (5.27) directly to find the flow

$$Q = \frac{(75 \text{ lb})(100 \text{ ft/min})}{5 \text{ ft}}$$

$$Q = 1500 \text{ lb/min}$$

Then, converting to lb/hr by multiplying by 60 min/hr

$$Q = \textbf{90,000 lb/hr}$$

5.6.2 Liquid Flow

The measurement of liquid flow is involved in nearly every facet of the process industry. The conditions under which the flow occurs and the vastly different

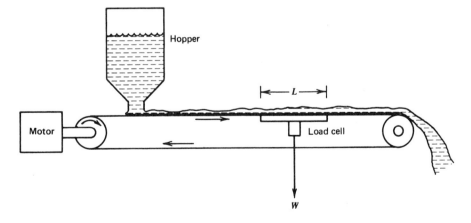

Figure 5.29   Conveyor system for illustrating solid flow measurement.

types of material that flow result in a great many types of flow measurement methods. Indeed, entire books are written devoted to the problems of how to measure liquid flow and how to interpret the results of flow measurements. It is impractical and is not within the scope of this book to present a comprehensive study of liquid flow, and only the basic ideas of liquid flow measurement will be presented.

Flow units

The units used to describe the flow measured can be of several types, depending on how the specific process needs the information. The most common descriptions are the following:

1. *Volume flow rate* Expressed as a volume delivered per unit time. Typical units are gallons/min, m^3/hr, ft^3/hr.
2. *Flow velocity* Expressed as the distance the liquid travels in the carrier per unit time. Typical units are m/min, ft/min. This is related to the volume flow rate by

$$V = \frac{Q}{A} \qquad (5.28)$$

where

V = flow velocity
Q = volume flow rate
A = cross-sectional area of flow carrier (pipe, and so on)

3. *Mass or weight flow rate* Expressed as mass or weight flowing per unit time. Typical units are kg/hr, lb/hr. This is related to the volume flow rate by

$$F = \rho Q \qquad (5.29)$$

where

F = mass or weight flow rate
ρ = mass density or weight density
Q = volume flow rate

Example 5.14

Water is pumped through a 1.5-in diameter pipe with a flow velocity of 2.5 ft per second. Find the volume flow rate and weight flow rate. The weight density is 62.4 lb/ft^3.

Solution   The flow velocity is given as 2.5 ft/s so the volume flow rate can be found from Equation (5.28), $Q = VA$. The area is given by

$$A = \pi d^2/4$$

where

the diameter $d = (1.5$ in$)(1/12$ ft/in$) = 0.125$ ft

so that $A = (3.14)(0.125)^2/4 = 0.0122$ ft^2. Then, the volume flow rate is

$$Q = (2.5 \text{ ft/s})(0.0122 \text{ ft}^2)(60 \text{ s/min})$$

$$Q = 1.8 \text{ ft}^3/\text{min}$$

The weight flow rate is found from Equation (5.29)

$$F = (62.4 \text{ lb/ft}^3)(1.8 \text{ ft}^3/\text{min})$$

$$F = \textbf{112 lb/min}$$

Pipe flow principles

The flow rate of liquids in pipes is determined primarily by the pressure that is forcing the liquid through the pipe. The concept of pressure head or simply *head* introduced in the previous sections often is used to describe this pressure because it is easy to relate the forcing pressure to that produced by a depth of liquid in a tank from which the pipe exits. In Figure 5.30, flow through pipe P is driven by the pressure in the pipe, but this pressure is caused by the weight of liquid in the tank of height h (head). The pressure is found by Equation (5.24) or (5.25). Many other factors affect the actual flow rate produced by this pressure, including liquid viscosity, pipe size, pipe roughness (friction), turbulence of flowing liquid, and others. It is beyond the scope of this book to detail exactly how these factors determine the flow. Instead, it is our objective to discuss how such flow is measured, regardless of those features that may determine exactly what the flow is relative to the conditions.

Restriction flow sensors

One of the most common methods of measuring the flow of liquids in pipes is by introducing a restriction in the pipe and measuring the pressure drop that results across the restriction. When such a restriction is placed in the pipe, the velocity of the fluid through the restriction *increases*, and the pressure in the restriction

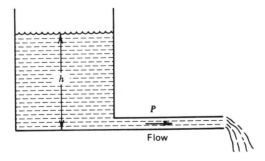

Figure 5.30   Flow through the pipe P is determined in part by the head pressure h.

decreases. We find that there is a relationship between the pressure drop and the rate of flow such that as the flow increases, the pressure drops. In particular, one can find an equation of the form

$$Q = K\sqrt{\Delta p} \tag{5.30}$$

where

Q = volume flow rate
K = a constant for the pipe and liquid type
Δp = drop in pressure across the restriction

The constant K depends on many factors, including the type of liquid, size of pipe, velocity of flow, temperature, and so on. The type of restriction employed also will change the value of the constant used in this equation. The flow rate is not linearly dependent on the pressure drop but on the square root. Thus, if the pressure drop in a pipe increased by a factor of 2 when the flow rate was increased, the flow rate will have only increased by a favor of 1.4 (the square root of 2). Certain standard types of restrictions are employed in exploiting the pressure-drop method of measuring flow.

Figure 5.31 shows the three most common methods. It is interesting to note that having converted flow information to pressure, we now employ one of the methods of measuring pressure, often by conversion to displacement, which is

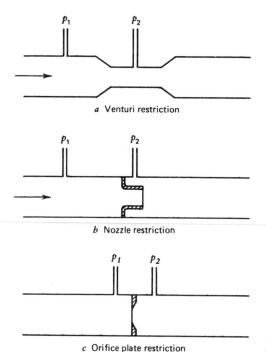

a Venturi restriction

b Nozzle restriction

c Orifice plate restriction

Figure 5.31 Three different types of restrictions commonly are used to convert flow rate to a pressure difference, $p_1 - p_2$.

measured by a displacement transducer before finally getting a signal that will be used in the process-control loop. The most common method of measuring the pressure drop is to use a differential pressure transducer similar to that shown in Figure 5.27. These are often described by the name *DP cell*.

Example 5.15

Flow is to be controlled from 20 to 150 gal/min. The flow is measured using an orifice plate system such as that shown in Figure 5.31c. The orifice plate is described by Equation (5.30) with $K = 119.5$ (gal/min)/psi$^{1/2}$. A bellows measures the pressure with an LVDT so that the output is 1.8 V/psi. Find the range of voltages that result from the given flow range.

Solution   From Equation (5.30), we find the pressures that result from the given flow

$$\Delta p = (Q/K)^2$$

For 20 gal/min

$$\Delta p = (20/119.5)^2 = 0.0280 \text{ psi}$$

and for 150 gal/min

$$\Delta p = (150/119.5)^2 = 1.5756 \text{ psi}$$

Because there are 1.8 V/psi, the voltage range is easily found

$$\text{For 20 gal/min, } V = 0.0280(1.8) = 0.0504 \text{ V}$$

$$\text{For 150 gal/min, } V = 1.5756(1.8) = 2.836 \text{ V}$$

Obstruction flow sensor

Another type of flow sensor operates by the effect of flow on an obstruction placed in the flow stream. In a *rotameter*, the obstruction is a float that rises in a vertical tapered column. The lifting force and thus the distance to which the float rises in the column is proportional to the flow rate. The lifting force is produced by the differential pressure that exists across the float, because it is a restriction in the flow. This type of sensor is used for both liquids and gases. A *moving vane* flow meter has a vane target immersed in the flow region, which will be rotated out of the flow as the flow velocity increases. The angle of the vane is a measure of the flow rate. If the rotating vane shaft is attached to an angle-measuring transducer, the flow rate can be measured for use in a process-control application. A *turbine* type of flow meter is composed of a freely spinning turbine blade assembly in the flow path. The rate of rotation of the turbine is proportional to the flow rate. If the turbine is attached to a tachometer, a convenient electrical signal can be produced. In all of these methods of flow measurement, it is necessary to present a substantial obstruction into the flow path to measure the flow. For this reason, these devices are used only when an obstruction does not cause any unwanted reaction on the flow system. These devices are illustrated in Figure 5.32.

Magnetic flow meter

It can be shown that if charged particles move across a magnetic field, a potential is established across the flow, perpendicular to the magnetic field. Thus, if the flowing liquid is also a conductor, although not necessarily a good conductor, of

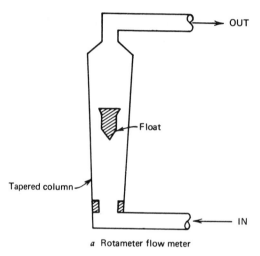

a Rotameter flow meter

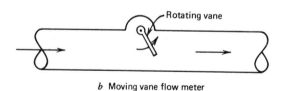

b Moving vane flow meter

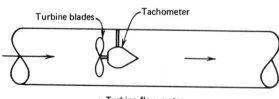

c Turbine flow meter

Figure 5.32 Three different types of obstruction flow meters.

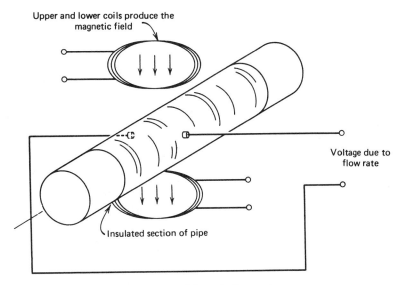

Figure 5.33 Magnetic flow meter.

electricity, the flow can be measured by allowing the liquid to flow through a magnetic field and measuring the transverse potential produced. The pipe section in which this measurement is made must be insulated and a nonconductor itself, or the potential produced will be cancelled by currents in the pipe. A diagram of this type of flow meter is presented in Figure 5.33. This type of transducer produces an electrical signal directly and is convenient for process-control applications involving conducting fluid flow.

SUMMARY

In this chapter, an assortment of measurement systems that fall under the general description of mechanical transducers have been studied. The objective was to gain familiarity with the essential features of the variables themselves and the typical measurement methods.

Topics covered were the following:

1. Position, location, and displacement sensors, including the potentiometric, capacitive, and LVDT. The LVDT converts displacement linearly into a voltage.

2. The strain gauge measures deformation of solid objects resulting from applied forces called stress. The strain gauge converts strain into a change of resistance.

3. Accelerometers are used to measure the acceleration of objects because of rectilinear motion, vibration, and shock. Most of them operate by the spring-mass principle, which converts acceleration information into a displacement.

4. Pressure is the force per unit area that a fluid exerts on the walls of a container. Pressure sensors often convert pressure information into a displacement. Examples include diaphragms, bellows, and the Bourdon tube. Electronic measures are often used for low pressures.

5. For gas pressures less than 1 atmosphere, purely electrical techniques are used. In some cases, the temperature of a heated wire is used to indicate pressure.

6. Flow sensors are very important in the manufacturing world. Typically, solid flow is mass or weight per unit time.

7. Fluid flow through pipes or channels typically is measured by converting the flow information into pressure by a restriction in the flow system.

PROBLEMS

Section 5.2

5.1 A 10-turn, 50-kΩ, wire-wound pot is used to measure displacement of a work piece. A linkage is employed so as the work piece moves over a distance of 12 cm, the pot rotates by the full 10 turns. The pot is wound on a 3 cm diameter form, 5 cm long and the distance between wires in the pot is 0.25 mm. What is the resolution in work-piece motion? What resistance change corresponds to this resolution?

5.2 Develop signal conditioning for Problem 5.1 so the output is -6 to $+6$ volts as the work piece moves over the 0- to 12-cm motion limit.

5.3 A capacitive displacement sensor is used to measure rotating shaft wobble, as shown in Figure 5.34. The capacity is 880 pF with no wobble. Find the change in capacity for a $+0.02$- to -0.02-mm shaft wobble.

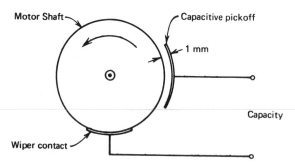

Figure 5.34   Figure for Problem 5.3.

5.4 Develop an ac bridge for Problem 5.3. Use 880 pF for all the bridge capacitors, and assume a 10-vrms, 10-kHz excitation. What is the maximum bridge offset voltage amplitude?

5.5 Design a linkage system so as a float for liquid level measurement moves from 0 to

1 meter, an LVDT core moves over its linear range of 3 cm. Suppose the LVDT output is interfaced to a 10-bit ADC. What is the resolution in level measurement?

5.6 For Example 5.2, what capacity change would need to be measured to have a resolution of 2 cm? If the measurement will ultimately go to a computer, how many bits must the ADC have to support this resolution?

5.7 Design an ac bridge like that of Figure 2.10 to convert the capacity change of Example 5.2 into an ac offset voltage. Plot the voltage versus level.

Section 5.3

5.8 An aluminum beam supports a 550-kg mass. If the beam diameter is 6.2 cm, calculate (a) the stress and (b) the strain of the beam.

5.9 A strain gauge has GF = 2.14 and a nominal resistance of 120 Ω. Calculate the resistance change resulting from a strain of 144 μin/in.

5.10 A strain gauge with GF = 2.03 and 120 Ω nominal resistance is to be used to measure strain with a resolution of 5 μs. Design a bridge and detector that provides this over five switched ranges of 1000-μ spans (i.e., 0 to 1000 μ, 1000 to 2000 μ, etc.). The idea is that a bridge null is found by a combination of switching to the appropriate range and then making a smooth null adjustment within that range. If the strain were 3390 μ, it would be necessary to switch to the 3000- to 4000-μ range and then adjust the smooth pot until a null occurred, which would be at 390 μ.

5.11 For Example 5.7, develop signal conditioning to provide input to a 10-bit ADC with a 5.000-volt reference. How many pounds does each LSB represent? Plot the ADC output in hex versus force. Evaluate the linearity. Do not use a gain of more than 500 for any single op amp circuit stage in the design.

5.12 We will weigh objects by a strain gauge of R = 120 Ω, GF = 2.02 mounted on a copper column of 6-in diameter. Find the change in resistance per pound placed on the column. Is this change an increase or decrease in resistance? Draw a diagram of the system showing how the active and dummy gauges should be mounted.

Section 5.4

5.13 Calculate the rotation rate of a 10,000-rpm motor in rad/s.

5.14 An object falls from rest near the earth's surface, accelerating downward at 1 **g**. After 5 seconds, what is the speed and the distance moved?

5.15 A force of 2.7 lb is applied to a 5.5-kg mass. Find the resulting acceleration in m/s^2.

5.16 Calculate the average shock in **g**s experienced by a transistor that falls 1.5 m from a tabletop, if it takes 2.7 ms to deaccelerate to zero when reaching the floor.

5.17 An automobile fender vibrates at 16 Hz with a peak-to-peak amplitude of 5 mm. Calculate the peak acceleration in **g**s.

5.18 A spring-mass system has a mass of 0.02 kg and a spring constant of 140 N/m. Calculate the natural frequency.

5.19 An LVDT is used in an accelerometer to measure seismic mass displacement. The LVDT and signal conditioning outputs 0.31 mV/mm with a ±2 cm maximum core displacement. The spring constant is 240 N/m, and the core mass is 0.05 kg. Find (a) the relation between acceleration in m/s^2 and output voltage, (b) the maximum acceleration which can be measured, and (c) the natural frequency.

5.20 For the accelerometer of Problem 5.19, design a signal conditioning system which provides velocity information at 2 mV/(m/s) and position information at 0.5 V/m.

5.21 A piezoelectric accelerometer has a transfer function of 61 mV/g and a natural frequency of 4.5 kHz. In a vibration test at 110 Hz, a reading of 3.6 volts peak results. Find the vibration peak displacement.

5.22 An accelerometer for shock is designed as shown in Figure 5.35. Find a relation between strain gauge resistance change and shock in **g** (i.e., the resistance change per **g**). The force rod cross-sectional area is 2.0×10^{-4} m².

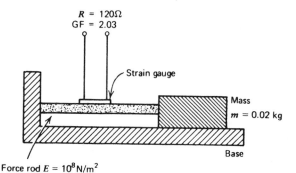

$R = 120\Omega$
$GF = 2.03$

Strain gauge

Mass
$m = 0.02$ kg

Base

Force rod $E = 10^8$ N/m²

Figure 5.35   Figure for Problem 5.22.

5.23 Design a signal conditioning scheme for the accelerometer of Problem 5.22 using a bridge circuit. Plot the bridge offset voltage versus shock in **g** from 0 to 5000 **gs**.

Section 5.5

5.24 Calculate (a) the pressure in atmospheres that a water column 3.3 m high exerts on its base and (b) the pressure if the liquid is mercury, and convert these results to pascals.

5.25 A welding tank holds oxygen at 1500 psi. What is the tank pressure expressed in Pa? What is the pressure in atmospheres?

5.26 A diaphragm has an effective area of 25 cm². If the pressure difference across the diaphragm is 5 psi, what force is exerted on the diaphragm?

Section 5.6

5.27 A grain conveyor system finds the weight on a 1.0-meter platform to be 258 N. What conveyor speed is needed to get a flow of 5200 kg/hr?

5.28 Convert water flow of 52.2 gal/hr into kg/hr and velocity in m/s through a 2-inch diameter pipe.

5.29 For an orifice plate in a system pumping alcohol, we find $K = 0.4$ m²/min(kPa)$^{1/2}$. Plot the pressure versus flow rate from 0 to 100 m³/hr.

CHAPTER 6

OPTICAL SENSORS

INSTRUCTIONAL OBJECTIVES

After you have read this chapter, you should be able to

1. Describe EM radiation in terms of frequency, wavelength, speed of propagation, and spectrum.
2. Define the energy of EM radiation in terms of power, intensity, and the effects of divergence.
3. Compare photoconductive, photovoltaic, and photoemissive-type photodetectors.
4. Describe the principles and structure of both total radiation and optical pyrometers
5. Distinguish incandescent, atomic, and laser light sources by the characteristics of their light.
6. Design the application of optical techniques to process-control measurement applications.

6.1 INTRODUCTION

A previously emphasized desired characteristic of all sensors is that a *linear* relation is desired between output and variable input. Such a linear characteristic simplifies both analysis and signal conditioning and thereby provides more confidence that the output signal represents the measured variable.

A second desirable characteristic of sensors is negligible effect on the measured environment, that is, the process. Thus, if a resistance temperature detector (RTD) *heats up* its own temperature environment, there is less confidence that the RTD resistance truly represents the environmental temperature. Much effort is made in sensor and transducer design to reduce backlash from the measuring instrument on its environment.

When electromagnetic (EM) radiation is used to perform process variable measurements, transducers that do not affect the system measured emerge. Such systems of measurement are called *nonlocal* or *noncontact* because no physical contact is made with the environment of the variable. Noncontact characteristic measurements often can be made from a distance.

In process control, EM radiation in either the visible or infrared light band is used in measurement applications. The techniques of such applications are called *optical* because such radiation is close to visible light.

A common example of optical transduction is measurement of an object's temperature by its emitted EM radiation. Another example involves radiation reflected off the surface to yield a level or displacement measurement.

In general, there are several levels of approach to the study of transducer technology. The first and most elementary level is the simple substitution of a black-box instrument for the measurement of some variable. A second level is the knowledge and training necessary to repair such an instrument using manuals. This requires rudimentary knowledge of the variable being measured and how such measurement is made. A third level might be the knowledge and understanding necessary to select an appropriate transducer for the specific measurement requirement. This involves even more comprehension of the variable, the transduction methods, and measurement limitations. The fourth and highest level requires the knowledge to *design* the actual transducer, given the variation of one physical variable compared to another. In this text, sufficient background is provided to reach the third level.

Optical technology is a vast subject covering a span from geometrical optics, including lenses, prisms, gratings, and the like, to physical optics with lasers, parametric frequency conversion, and nonlinear phenomena. These subjects are all very interesting, but all that is required for our purposes is a familiarity with optical principles and a knowledge of specific transduction and measurement methods.

6.2 FUNDAMENTALS OF EM RADIATION

We are all familiar with EM radiation as *visible light*. Visible light is all around us. EM radiation is also familiar in other forms, such as radio or TV signals and ultraviolet or infrared light. Most of us falter, however, if asked to give a general technical description of such radiation including criteria for measurement and units.

This section covers a general method of characterizing EM radiation. Although much of what follows is valid for the complete range of radiation, particular attention is given to the infrared, visible, and ultraviolet because most sensor applications are concerned with these ranges.

6.2.1 Nature of EM Radiation

EM radiation is a form of *energy* that is always in motion, that is, propagates through space. An object that releases or *emits* such radiation *loses* energy. One that *absorbs* radiation *gains* energy. Thus, we must describe how this energy appears as EM radiation.

Frequency and wavelength

Because we use the term *electromagnetic* radiation to name this form of energy, it is no surprise that it is intimately tied to electricity and magnetism. Indeed, careful study shows that electrical and magnetic phenomena produce EM radiation. The radiation propagates through space in a manner similar to waves in water propagating from some disturbance. As such, we can define both a *frequency* and *wavelength* of the radiation. The *frequency* represents the oscillation per second as the radiation passes some fixed point in space. The *wavelength* is the spatial distance between two successive maxima or minima of the wave in the direction of propagation.

Speed of propagation

EM radiation propagates through a vacuum at a speed *independent* of both the wavelength and frequency. In this case, the velocity is

$$c = \lambda f \tag{6.1}$$

where

$c = 2.998 \times 10^8$ m/s $\approx 3 \times 10^8$ m/s = speed of EM radiation in a vacuum
λ = wavelength in meters
f = frequency in hertz (Hz) or cycles per second (s^{-1})

Example 6.1

Given an EM radiation frequency of 10^6 Hz, find the wavelength.

Solution   We have

$$c = \lambda f \tag{6.1}$$

and

$$\lambda = \frac{c}{f} = \frac{3 \times 10^8 \text{ m/s}}{10^6 \text{ s}^{-1}}$$

$$\lambda = \mathbf{300 \ m}$$

Note: This EM radiation is used to carry AM radio signals.

When such radiation moves through a nonvacuum environment, the propagation velocity is *reduced* to a value *less* than *c*. In general, the new velocity is indicated by the *index of refraction* of the medium. The index of refraction is a ratio defined by

$$n = \frac{c}{v} \tag{6.2}$$

where

n = index of refraction
v = velocity of EM radiation in the material (m/s)

The index of refraction often varies with the radiation wavelength for some sample of material.

Example 6.2

Find the velocity of EM radiation in glass that has an index of refraction of $n = 1.57$.

Solution   We know that

$$n = \frac{c}{v} \tag{6.2}$$

so that

$$v = \frac{c}{n} = \frac{3 \times 10^8 \text{ m/s}}{1.57}$$

$$v = 1.91 \times 10^8 \text{ m/s}$$

Wavelength units

The most consistent description of EM radiation is via the *frequency* or *wavelength*. For many applications, this specification is made through the frequency of the radiation, as in a 1-MHz radio signal or a 1-GHz microwave signal. By convention, however, it has become more common to describe EM radiation by the wavelength, that is, by Equation (6.1). This is particularly true near the visible light band. The proper unit of measurement is, of course, the length in meters with associated prefixes. Thus, a 10-GHz signal is described by a 30-mm wavelength. Red light is emitted at about 0.7 μm wavelength.

Another unit often employed is the *Angstrom* (Å), defined as 10^{-10} m or 10^{-10} m/Å. Thus, the red light previously described has a wavelength of 7000 Å. The conversion is left as an exercise for the reader. (See also Problem 6.3.)

EM Radiation spectrum

We have seen that EM radiation is a type of energy the propagates through space at a constant speed or velocity if we specify the direction. The oscillating nature of this radiation gives rise to a different interpretation of this radiation in relation to our environment, however. In categorizing radiation by wavelength or frequency, we are describing its position in the *spectrum* of radiation. Figure 6.1 shows the range of EM radiation from very low frequency to very high frequency, together with the associated wavelength in meters from Equation (6.1) and how the bands of frequency relate to our world.

This one type of energy ranges from radio signals and visible light to X-rays and penetrating cosmic rays and all through the smooth variation of frequency. In process-control instrumentation we are particularly interested in two of the bands, infrared and visible light.

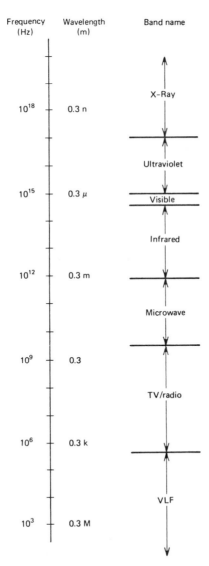

Figure 6.1 The electromagnetic radiation spectrum covers everything from very low frequency (VLF) radio to X-rays and beyond.

Visible light

The small band of radiation between approximately 400 nm and 760 nm represents *visible light* (Figure 6.1). This radiation band covers those wavelengths to which our eyes (or radiation detectors in our heads) are sensitive.

Infrared light

The longer-wave radiation band that extends from the limit of eye sensitivity at 0.76 μm to approximately 100 μm is called *infrared* (IR) radiation. In some cases, the band is farther subdivided so that radiation of wavelength 3 to 100 μm is called *far-infrared*.

   The limits of these bands are not distinct and serve only to separate roughly the described radiation into broad categories. Our treatment for the rest of this chapter refers simply to *light*, meaning either IR or visible, because our concern is with these bands, exclusively.

Example 6.3

   Describe (a) the wavelength (in m and Å units), and (b) the nature of EM radiation of 5.4×10^{13} Hz frequency.

   Solution   The wavelength is given by

(a)

$$\lambda = \frac{c}{f}$$

$$\lambda = (3 \times 10^8 \text{ m/s})/(5.4 \times 10^{+13} \text{ Hz}) \qquad (6.1)$$

$$\lambda = \textbf{5.56 μm}$$

$$\lambda = 5.56 \times 10^{-6} \text{ m} \times \frac{1 \text{ Å}}{10^{-10} \text{ m}} = \textbf{55600 Å}$$

(b) From Figure 6.1 we see that this radiation lies in the *infrared* band, generally, and is designated as far-infrared, specifically.

6.2.2 Characteristics of Light

Because light has been described as a source of energy, it is natural to inquire about the energy content and its relation to the spectrum.

Photon

No description of EM radiation is complete without a discussion of the photon. EM radiation at a particular frequency can propagate only in *discrete* quantities of energy. Thus, if some source is emitting radiation of one frequency, then in fact it is emitting this energy as a large number of discrete units or *quanta*. These quanta are called *photons*. The actual energy of one photon is related to the frequency by

$$W_p = hf = \frac{hc}{\lambda} \tag{6.3}$$

where

W_p = photon energy (J)
h = 6.63×10^{-34} J $-$ s (known as Planck's constant)
f = frequency (s^{-1})
λ = wavelength (m)

The energy of one photon is very small compared to electric energy, as shown by Example 6.4.

Example 6.4

A microwave source emits a pulse of radiation at 1 GHz with a total energy of 1 joule. Find (a) the energy per photon and (b) the number of photons in the pulse.

Solution   We find the energy per photon where 1 GHz = 10^9 s^{-1}

(a)

$$W_p = hf$$

$$W_p = (6.63 \times 10^{-34} \text{ J} - \text{s})(10^9 \text{ s}^{-1}) \tag{6.3}$$

$$W_p = \mathbf{6.63 \times 10^{-25} \text{ J}}$$

(b) The number of photons is

$$N = \frac{W}{W_p} = \frac{1 \text{ J}}{6.63 \times 10^{-25} \text{ J/photon}}$$

$$N = \mathbf{1.5 \times 10^{24} \text{ photons}}$$

Figure 6.2 shows the energy carried by a single photon at various wavelengths. This energy is expressed in electron volts where (1 eV = 1.602×10^{-19} J). This unit is conventionally employed and provides convenient magnitude for photon energy discussion.

Energy

When dealing with typical sources and detectors of light, it is impractical and unnecessary to consider the discrete nature of the radiation. Instead, we deal with so-called *macroscopic* properties that result from the collective behavior of a vast number of photons moving together. The general energy principles involve a de-

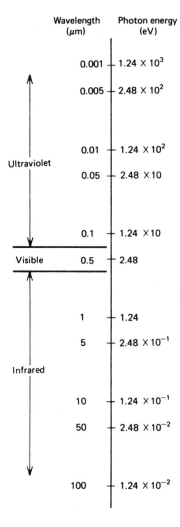

Figure 6.2 The energy carried by one photon varies inversely with the wavelength of the EM radiation.

scription of the net energy of the radiation as it propagates through a region of space. This description is then given in *joules* of energy in the propagating light. A simple statement of energy is insufficient, however, because of the motion of the energy and the spatial distribution of the energy.

Power

Because EM radiation is energy in motion, a more complete description is the joules per second or *watts* of power carried. Thus, one might describe a situation where a source emits 10 watts of light; that is, 10 joules of energy in the form of light radiation are emitted every second. Even this description is incomplete without specifying how the power is spatially distributed.

Intensity

A more complete picture of the radiation emerges if we also specify the spatial distribution of the power *transverse* to the *direction* of propagation. Thus, if the 10-watt source just discussed was concentrated in a beam with a cross-sectional area of 0.2 m², then we can specify the *intensity* as the watts per unit area, in this case as (10 W)/(0.2 m²) or 50 W/m². In general, the intensity is

$$I = \frac{P}{A} \tag{6.4}$$

where
  I = intensity in W/m²
  P = power in W
  A = beam cross-sectional area in m²

Divergence

We still have not quite exhausted the necessary descriptors of the energy because of the tendency of light to travel in straight lines. Because the radiation travels in straight lines, it is possible for the intensity of the light to *change* even though the power remains *constant*. This is best seen in Figure 6.3. We have a 10-watt source with an area A_1 at the source. Because of the nature of the source and the straight-line propagation, however, we see that some distance away the same 10 watts is distributed over a larger area A_2 and hence the *intensity* is *diminished*. Such spreading of radiation is called *divergence* and specified as the angle θ, which is made by the outermost edge of the beam to the central direction of propagation, as shown in Figure 6.3.

A complete description of the energy state of light propagating through a region of space demands knowledge of the power carried, the cross-sectional area over which the power is distributed, and the divergence.

Example 6.5

Find the intensity of a 10-watt source (a) at the source and (b) 1 meter away for the case shown in Figure 6.3 if the radius at A_1 is 0.05 m and the divergence is 2°.

Solution   The intensity at the source is simply the power over the area. Thus, for

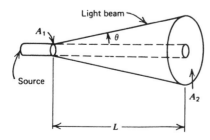

Figure 6.3   Sources of EM radiation exhibit divergence through the spreading of the beam with distance from the source.

a circular cross section

$$A_1 = \pi R_1^2 = (3.14)(0.05 \text{ m})^2$$

$$A_1 = 7.85 \times 10^{-3} \text{ m}^2$$

(a) The intensity at the source is

$$I_1 = \frac{P}{A_1} \qquad (6.4)$$

$$I_1 = \frac{10 \text{ W}}{7.85 \times 10^{-3} \text{ m}^2} = \mathbf{1273 \text{ W/m}^2}$$

(b) We first must find the beam area at 1 meter to get A_2 and then the intensity at
1 meter. We can find the radius using Figure 6.4 as a guide. From elementary
trigonometry we see that, for the right triangle formed

$$R_2 = R_1 + L \tan(\theta) \qquad (6.5)$$

then

$$R_2 = 0.05 \text{ m} + (1 \text{ m}) \tan(2°)$$

$$R_2 = 0.085 \text{ m}$$

$$A_2 = \pi R_2^2 = (3.14)(0.085 \text{ m})^2$$

$$A_2 = 0.0227 \text{ m}^2$$

$$I_2 = \frac{10 \text{ W}}{0.0227 \text{ m}^2}$$

$$I_2 = 440.53 \text{ W/m}^2 \text{ or } \mathbf{441 \text{ W/m}^2}$$

When dealing with sources originating from a very small point but propagating
in *all* directions, we have a condition of *maximum divergence*. Such *point sources*
have an intensity that decreases as the *inverse square* of the distance from the
point. This can be seen from a consideration of the divergence. Suppose a source
delivers a power P of light as shown in Figure 6.5. The intensity at a distance R
is found by dividing the total power by the surface area of a sphere of radius R
from the source.

$$I = \frac{P}{A} \text{ in W/m}^2 \qquad (6.4)$$

The surface area of a sphere of radius R is $A = 4\pi R^2$, thus over the entire surface

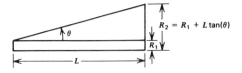

$R_2 = R_1 + L \tan(\theta)$

R_1

L

Figure 6.4   Diagram to aid in solving
divergence problems such as Example
6.5.

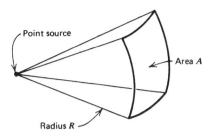

Figure 6.5 The intensity of light from a point source depends on the distance from the source R and the area considered A.

surrounding a point source

$$I = \frac{P}{4\pi R^2} \text{ in } W/m^2 \tag{6.6}$$

This equation shows that the intensity of a point source decreases as the inverse square of the distance from the point.

Chromaticity

Another factor of significance in the description of light includes the *spectral content* of the radiation. A source such as a laser beam, which delivers light of a *single* wavelength (or very nearly), is called a *monochromatic* source. A source such as an incandescent bulb may deliver a very *broad* spectrum of radiation and is referred to as a *polychromatic* source.

Coherency

A less familiar characteristic of the radiation is its coherency. We have seen that light is described through electric and magnetic effects that oscillate in time and space. Whenever we consider oscillating phenomena, it is of interest to determine the phase relation of oscillations of different parts of the beam. When all points along some cross section of a radiation beam are in phase, the beam has *spatial coherence*. When the radiation in a line along the beam has a fixed-phase relation, the beam has *temporal coherence*. In general, conventional sources of light, such as incandescent and fluorescent light bulbs, produce beams with no coherence. A laser is the only convenient source of coherent radiation available.

6.2.3 Photometry

The conventional units described previously would seem to be satisfactory for a complete description of optical processes. For designs in human engineering, however, these traditional units are insufficient. This is because the human eye responds not only to the intensity of light but also to its spectral content.

Suppose it is known that a human needs an intensity of 1 W/m^2 to read. If an infrared source at that intensity were used, the human would still not be able to read, because the eye does not respond to infrared.

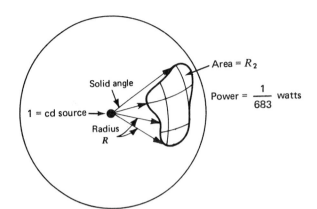

Figure 6.6 The candela is defined in terms of uniform monochromatic radiation power.

Because of these problems and others, special sets of units for EM radiation have been developed to be used in such human-related design problems. The most basic unit is the SI candela (cd). A 1-candela source is defined as one which emits monochromatic radiation at 340×10^{12} Hz ($\approx$555 nm wavelength) at such an intensity that there is 1/683 watts passing through one steradian of solid angle. This radiation is roughly in the middle of the visible spectrum.

Figure 6.6 shows how a 1-candela source is pictorially defined. The 1-candela source at the center of a sphere of radius R will emit radiation of the aforementioned frequency with an energy of 1/683 watts through any part of the surface with an area of R^2.

There are many other units employed to describe visible or luminous energy. These units do not find significant application in process measurement applications using optical technology, so they will not be covered.

6.3 PHOTODETECTORS

An important part of any application of light to an instrumentation problem is how to measure or detect radiation. In most process-control-related applications, the radiation lies in the range from IR through visible and sometimes UV bands. The measurement transducers generally used are called *photodetectors* to distinguish them from other spectral ranges of radiation such as RF detectors in radio frequency (RF) applications.

In this section, we will study the principal types of photodetectors through a description of their operation and specifications.

6.3.1 Photodetector Characteristics

Several characteristics of photodetectors are particularly important in typical applications of these devices in instrumentation. In the following discussions, the various types of detectors are described in terms of these characteristics.

Spectral response

Most detectors are able to function over some specified range of radiation wavelengths that defines the *spectral* response of the device. In most cases, the response is flat within some allowed deviation within this band of radiation.

Time constant

Much of the instrumentation associated with optical or EM radiation techniques is associated with time-varying phenomena. For this reason, the response time, as specified through the time constant defined in Chapter 2, becomes an important part of the specifications.

Detectivity

Various photodetectors vary remarkably so that no general contrast in detectivity or sensitivity can be made. An important part of the specification, however, is to note the type of transduction and sensitivity to the optical variable, which is usually given as intensity of the radiation at a particular wavelength.

6.3.2 Photoconductive Detectors

One of the most common photodetectors is based on the change in *conductivity* of a semiconductor material with radiation *intensity*. The change in conductivity appears as a change in *resistance* so that these devices also are called *photoresistive* cells. Because resistance is the parameter used as the transduced variable, we describe the device from the point of view of *resistance changes* versus *light intensity*.

Principle

In Section 4.4.1, we noted that a semiconductor is a material in which an energy gap exists between conduction electrons and valence electrons. In a semiconductor photodetector, a photon is absorbed and thereby excites an electron from the valence to the conduction band. As many electrons are excited into the conduction band, the semiconductor resistance *decreases*, making the resistance an inverse function of radiation intensity. For the photon to provide such an excitation it must carry at least as much energy as the gap. From Equation (6.3) this indicates a maximum wavelength

$$E_p = \frac{hc}{\lambda_{\max}} = \Delta W_g$$

$$\lambda_{\max} = \frac{hc}{\Delta W_g}$$

(6.7)

where

$$h = \text{Planck's constant, } 6.63 \times 10^{-34} \text{ J} - \text{s}$$
$$\Delta W_g = \text{semiconductor energy gap (J)}$$
$$\lambda_{max} = \text{maximum detectable radiation wavelength (m)}$$

Any radiation with a wavelength greater than that predicted by Equation (6.7) *cannot* cause any resistance change in the semiconductor.

Example 6.6

Germanium has a band gap of 0.67 eV. Find the maximum wavelength for resistance change by photon absorption. Note 1.6×10^{-19} J = 1 eV.

Solution   We find the maximum wavelength from

$$\lambda_{max} = \frac{hc}{\Delta W_g}$$

$$\lambda_{max} = \frac{(6.63 \times 10^{-34} \text{ J} - \text{s})(3 \times 10^8 \text{ m/s})}{(0.67 \text{ eV})(1.6 \times 10^{-19} \text{ J/eV})} \qquad (6.7)$$

$$\lambda_{max} = 1.86 \ \mu\text{m}$$

which lies in the IR.

It is important to note that the operation of a thermistor involves *thermal-energy-exciting* electrons into the conduction band. To prevent the photoconductor from showing similar thermal effects, it is necessary either to operate the devices at a controlled temperature or to make the gap too large for thermal effects to produce conduction electrons. Both approaches are employed in practice. The upper limit of the cell spectral response is determined by many other factors, such as reflectivity and transparency to certain wavelengths.

Cell structure

The two most common photoconductive semiconductor materials are cadmium sulfide (CdS) with a band gap of 2.42 eV and cadmium selenide (CdSe) with a 1.74 eV gap. Because of these large gap energies, both materials have a very high resistivity at room temperature. This gives bulk samples a resistance much too large for practical applications. To overcome this, a special configuration is used, as shown in Figure 6.7, that minimizes resistance geometrically and provides maximum surface area for the detector. This result is based on Equation (4.7).

$$R = \rho \, l/A \qquad (4.7)$$

where

$$R = \text{resistance } (\Omega)$$
$$\rho = \text{resistivity } (\Omega - \text{m})$$

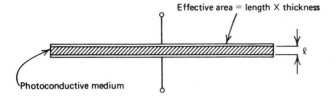

a) A long, narrow, thin sample gives optimum response and resistance

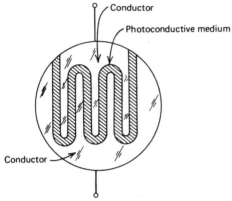

b) By folding the above pattern back
and forth we concentrate the sensitive
material in space

Figure 6.7 Photoconductive cell
structure.

l = length (m)
A = cross-sectional area (m^2)

The pattern of Figure 6.7d gives a minimum l and a maximum A. By using
a thin, narrow strip and winding this arrangement back and forth as in Figure
6.7b, we get a maximum surface area.

Cell characteristics

The characteristics of photoconductive detectors vary considerably when different semiconductor materials are used as the active element. These characteristics
are summarized for typical values in Table 6.1.

The nominal dark resistance and variation of resistance with intensity are
usually provided in terms of a graph or table of resistance versus intensity at a
particular wavelength within the spectral band. Typical values at dark resistance
vary from hundreds of ohms to several MΩ for various types of photoconductors.

TABLE 6.1   Photoconductor Characteristics

Photoconductor	Time Constant	Spectral Band
CdS	~100 ms	0.47 to 0.71 μm
CdSe	~10 ms	0.6 to 0.77 μm
PbS	~400 μs	1 to 3 μm
PbSe	~10 μs	1.5 to 4 μm

The variation with radiation intensity is usually nonlinear, with resistance decreasing as the radiation intensity increases.

Signal conditioning

Like the thermistor, a photoconductive cell exhibits a resistance that decreases nonlinearly with the dynamic variable, in this case, radiation intensity. Generally, the change in resistance is very pronounced where a resistance can change by several hundred orders of magnitude from dark to normal room daylight.

If an absolute intensity measurement is desired, calibration data is used in conjunction with any accurate resistance measurement method.

Sensitive control about some ambient radiation intensity is obtained using the cell in a bridge circuit adjusted for a null at the ambient level.

Various op amp circuits using the photoconductor as a circuit element are used to convert the resistance change to a current or voltage change.

It is important to note that the cell is a variable resistor and therefore has some maximum power dissipation that cannot be exceeded. Most cells have a dissipation from 50 to 500 mW, depending on size and construction.

Example 6.7

A CdS cell has a dark resistance of 100 kΩ and a resistance in a light beam of 30 kΩ. The cell time constant is 72 ms. Devise a system to trigger a 3-volt comparator within 10 ms of the beam interruption.

Solution   There are many possible solutions to this problem. Let us first find the cell resistance at 10 ms using

$$R(t) = R_i + (R_f - R_i)[1 - e^{-t/r}] \tag{1.7}$$

$$R(10 \text{ ms}) = 30 \text{ k} + 70 \text{ k} (0.1296) = 39.077 \text{ k}\Omega$$

so that we must have a +3-volt signal to the comparator when the cell resistance is 39.077 kΩ. The circuit of Figure 6.8 will accomplish this. The cell is R_2 in the feedback of an inverting amplifier with a −1.0-volt constant input. The output is

$$V_{\text{out}} = - \left(\frac{R_2}{R_1}\right) (-1 \text{ V}) = \frac{R_2}{R_1}$$

when $R_2 = 39.077$ kΩ, we make $V_{\text{out}} = 3$ volts so that

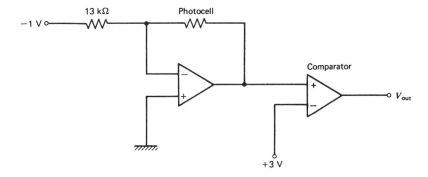

Figure 6.8   This circuit is the solution of Example 6.7.

$$R_2 = 39.077 \text{ k}\Omega$$

$$R_1 \simeq \textbf{13 k}\Omega$$

which assures that the comparator will trigger at 10 ms from beam interruption. To see that the comparator will *not* trigger with the beam present, set $R_2 = 30$ kΩ and the amplifier output is

$$V_{\text{out}} = -\frac{30 \text{ k}\Omega}{13 \text{ k}\Omega}(-1 \text{ V})$$

$$V_{\text{out}} = 2.3 \text{ V}$$

which is insufficient to trigger the comparator.

6.3.3 Photovoltaic Detectors

An important class of photodetectors generates a voltage that is proportional to incident EM radiation intensity. These devices are called *photovoltaic* cells because of their *voltage-generating* characteristics. They actually convert the EM energy into electrical energy. Applications are found as both EM radiation detectors and power sources converting solar radiation into electrical power. The emphasis of our consideration is on instrumentation-type applications.

Principle

Operating principles of the photovoltaic cell are best described by Figure 6.9. We see that the cell is actually a giant diode that is constructed using a *pn* junction between appropriately doped semiconductors. Photons striking the cell pass through the thin *p*-doped upper layer and are absorbed by electrons in the *n* layer, which causes formation of conduction electrons and holes. The depletion zone potential of the *pn* junction then separates these conduction electrons and holes, which causes a difference of potential to develop across the junction. The upper

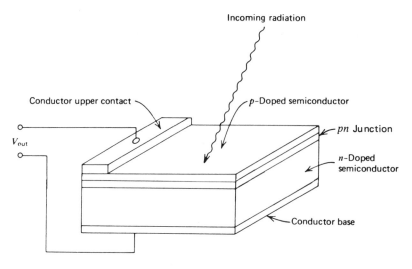

Figure 6.9 Structure of a photovoltaic cell.

terminal is positive and the lower negative. It is also possible to build a cell with a thin n-doped layer on top so that all polarities are opposite. In general, the open circuit voltage V developed in these cells varies logarithmically with the radiation intensity by

$$V = V_0 \log_e(I) \qquad (6.8)$$

where

I = intensity in W/m^2
V_0 = calibration voltage, constant
V = output voltage, unloaded

All photovoltaic devices have low but finite internal resistance. When connected in a circuit with some load resistance, the cell voltage is reduced somewhat from the value predicted by Equation (6.8).

The constant voltage amplitude V_0 is a function of the cell material only and shows that the voltage produced is independent of the cell geometry. The current delivered from such a cell into a load depends on the intensity via Equation (6.8) and increases with cell surface area. Cells are often arranged in series and parallel combinations to obtain desired voltage and current output levels.

The spectral range of the photovoltaic cell depends on the photon energy required to excite an electron-hole pair in the doped semiconductor and other properties of the materials.

Example 6.8

A photovoltaic cell produces 0.33 V open circuit when illuminated by 10 W/m^2 radiation intensity. A current 2.2 mA is delivered into a 100-Ω load at that intensity. Calculate (a) the internal resistance and (b) the open circuit voltage at 25 W/m^2.

Solution

(a) We find the internal resistance by noting that

$$I = \frac{V}{R_{in} + R_L}$$

where V is the open circuit cell voltage, R_L the load resistance, I the current, and R_{in} the internal resistance. We find then

$$R_{in} = \frac{0.33 \text{ V} - (2.2 \text{ mA})(100 \text{ } \Omega)}{2.2 \text{ mA}}$$

$$R_{in} = \mathbf{50 \text{ } \Omega}$$

(b) The open circuit voltage at 25 W/m^2 requires knowledge of V_0 in Equation (6.8). This can be found from

$$V_0 = \frac{V}{\log_e(I)}$$

$$V_0 = \frac{0.33 \text{ V}}{\log_e(10)} = 0.143 \text{ V}$$

so that 25 W/m^2 we have

$$V = (0.143 \text{ V}) \log_e(25)$$

$$V = \mathbf{0.46 \text{ V}}$$

Cell characteristics

The properties of photovoltaic cells depend on the materials employed for the cell and the nature of the doping used to provide the n and p layers. Some cells are used only at low temperature to prevent thermal effects from obscuring radiation detection. The silicon photovoltaic cell is probably the most common. Table 6.2 lists several types of cells and their typical specifications.

Thermal effects can be quite pronounced, producing changes of the order of mV/°C in output voltage of fixed intensity.

Signal conditioning

Generally, signal conditioning depends on the application where the cell is used. Simple op amp configuration provides a measure of either open circuit voltage

TABLE 6.2 Typical Photovoltaic Cell Characteristics

Cell Material	Time Constant	Spectral Band
Silicon (Si)	~20 μs	0.44 μm to 1 μm
Selenium (Se)	~2 ms	0.3 μm to 0.62 μm
Germanium (Ge)	~50 μs	0.79 μm to 1.8 μm
Indium arsenide (InAs)	~1 μs	1.5 μm to 3.6 μm (cooled)
Indium antimonide (InSb)	~10 μs	2.3 μm to 7 μm (cooled)

or current at any specified load impedance. Fast measurements require circuits that account for the internal capacity of the cell and the inherent time constant of the cell.

6.3.4 Photodiode Detectors

The *pn* junction of any diode is sensitive to EM radiation that may strike the junction. This sensitivity is usually in the form of an alteration of the I versus V characteristic of the junction because of a change in current carriers. Special diodes that allow the junction to be exposed to incident EM radiation frequently are used for photodetectors. Generally, the junction is very small, requiring the use of lenses to focus the radiation on the junction. The most significant advantage of these detectors is their very fast response times. Most photodiodes have a time constant near 1 μs, but units are common with time constants less than 1 ns. These latter devices are employed in high-speed measurement or communication applications.

The most common diodes are silicon, which are used between 0.82 μm and 1.1 μm, and germanium, which are used between 1.4 μm and 1.9 μm radiation wavelength. Signal conditioning usually involves standard diode circuits where incident radiation will cause a shift in the diode operating point. High-speed measurements require special, frequency-compensated circuits.

6.3.5 Photoemissive Detectors

This type of photodetector was developed many years ago, but it still is one of the most sensitive types. A wide variety of spectral ranges and sensitivities can be selected from the many types of photoemissive detectors available.

Principles

To understand the basic operational mechanism of photoemissive devices, let us consider the two-element vacuum phototube shown in Figure 6.10. Such photodetectors have been largely replaced by other detectors in modern measurements. In Figure 6.10, we note that the cathode is maintained at some negative voltage with respect to the wire anode that is grounded through resistor R. The inner surface of the cathode has been coated with a *photoemissive* agent. This material is a metal for which electrons are easily detached from the metal surface. "Easily" means it does not take much energy to cause an electron to leave the material. In particular, then, a photon can strike the surface and impart sufficient energy to an electron to eject it from this coating. The electron will then be driven from the cathode to anode and thence through resistor R. Thus, we have a current that depends on the intensity of light striking the cathode.

Photomultiplier Tube

The simple diode described previously is the basis of one of the most sensitive photodetectors available, as shown in Figure 6.11a. As previously, a cathode is

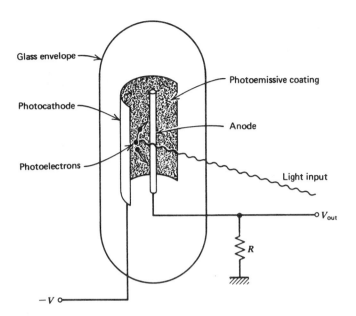

Figure 6.10   Structure of the basic photoemissive diode.

maintained at a large negative voltage and coated with a photoemissive material. In this case, however, we have many following electrodes, called *dynodes*, maintained at successively more positive voltages. The final electrode is the *anode* that is grounded through a resistor R. A photoelectron from the cathode strikes the first dynode with sufficient energy to eject several electrons. All of these electrons are accelerated to the second dynode where *each* strikes the surface with sufficient energy to again eject several electrons. This process is repeated for each dynode until the electrons that reach the anode are greatly multiplied in number, where they constitute a current through R. Thus, the photomultiplier (unlike some other transducers) has a *gain* associated with its detection. One single photon striking the cathode may result in a million electrons at the anode! It is this effect that gives the photomultiplier its excellent *sensitivity*. Many other electrode design arrangements are used with the same principle of operation shown in Figure 6.11a.

Specifications

The specifications of photomultiplier tubes depend on several features:

1. The *number* of dynodes and material from which they are constructed determines the amplification or current gain. Gains of 10^5 to 10^7 (relating direct photoelectrons from the cathode to electrons at the anode) are typical.
2. The spectral response is determined by two factors. The first is the spectral response of the photoemissive material coated on the cathode. The second

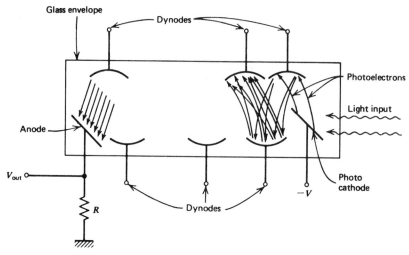

a) Basic structure of the photomultiplier tube

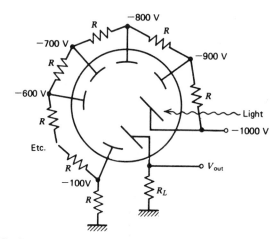

b) Typical voltage divider used for dynode potentials

Figure 6.11   The photomultiplier tube.

is the transparency of the glass envelope or window through which the EM radiation must pass. Using various materials, it is possible to build different types of photomultipliers that, taken together, span wavelengths from 0.12 to 0.95 μm.

The combination of cathode coating and window material is described by a standard system to indicate the spectral response. Thus, a designation of S and a number identifies a particular band. For example, S-3 designates flat response from about 0.35 to 0.7 μm.

The time constant for photomultiplier tubes ranges from 20 μs down to 0.1 μs, typically.

Signal conditioning

The signal conditioning is usually a high-voltage negative supply directly connected to the photocathode. A resistive voltage divider (Figure 6.11b) provides the respective dynode voltages. Usually, the anode is grounded through a resistor, and a voltage drop across this resistor is measured. The cathode typically requires −1000 to −2000 V, and each dynode is then divided evenly from that. A 10-dynode tube with a −1000-V cathode thus has −900 V on the first dynode, −800 V on the second, and so on, as shown in Figure 6.11b.

6.4 PYROMETRY

One of the most significant applications of optoelectronic transducers is in the noncontact measurement of temperature. The early term *pyrometry* has been extended to include any of several methods of temperature measurement that rely on EM radiation. These methods depend on a direct relation between an object's temperature and the EM radiation emitted. In this section, we consider the mechanism by which such radiation and temperature are related and how it is used for temperature measurement.

6.4.1 Thermal Radiation

All objects having a finite absolute temperature emit EM radiation. The nature and extent of such radiation depend on the temperature of the object. The understanding and description of this phenomenon occupied the interest and attention of physicists for many years, but we can briefly note the results of these researches through the following argument. It is well known that EM radiation is generated by the acceleration of electrical charges. We also have seen that the addition of thermal energy to an object results in vibratory motion of the molecules of the object. A simple marriage of these concepts, coupled with the fact that molecules consist of electrical charges, led to the conclusion that an object with finite thermal energy emits EM radiation because of *charge* motion.

Because an object is *emitting* EM radiation and such radiation is a form of energy, the object must be *losing* energy, but if this were true its temperature would decrease as the energy radiates away. (In fact, for an isolated small sample, this does occur.) In general, however, a state of equilibrium is reached where the object gains as much energy as it radiates, and so remains at a fixed temperature. This energy gain may be because of thermal contact with another object or the absorption of EM radiation from surrounding objects. It is most important to note this interplay between EM radiation *emission* and heat *absorption* by an object in achieving *thermal equilibrium.*

Blackbody radiation

To develop a quantitative description of thermal radiation, consider first an *idealized* object. An idealized object absorbs *all* radiation impinging on it, regardless of wavelength, and therefore becomes an *ideal absorber*. This object also emits radiation without regard to any special peculiarities of particular wavelength and therefore becomes an *ideal emitter*.

Assume this ideal object is now placed in thermal equilibrium so that its temperature is controlled. In Figure 6.12, the EM radiation emitted from such an ideal object is plotted to show the intensity and spectral content of the radiation for several temperatures. The abscissa is the radiation wavelength, and the ordinate is the energy emitted per second (power dissipated) per unit area at a particular wavelength. The area under the curve indicates the total energy per second (power dissipated) per unit area emitted by the object. Several curves are shown on the plot for different temperatures. We see that at low temperatures, the radiation emitted is predominately in the long wavelength (far-infrared to microwave) region. As the temperature is increased, the maximum emitted radiation is in the shorter wavelengths and finally, at very *high* temperatures, the maximum emitted radiation is near the *visible* band. Because of this shift in emission peak with temperature, an object begins to glow as its temperature is in-

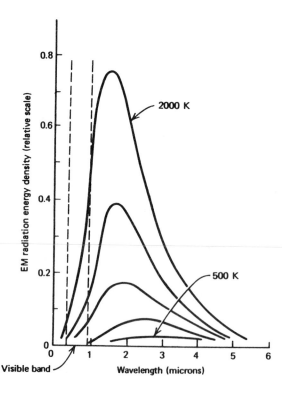

Figure 6.12 Idealization of the EM radiation emitted by a perfect blackbody as a function of temperature.

creased. For the blackbody, the temperature and emitted radiation are in one-to-one correspondence in the following respects:

1. *Total radiation*   A study of blackbody radiation shows that the total emitted radiation energy per second for all wavelengths increases with the fourth power of the temperature, or

$$E \propto T^4 \tag{6.9}$$

where

E = radiation emission in J/s per unit area or W/m^2
T = temperature (K) of object

2. *Monochromatic radiation*   It is also clear from Figure 6.12 that the radiation energy emitted at any particular wavelength increases as a function of temperature. Thus, the J/s per area at some given wavelength increases with temperature. This is manifested by the object getting brighter at the (same) wavelength as its temperature increases.

Blackbody approximation

Most materials emit and absorb radiation at *preferred* wavelengths, giving rise to *color*, for example. Thus, these objects cannot display a radiation energy versus wavelength curve like that of an ideal blackbody. Correction factors are applied to relate the radiation curves of *real* objects to that of an ideal blackbody. For calibration purposes, a blackbody is constructed as shown in Figure 6.13. The radiation emitted from a small hole in a metal enclosure is close to an ideal blackbody.

6.4.2 Broadband Pyrometers

One type of temperature measurement system based on emitted EM radiation uses the relation between total emitted radiation energy and given temperature. This shows that the total EM energy emitted for all wavelengths, expressed as joules per second per unit area, varies as the fourth power of the temperature. A system that responded to this energy could thus measure the temperature of the emitting object. In practice, it is virtually impossible to build a detection system

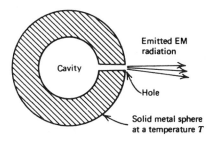

Figure 6.13   The EM radiation emitted by a small hole in a metal sphere at a temperature T with a cavity, as shown, simulates blackbody radiation.

to respond to radiation of all wavelengths. A study of the curves in Figure 6.12, however, shows that most of the energy is carried in the IR and visible bands of radiation. Collection of radiation energy in these bands provides a good approximation of the *total* radiated energy.

Total radiation pyrometer

One type of broadband pyrometer is designed to collect radiation extending from the visible through the infrared wavelengths and is referred to as a *total radiation pyrometer*. One form of this device is shown in Figure 6.14. The radiation from an object is collected by the spherical mirror *S* and focused on a broadband detector *D*. The signal from this detector then is a representation of the incoming radiation intensity and thus the object's temperature. In these devices, the detector is often a series of microthermocouples attached to a blackened platinum disc. The radiation is absorbed by the disc, which heats up, and results in an emf developed by the thermocouples. The advantage of such a detector is that it responds to visible and IR radiation with little regard to wavelength.

IR pyrometer

Another popular version of the broadband pyrometer is one that is mostly sensitive to IR wavelengths. This device often uses a lens formed from silicon or germanium to focus the IR radiation on a suitable detector. IR pyrometers are often hand-held, "pistol-shaped" devices that read the temperature of an object toward which they are pointed.

Characteristics

Broadband pyrometers often have a readout directly in temperature, either analog or digital. Generally, the switchable range is 0°C to 1000°C with an accuracy of ±5°C to ±0.5C°, depending on cost. Accurate measurements require the input of emissivity information by variation in a reading scale factor. Such a correction

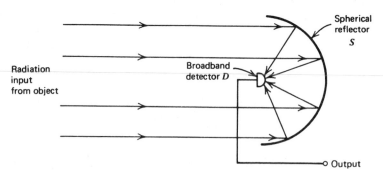

Figure 6.14   A total radiation pyrometer attempts to determine an object's temperature by input of radiation of all wavelengths emitted by the object.

factor accounts for the fact that an object is not an ideal blackbody and does not conform exactly to the radiation curves.

Applications

As the technology of IR pyrometers has advanced, these devices have experienced a vast growth in industrial applications. Some applications are as follows:

1. *Metal production facilities*   In the numerous industries associated with the production and working of metals, temperatures in excess of 500°C are common. Contact temperature measurements usually involve a very limited lifetime of the measurement element. With broadband pyrometers, however, a noncontact measurement can be made and the result converted into a process-control loop signal.

2. *Glass industries*   Another area where high temperatures must be controlled is in the production, working, and annealing of glasses. The broadband pyrometers find ready application in process-control loop situations. In carefully designed control systems, glass furnace temperatures have been regulated to within ±0.1 K.

3. *Semiconductor processes*   The extensive use of semiconductor materials in electronics has resulted in a need for carefully regulated, high-temperature processes producing pure crystals. In these applications, the pyrometer measurements are used to regulate induction heating equipment, crystal pull rates, and other related parameters.

As the accuracy of IR pyrometers is improved in the temperature ranges below 500 K, many applications that have historically used contact measurements will change to IR pyrometers.

6.4.3 Narrowband Pyrometers

Another class of pyrometer depends on the variation in *monochromatic* radiation energy emission with temperature. These devices often are called *optical pyrometers* because they generally involve wavelengths only in the *visible* part of the spectrum. We know that the intensity at any particular wavelength is proportional to temperature. If the intensity of one object is *matched* to another, the temperatures are the same. In the optical pyrometer, the intensity of a heated platinum filament is varied until it matches an object whose temperature is to be determined. Because the temperatures are now the same and filament temperature is calibrated versus a heat setting, the temperature of the object is determined.

Figure 6.15 shows a typical system for implementation of an optical pyrometer. The system is focused on the object whose temperature is to be determined, and the filter picks out only the desired wavelength, which is usually in the red. The viewer also sees the platinum filament superimposed on an image of the

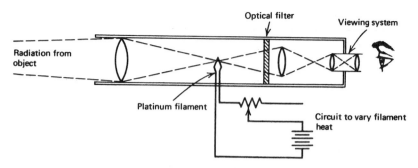

Figure 6.15 An optical pyrometer matches the intensity of the object to a heated, calibrated filament, usually for radiation in the red wavelengths.

object. At low heating the filament appears dark against the background object, as in Figure 6.16a. As the filament is heated, it eventually appears as a bright filament against the background object, as in Figure 6.16c. Somewhere in between is the point when the brightness of the filament and the measured object match. At this setting, the filament disappears with respect to the background object and the object temperature is read from the filament heating dial.

The range of optical pyrometer devices is determined in the low end at the point where an object becomes visible in the red (~500 K) and is virtually limited by the melting point of platinum at the upper end (~3000 K). Accuracy is typically ±5 K to ±10 K and is a function of operator error in matching intensities and

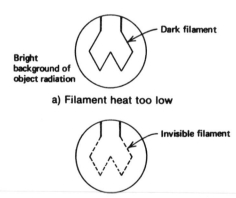

a) Filament heat too low

b) Filament heat adjusted correctly

c) Filament heat too high

Figure 6.16 Examples of the appearance of the filament during optical pyrometer measurement.

emissivity corrections for the object. These devices are not easily adapted to control processes because they require acute optical comparisons, usually by a human operator. Applications are predominantly in spot measurements where constant monitoring or control of temperature is *not* required.

6.5 OPTICAL SOURCES

One limitation in the application of EM radiation devices to process control has been the lack of convenient characteristics of available optical sources. Often, complicated collimating lens systems are required, heat dissipation may be excessive, wavelength characteristics may be undesirable, or a of host other problems may arise. The development of sources relying on Light Amplification by Stimulated Emission of Radiation (LASER) has provided EM radiation sources having good characteristics for application to process-control measurements. In this section, we will consider the general characteristics of both conventional and laser light sources and their applications to measurement problems. Our discussion is confined to sources in the visible or IR wavelength bands, although it should be noted that many applications exist in other regions of the EM radiation spectrum.

6.5.1 Conventional Light Sources

Before the development of the laser, two primary types of light sources were employed. Both of these are fundamentally, *distributed* because radiation emerges from a physically distributed source. They also are both *divergent*, incoherent, and often not particularly monochromatic.

Incandescent sources

A common light source is based on the principle of thermal radiation discussed in Section 6.4. Thus, if a fine current-carrying wire is heated to a very high temperature by I^2R losses, it emits considerable EM radiation in the visible band. A standard lamp is an example of this type of source, as are flashlight lamps, automobile headlights, and so on. Because the light is distributed in a very broad wavelength spectrum (Figure 6.12), it clearly is *not* monochromatic. Such light actually results from molecular vibrations induced by heat, and light from one section of the wire is not associated with the light from another section. From this argument we see that the light is incoherent. The divergent nature of the light is inherent in the observation that no direction of emission is preferential. In fact, the employment of lenses of mirrors to collimate light is familiar to anyone who uses a flashlight. A large fraction of the emitted radiation lies in the IR spectrum, which shows up as a radiant heat loss rather than effective lighting. In fact, to a great extent, the elevated temperature of the glass bulb of an incandescent lamp is caused through absorption by the glass of the IR radiation emitted by the filament.

From this we see that an incandescent source is *polychromatic, divergent, incoherent*, and *inefficient* for visible light production. Yet this source has been a workhorse for lighting for many years. It is most deserving, but for use as a measurement transducer its limitations are severely restrictive.

Atomic sources

The light sources that provide the red neon signs used in advertisements and the familiar fluorescent lighting are examples of another type of light source. Such light sources are atomic in that they depend on rearrangements of electrons within atoms of the material from which the light originates. Figure 6.17 shows a schematic representation of an atom with the nucleus and associated electrons. If one of these electrons is excited from its normal energy level to a different position as indicated by an $a \rightarrow b$ transition, energy must be provided to the atom. This electron returns to its normal level in a very short time ($\sim 10^{-8}$ s for most atoms). In doing so, it gives up energy in the form of emitted EM radiation, as indicated by the $b' \rightarrow a'$ transition in Figure 6.18. This process is often represented by an *energy diagram* as in Figure 6.18. Here the lines represent all possible positions or energies of electron-excited states. The excitation $a \rightarrow b$ of the previous example is indicated by the upward arrow. The de-excitation and resulting radiation is shown as $b' \rightarrow a'$. The other arrows show that a multitude of possible excitations and de-excitations may occur. The wavelength of the emitted radiation is *inversely* proportional to the energy of the transition (Equation 6.3). Thus, $b' \rightarrow a'$ will have a shorter wavelength than $c' \rightarrow d'$.

In *atomic* sources, a mechanism is provided to cause excitation of electron

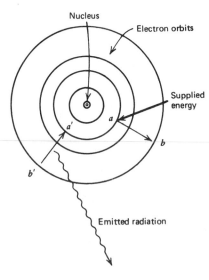

Figure 6.17 A schematic representation of electron transitions in an atom which can cause the emission of EM radiation.

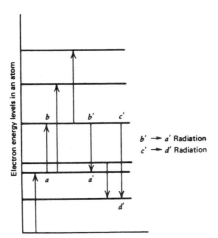

Figure 6.18 An energy-level diagram schematically shows the electron orbit energies and possible transitions.

states, so that the EM radiation emitted by the resulting de-excitation appears as visible light.

In a neon light, the excitation is provided by collision between electrons and ions in the gas, where the electrons are provided by an electric current through the gas. In the case of neon gas, the de-excitation results in light emission whose wavelength is predominantly in the red-orange part of the visible spectrum. Thus, the light is nearly monochromatic, although light of other wavelengths representing different de-excitation modes is still present.

In a fluorescent light, a two-step process occurs. The initial de-excitation produces light predominantly in the *ultraviolet* (UV) part of the spectrum that is absorbed on an inner coating of the bulb. Electrons of atoms in the coating material are excited by UV radiation and then de-excite by many level transitions, producing radiation of a broad band of wavelengths in the *visible* region. Thus, the radiation emitted is *polychromatic*. The radiation in these sources is also divergent and incoherent.

Fluorescence

Certain materials exhibit a peculiar characteristic with regard to the de-excitation transition time of electrons; that is, a transition may take much longer on the average than the normal 10^{-8} s. In some cases, the average transition time may be even hours or days. Such levels are called *long-lived states* and show up in materials that fluoresce or "glow in the dark" following exposure to an intense light source. What actually happens is that the material is excited by exposure to the light source and fills electrons into some long-lived excited states. Because the transition time may be several minutes, the object continues to emit light when taken into a dark room until the excited levels are finally depleted. Such long-lived state materials actually form a basis for development of a laser.

6.5.2 Laser Principles

Stimulation emission

The basic operation of the laser depends on a principle formulated by Albert Einstein regarding the emission of radiation by excited atoms. He found that if several atoms in a material are excited to the *same* level and one of the atoms emits its radiation before the others, then the passage of this radiation by such excited atoms can also *stimulate* them to de-excite. It is significant that when stimulated to de-excite, the emitted radiation will be *in-phase* and *in the same direction* as the stimulating radiation. This effect is shown in Figure 6.19, where atom *a* emits radiation spontaneously. When this radiation passes by atoms indicated by *b, c*, and so on, they also are stimulated to emit in the *same* direction and *in-phase* (coherently). Such radiation is also *monochromatic* because only a single transition energy is involved. Such *stimulated* emission is the first requirement in the realization of a laser.

Laser structure

To see how the concept of stimulated emission is employed in a laser, consider Figure 6.20. We have a host material that also contains atoms having long-lived states described earlier. If some of these atoms spontaneously de-excite, their radiation stimulates other atoms in the radiation path to de-excite, giving rise to pulses of radiation indicated by P_1, P_2, and so on. Now consider one of these pulses directed perpendicularly to mirrors M_1 and M_2. This pulse reflects between the mirrors at the speed of light, stimulating atoms in its paths to emit. The majority of excited atoms is quickly de-excited in this fashion. If mirror M_2 is only 60% reflecting, some of this pulse in each reflection will be passed. The overall result

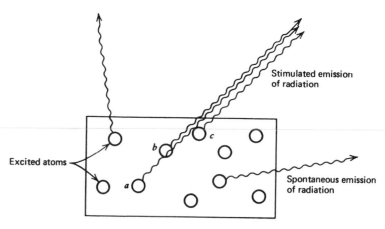

Figure 6.19 Stimulated emission of radiation gives rise to monochromatic coherent radiation pulses moving in various random directions.

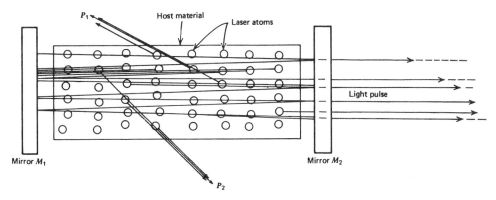

Figure 6.20 In the most elementary sense, the laser simply collects the stimulated radiation emitted perpendicular to reflecting surfaces.

is that following excitation, a pulse of light emerges from M_2 that is *monochromatic, coherent*, and has very little divergence. This system can also be made to operate continuously by providing a continuous excitation of the atoms to replenish those de-excited by stimulated emission.

Properties of laser light

The light that comes from the laser is characterized by the following properties:

1. *Monochromatic*   Laser light comes predominantly from a particular energy level transition and is therefore almost monochromatic. (Thermal vibration of the atoms and the presence of impurities cause some other wavelengths to be present.)

2. *Coherent*   Laser light is coherent as it emerges from the laser output mirror and remains so for a certain distance from the laser; this is called the *coherence length*. (Slight variations in coherency induced by thermal vibrations and other effects cause the beam eventually to lose coherency.)

3. *Divergence*   Because the laser light emerges perpendicular to the output mirror, the beam has very little divergence. Typical divergency may be 0.001 radians.

4. *Power   Continuous-operation* lasers may have power outputs of 0.5 mW to 100 W or more. *Pulse-type* lasers have power levels up to terrawatts, but for only very short time pulses—microsecond or even nanoseconds in duration.

Table 6.3 summarizes some characteristics of typical industrial lasers. The workhorses of measurement applications are the He-Ne continuous-operation lasers. These lasers are relatively cheap and operate in the visible (red) wavelength regions.

TABLE 6.3 Laser Characteristics

Material	Wavelength (μm)	Power	Applications
Helium-neon (gas)	0.6328 (red)	0.5 to 100 mW (cw)	General purpose, ranging, alignment communication, and so on
Argon (gas)	0.4880 (green)	0.1 to 5 W (cw) 10 to 100 W (pulsed)	Heating, small part welding, communication
Carbon dioxide (gas)	10.6 (IR)	0 to 1 kW (cw) 0 to 100 kW (pulse)	Cutting, welding, communication, vaporization, drilling
Ruby (solid)	0.6943 (red)	0 to 1 GW (pulse)	Cutting, welding, vaporization, drilling ranging
Neodymium (solid)	1.06 (IR)	0 to 1 GW (pulse)	Cutting, welding, vaporization, drilling, communication

Example 6.9

An He-Ne laser with an exit diameter of 0.2 cm, power of 7.5 mW, and divergence of 1.7×10^{-3} rad is to be used with a detector 150 m away. If the detector has an area of 5 cm^2, find the power of the laser light to which the detector must respond.

Solution   To solve this problem, we must ultimately find the intensity of the laser beam 150 m away. We first find the beam area at 150 m using the known divergence. From Figure 6.4, we find (see Example 6.5) the radius at 150 m by

$$R_2 = R_1 + L \tan(\theta)$$

$$R_2 = 10^{-3} + (150 \text{ m}) \tan(1.7 \times 10^{-3} \text{ rad}) \qquad (6.5)$$

$$R_2 = 0.256 \text{ m}$$

Thus, the area is

$$A_2 = \pi R_2^2 = (3.14)(0.256 \text{ m})^2$$

$$A_2 = 0.206 \text{ m}^2$$

The intensity at 150 m is

$$I_2 = P/A$$

$$I = \frac{7.5 \times 10^{-3} \text{ W}}{0.206 \text{ m}^2} \qquad (6.4)$$

$$I = 0.036 \text{ W/m}^2$$

Then we find the power intercepted by the detector as

$$P_{\text{det}} = IA_{\text{det}} = (0.036 \text{ W/m}^2)(5 \times 10^{-4} \text{ m}^2)$$

$$P_{\text{det}} = \mathbf{1.82 \times 10^{-5} \text{ W}}$$

Thus, the detector must respond to a power of 18.2 μW of power. Although small, such low power is measured by many detectors. The reader can show (by similar calculation) that a source such as a flashlight with a divergence of 2° or 34.9×10^{-3} rad would have resulted in a much lower power at the detector of $\mathbf{4.35 \times 10^{-8}\ W}$.

6.6 APPLICATIONS

Several applications of optical transduction techniques in process control will be discussed. The intention is to demonstrate only the typical nature of such application and not to design details. Pyrometry for temperature measurement has already been discussed and is not considered in these examples.

6.6.1 Label Inspection

In many manufacturing processes, a large number of items are produced in batch runs where an automatic process attaches labels to the items. Inevitably, some items are either missing labels or the labels are incorrectly attached. The system of Figure 6.21 examines the presence and alignment of labels on boxes moving on a conveyer belt system. If the label is missing or improperly aligned, the photodetector signals are incorrect in terms of light reflected from the sources and a solenoid pushout rejects the item from the conveyer. The detectors in this case could be CdS cells and the sources either focused incandescent lamps or a small He-Ne laser. Source/detector system *A* detects the presence of a box and initiates measurements by source/detector systems *B* and *C*. If the signals received by detectors *B* and *C* are identical and at a preset level, the label is correct, and the box moves on to the accept conveyer. In any other instance, a misalignment or missing label is indicated and the box is rejected onto the reject conveyer.

Example 6.10

Devise signal conditioning circuitry for the application of Figure 6.22 using CdS cells as the detectors. If *both* cells have resistance of $1000 \pm 100\ \Omega$ or less, the label is considered *correct*. No label produces $2000\ \Omega$ or more.

Solution One of the many possible methods for implementing this solution is shown in Figure 6.22. If the label is misaligned or missing, one or both comparator outputs is high, thus driving the summing amplifier to close the *reject* relay. If the box is present, then power is applied to the reject solenoid and the box is ejected. The resistors are chosen so that if the cell resistance exceeds $1100\ \Omega$, the comparator outputs go high. The relay is chosen to close if either (or both) comparator signal is present.

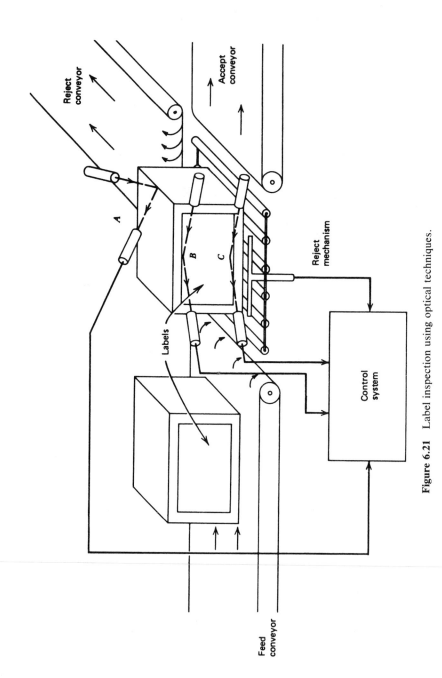

Figure 6.21 Label inspection using optical techniques.

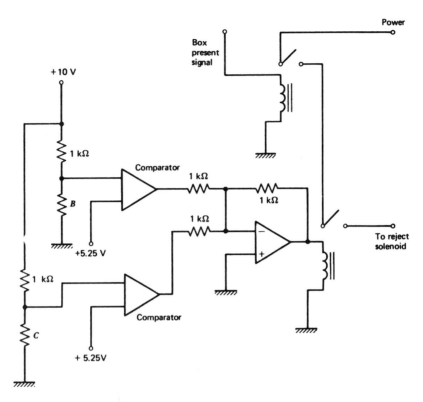

Figure 6.22   One possible circuit to implement the requirements of Example 6.10.

6.6.2 Turbidity

One of the many characteristics of liquids involved in process industries is called *turbidity*. Turbidity refers to the lack of clarity of the liquid, which can be caused by suspended particulate material. Turbidity can be an indication of a problematic condition because of impurities or improperly dissolved products. It also can be intentional as, for example, when some material is suspended in a liquid for ease of transport through pipes. Turbidity can be measured optically because it affects the propagation of light through the liquid.

It is also possible to measure the turbidity of liquids in a process in-line, that is, without taking periodic samples, by a method similar to that in Figure 6.23. In this case, a laser beam is split and passed through two samples to matched photodetectors. One sample is a carefully selected *standard* of allowed (acceptable) turbidity. The other is an in-line sample of the process liquid itself. If the in-line sample attenuates the light more than the standard, the signal conditioning system triggers an alarm or takes other appropriate action to reduce turbidity.

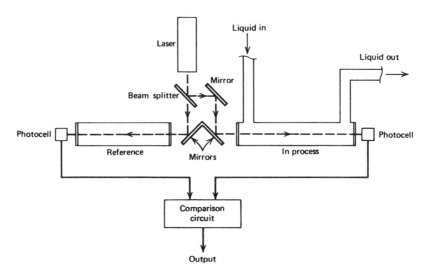

Figure 6.23   Turbidity measurements can be made in-line with the optical system shown.

6.6.3 Ranging

The development of the laser and fast (small τ) photodetectors has introduced a number of methods for measuring distances and the rate of travel of objects by noncontact means. Distances can be measured by measuring the time of flight of light pulses scattered off a distant object. Because the speed of light is constant, we use a simple equation to find distance providing the time of flight T is known. Thus, if a pulse of light is directed at a distant object and the reflection is detected a time T later, then the distance is

$$D = cT/2 \qquad (6.10)$$

where

D = distance to the object (m)
c = speed of light (m/s)
T = time for light round trip (s)

Example 6.11

An object is approximately 300 m away. Find the approximate time difference to calculate the distance using a light pulse reflected from the object.

Solution   From

$$D = \frac{cT}{2} \qquad (6.10)$$

we have

$$T = \frac{2D}{c} = \frac{(2)(3 \times 10^2 \text{ m})}{3 \times 10^8 \text{ m/s}}$$

$$T = \textbf{2 } \boldsymbol{\mu}\textbf{s}$$

This ranging method can be employed for measuring shorter distances limited by time measurement capability and detecting the reflected signal for longer distances. Surveying instruments for measuring distance have been developed by this method. Velocity or rate of motion can be measured by an electronic computing system that records the changing reflected pulse travel time and computes velocity. These and the interferometric methods such as those used in Doppler radar are beyond the scope of this text.

SUMMARY

EM radiation measurement allows noncontact measurement techniques for many variables, such as temperature, level, and others. This chapter presented the essential elements of EM radiation and its application to process-control measurement. This included the following:

1. EM radiation is defined as a form of energy characterized by constant propagation speed. The wavelength and frequency are related by

$$c = \lambda f \tag{6.1}$$

2. As a form of energy, EM radiation is best described by the intensity in watts per unit area, divergence (spreading) in radians, and spectral content.

3. When associated with human eye detection, light is described by a set of luminous units that are all relative to a standard, visible source.

4. There are four basic photodetectors: photoconductive, photovoltaic, photoemissive, and photodiode. Each has its special characteristics relative to spectral sensitivity, detectable power, and response time.

5. Pyrometry relates to measurement of temperature by measuring the intensity of EM radiation from an object as a function of its temperature.

6. Total radiation and IR pyrometers may be used in process-control applications for measurement of temperature from about 300 K to almost no limit. Accuracy varies from ± 10 to ± 0.5 K or better in some cases.

7. Conventional light sources are usually divergent and incoherent. Incandescent types are polychromatic, but atomic sources may be almost monochromatic.

8. A laser is based on a concept of stimulated emission where one photon can stimulate excited atoms to emit many more photons.

9. The light from a laser is nearly nondivergent, coherent, and monochromatic.

10. Applications of optical techniques are particularly useful where contact measurement is difficult.

PROBLEMS

Section 6.2

6.1 Find the frequency of 3-cm wavelength EM radiation. What band of phenomena does this radiation represent?

6.2 A source of green light has a frequency of 6.5×10^{14} Hz. What is its wavelength in nanometers (nm) and Å?

6.3 A light beam is passed through 100 m of liquid with an index of refraction $n = 1.7$. How long does it take the light to traverse the 100 m with and without the liquid?

6.4 A flashlight beam has an exit power of 100 mW, an exit diameter of 4 cm, and a divergence of 1.2°. Calculate the intensity at 60 m and the size of the beam at that distance.

6.5 A detector can just resolve a light intensity of 25 mW/m². How far away can it be placed from a 10-watt point source?

6.6 How many watts of power are emitted by a one-candela source?

Section 6.3

6.7 A CdS cell has a time constant of 73 ms and a dark resistance of 150 kΩ. A light pulse which is only 20 ms in length strikes the cell. If the intensity of the light pulse is such that the final resistance would be 45 kΩ, plot the resistance versus time for 100 ms. What is the resistance at 20 ms?

6.8 Suppose there is another 20-ms light pulse which may strike the cell of Problem 6.7, but it is of lower intensity so that the final resistance would be 85 kΩ. Call this one P_1 and that of Problem 6.7 P_2. Devise a system which will latch on a red LED if P_1 strikes the cell or a green LED if P_2 strikes the cell. Each time a pulse strikes the cell, the latches are cleared and the appropriate LED turned on.

6.9 A silicon photovoltaic cell is employed in the circuit shown in Figure 6.24. The cell

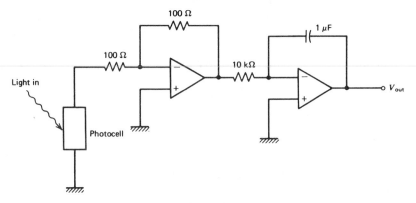

Figure 6.24   Circuit for Problem 6.9.

has an internal resistance of 65 Ω. A light pulse of 20 ms duration and 2 W/m^2 intensity strikes the detector. Sketch the output voltage versus time and find the maximum voltage produced. The cell calibration voltage is $V_0 = 0.60$ volts.

6.10 A photomultiplier has a current gain of 3×10^6. A weak light beam produces 50 electrons/s at the photocathode. What anode to ground resistance must be used to get a 3-μV voltage from this light pulse? The charge of an electron is 1.6×10^{-19} coulomb.

Section 6.5

6.11 A 10-W argon laser has a beam diameter of 1.5 cm. If focused to a 1-mm diameter, find the intensity of the radiation beam at the focus.

6.12 In a turbidity system such as that of Figure 6.23, the tanks are 2 m in length. A laser with a 2.2 mrad divergence, 2.1 mW power, and 1 mm exit radius is employed. If the sample nominally detracts from the beam power by 12% per meter, find the intensity of the beam at the detector. The laser is 1.5 m from the beam splitter. Note that the splitter halves the power and does not affect the divergence.

6.13 A special timing circuit can resolve 2.4 ns time difference. Find the closest distance that could be measured using laser ranging.

6.14 For the turbidity system of Figure 6.23, two matched photoconductive cells are used with R vs. I_L as given in Figure 6.25. Design a signal conditioning system that outputs the deviation of the flowing system turbidity in volts and triggers an alarm if the intensity is reduced by 10% from the nominal of 15 mW/cm^2.

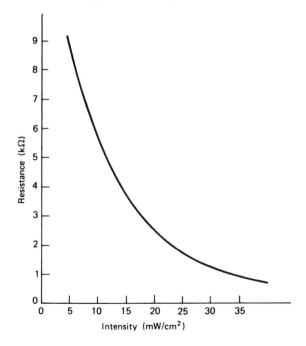

Figure 6.25 Figure for Problem 6.14.

CHAPTER 7

FINAL CONTROL

INSTRUCTIONAL OBJECTIVES

In this chapter, the general techniques used to implement the final control element function are presented. After you have read this chapter, you should be able to

1. **Define the three parts of final control operation.**
2. **Give two examples of electrical signal conversion.**
3. **Make a diagram and describe the operating principles of the flapper/nozzle pneumatic system.**
4. **Describe the operating principle of ac, dc, and stepping motors.**
5. **Explain how a pneumatic positioning actuator functions in both the direct and reverse modes.**
6. **Contrast quick-opening, linear, and equal percentage control valves in terms of the flow versus stem position.**
7. **Explain how control valve sizing techniques allow selection of the proper size of control valve.**

7.1 INTRODUCTION

In a typical process-control application, the measurement and evaluation of some process variable are carried out by a low-energy analog or digital representation of the variable. The control signal that carries feedback information back to the

272

process for necessary corrective action is expressed by the same low level of representation. In general, the controlled process itself may involve a high-energy condition, such as the flow of thousands of cubic meters of liquid or several hundred thousand newton hydraulic forces, as in a steel rolling mill. The function of the final control element is to translate low-energy control signals into a level of action commensurate with the process under control. This can be considered an amplification of the control signal, although in many cases the signal is also converted into an entirely *different* form.

In this chapter, the general concepts to implement the final control element function are presented together with specific examples in several areas of process control.

A sensor used to measure some variable in a process-control application should have negligible effect on the process itself. It follows that sensor selection is based mainly on required measurement specifications and necessary protections (of the sensor) from harmful effects of the process environment. In sensor selection, the process-control technologist need not have intimate knowledge of the mechanisms of the process itself.

These arguments do not apply, however, when considering the *final control element*. By necessity, the final control element has a profound effect on the process and therefore must be selected after detailed considerations of the process operational mechanisms. Such a selection, therefore, cannot be the responsibility of the process-control technologist alone. In this view, the process-control technologist should have sufficient background on the final control element and its associated signal conditioning to know how such devices interface with preceding process controllers and transducers. The technologist should be able to communicate and work closely with process engineers on these subjects. The objectives of this chapter were selected to fulfill such responsibility.

7.2 FINAL CONTROL OPERATION

Final control element operations involve the steps necessary to convert the control signal (generated by a process controller) into proportional action on the process itself. Thus, to use a typical 4–20 mA control signal to vary a large flow rate from, say, 10.0 m³/min to 50.0 m³/min certainly requires some intermediate operations. The specific intermediate operations vary considerably depending on the process-control design, but certain generalizations can be made regarding the steps leading from the control signal to the final control element. For a typical process-control application the conversion of a process-controller signal to a control function can be represented by the steps shown in Figure 7.1. The input control signal may take many forms, including an electric current, digital signal, or pneumatic pressure.

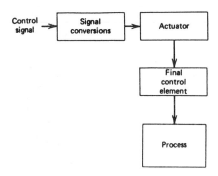

Figure 7.1 Elements of the final control operation.

7.2.1 Signal Conversions

This step refers to the modifications that must be made to the control signal to properly interface with the next stage of control, that is, the actuator. Thus, if a valve control element is to be operated by an electric motor actuator, then a 4–20 mA dc control signal must be modified to operate the motor. If a dc motor is used, modification might be current-to-voltage conversion and amplification.

Standard types of signal modification are discussed in Section 7.3. The devices that perform such signal conversions are often called *transducers* because they convert control signals from one form to another, such as current-to-pressure, current-to-voltage, and so on.

7.2.2 Actuators

(See Figure 7.1.) The results of signal conversions provide an amplified and/or converted signal that is designed to operate (actuate) a mechanism that changes a controlling variable in the process. The direct effect is usually implemented by something in the process such as a valve or heater that must be operated by some device. The *actuator* is a translation of the (converted) control signal into action *on* the control element. Thus, if a valve is to be operated, then the actuator is a device that converts the control signal into physical action of opening or closing the valve. Several examples of actuators in common process-control use are discussed in Section 7.4.

7.2.3 Control Element

Finally (see Figure 7.1), we get to the final control element itself. This device has direct influence on the process dynamic variable and is designed as an integral part of the process. Thus, if flow is to be controlled, then the control element, a valve, must be built directly into the flow system. Similarly, if temperature is to be controlled, then some mechanism or control element that has a direct influence on temperature must be involved in the process. This could be a heater/cooler

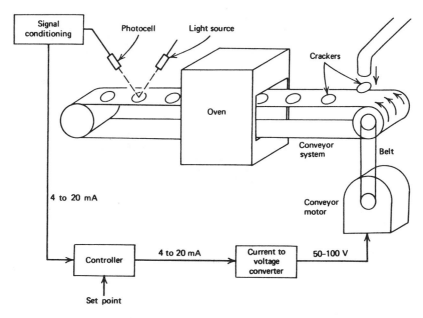

Figure 7.2   A process-control system showing the final control operations.

combination that is electrically actuated by relays or a pneumatic valve to control influx of reactants.

In Figure 7.2, a control system is shown to control the degree of baking of, say, crackers, as determined by the *cracker color*. The optical measurement system produces a 4–20 mA conditioned signal that is an analog representation of cracker color (and, therefore, proper baking). The controller compares the measurement to a setpoint and outputs a 4–20 mA signal that regulates the conveyer belt feed motor speed to adjust baking time. The final control operation is then represented by a *signal conversion* that transforms the 4–20 mA signal into 50–100 volt signal as required for motor speed control. The motor itself is the *actuator*, and the conveyer belt assembly is the *control element*.

Because applications of process-control techniques in industry are as varied as the industry itself, it is impractical to consider more than a few final control techniques. By studying some examples, the reader should be prepared to analyze and understand many other techniques that arise in industry.

7.3 SIGNAL CONVERSIONS

The principal objective of signal conversion is to convert the low-energy control signal to a high-energy signal to drive the actuator. Controller output signals are typically in one of three forms: (1) electrical current, usually 4–20 mA, (2) pneumatic pressure, usually 3–15 psi, and (3) digital signals, usually TTL level voltages in serial or parallel format. There are many different schemes for conversion of these signals to other forms depending on the desired final form and evolving technology. In the following sections, a number of the more common conversion schemes are presented. You should always be receptive to the advances of technology and the new subsequent methods of signal conditioning and conversion.

7.3.1 Analog Electrical Signals

Many of the methods of analog signal conditioning discussed in Chapter 2 are used in conversions necessary for final control. The following paragraphs summarize some of the more common approaches.

Relays

A common conversion is to use the controller signal to activate a relay when simple on/off or two-position is sufficient. In some cases, the low-current signal is insufficient to drive a heavy industrial relay, and an amplifier is used to boost the control signal to a level sufficient to do the job.

Amplifiers

High-power ac or dc amplifiers often can provide the necessary conversion of the low-energy control signal to a high-energy form. Such amplifiers may serve for motor control, heat control, light-level control, and a host of other industrial needs.

Example 7.1

A magnetic amplifier requires a 5–10 volt input signal from 4–20 mA control signal. Design a signal conversion system to provide this relationship.

Solution   We first must convert the current to a voltage, and then provide the required gain and bias. We can get a voltage using a resistor in the current line of, say, 100 Ω. Then the 4–20 mA becomes 0.4–2.0 volts. The amplifier system must provide an output given by

$$V_{out} = KV_{in} + V_B$$

where K is the gain and V_B is an appropriate bias voltage. We know that 0.4 volts input must provide 5 volts output, and 2 volts input must provide 10 volts output. This allows us to find K and V_B, using simultaneous equations as

$$5 = 0.4K + V_B$$

$$10 = 2K + V_B$$

Subtracting, we get

$$5 = 1.6K$$

$$K = \mathbf{3.125}$$

that we use in either equation to find

$$V_B = \mathbf{3.75}$$

Thus, the result is

$$V_{\text{out}} = 3.125V_{\text{in}} + 3.75$$

The circuit of Figure 7.3 shows how this can be implemented using an op amp configuration.

Motor control

Many motor control circuits are designed as packaged units that accept a low-level dc signal directly to control motor speed. If such a system is not available, it is possible to build circuits using amplifiers and SCRs or TRIACs to perform this control. The details of such a control system are beyond the intent of this text, however. The basic elements of electrical motors and some words about their control are discussed later in this chapter.

7.3.2 Digital Electrical Signals

Conversions of digital signals to forms required by final control operations usually are carried out using systems already discussed in Chapter 3. We mention again,

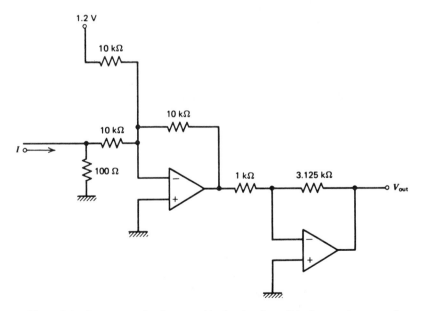

Figure 7.3 An op amp circuit to provide the signal conditioning requirements of Example 7.1.

however, the basic elements of the output interface between computer and final control.

On/off control

There are many cases in process control where the control algorithm is accomplished by simple commands to outside equipment to change speed, turn on (or off), move up, and so on. In such cases, the computer can simply load a latched output line with a **1** or **0** as appropriate. Then it is a simple matter to use this signal to close a relay or activate some other outside circuit.

DAC

When the digital output must provide a smooth control, as it does in valve positioning, the computer must provide an input to a DAC that then determines an appropriate analog output. When a computer must provide outputs to many final control elements, a data output module or system such as that described in Chapter 3 can be employed. These integrated modules contain channel addressing, DAC, and other required elements of a self-contained output interface system.

Example 7.2

A 4-bit digital word is intended to control the setting of a 2-Ω dc resistive heater. Heat output varies as a 0–24 volt input to the heater. Using a 10-volt DAC followed by an amplifier and a unity gain high-current amplifier, calculate (a) the settings from minimum to maximum heat dissipation, and (b) how the power varies with LSB changes.

Solution

(a) The 4-bit word has a total of 16 states; thus, the DAC outputs voltage from 0 V for a 0000_2 input to 9.375 V for a 1111_2 input. If we use a gain of 2.56, then the heater input will be 0–24 volts in 1.6-V steps. The heat dissipation is found from

$$P = \frac{V^2}{R} \tag{7.1}$$

Then, for the minimum $P = 0$ because $V = 0$ and for the maximum

$$P_{max} = \frac{(24V)^2}{2\ \Omega}$$

$$P_{max} = 288 \text{ W}$$

(b) The variation in heating with voltage is *not* linear because the power varies as the square of the voltage. The increase in power for a step change in voltage can be found by taking the difference in power between the two cases

$$\Delta P = \frac{V^2}{R} - \frac{(V - \Delta V)^2}{R}$$

Because $\Delta V = 1.6$ volts, we have

$$\Delta P = \frac{1.6(2V - 1.6)}{R} = 0.8(2V - 1.6)$$

The first bit change produces a power change of 1.28 W and the last bit a power change of 37.12 W.

Direct action

As the use of digital and computer techniques in process control becomes more widespread, new methods of final control have been developed that can be actuated directly by the computer. Thus, a stepping motor, to be discussed later, interfaces very easily to the digital signals that a computer outputs. In another development, special integrated circuits are made that reside within the final control element and allow the digital signal to be connected directly.

7.3.3 Pneumatic Signals

The general field of pneumatics covers a broad spectrum of applications of gas pressure to industrial needs. One of the most common applications is to provide a force by the gas pressure acting on a piston or diaphragm. Later in this chapter, we will deal with this application in process control. In this section, however, we are interested in pneumatics as a means of propagating information, that is, as a signal carrier, and how that signal can be converted to other forms of signals.

Principles

In a pneumatic system, information is carried by the pressure of gas in a pipe. If we have a pipe of any length and raise the pressure of gas in one end, this increase in pressure will propagate down the pipe until the pressure throughout is raised to the new value. The pressure signal travels down the pipe at a speed in the range of the speed of sound in the gas (say, air), which is about 330 m/s (1082 ft/ s). Thus, if a transducer varies gas pressure at one end of a 330-meter pipe (about 360 yards) in response to some controlled variable, then that same pressure occurs at the other end of the pipe after a delay of approximately 1 second. For many process-control installations, this delay time is of no consequence, although it is very slow compared to an electrical signal. This type of signal propagation was used for many years in process control before electrical/electronic technology advanced to a level of reliability and safety to enable its use with confidence. Pneumatics is still employed in many installations either because of danger to electrical equipment or as a carryover from previous years, where conversion to electrical methods would not be cost effective. In general, pneumatic signals are carried with dry air as the gas where signal information has been adjusted to lie within the range of 3–15 psi. In SI unit systems, the range of 20–100 kPa is used. There are three types of signal conversion of primary interest:

Amplification

A pneumatic amplifier, also called a *booster* or *relay*, raises the pressure and/or air flow volume by some linearly proportional amount from the input signal. Thus, if the booster has a pressure gain of 10, the output would be 30–150 psi for an input of 3–15 psi. This is accomplished via a regulator that is activated by the control signal. A schematic diagram of one type of pressure booster is shown in Figure 7.4. As the signal pressure varies, the diaphragm motion will move the plug in the body block of the booster. If motion is down, the gas leak is reduced and pressure in the output line is increased. The device shown is reverse acting because a high-signal pressure will cause output pressure to decrease. Many other designs are used.

Nozzle/flapper system

A very important signal conversion is from pressure to mechanical motion and vice versa. This conversion can be provided by a nozzle/flapper system (sometimes called a nozzle/baffle system). A diagram of this device is shown in Figure 7.5a. A regulated supply of pressure, usually over 20 psig, provides a source of air through the restriction. The nozzle is open at the end where the gap exists between the nozzle and flapper, and air escapes in this region. If the flapper moves down and closes off the nozzle opening so that no air leaks, the signal pressure will rise to the supply pressure. As the flapper moves away, the signal pressure will drop because of the leaking gas. Finally, when the flapper is far away, the pressure will stabilize at some value determined by the maximum leak through

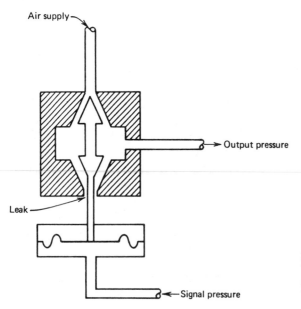

Figure 7.4 A pneumatic amplifier or booster converts the signal pressure to a higher pressure or the same pressure but with greater gas volume.

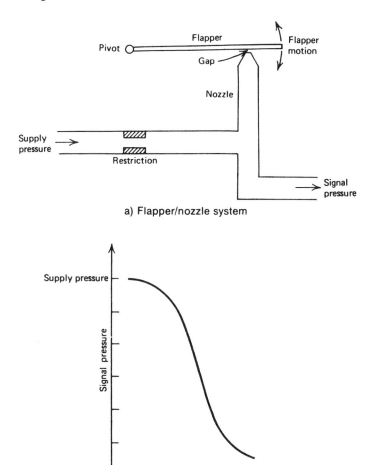

a) Flapper/nozzle system

b) Signal pressure versus gap distance

Figure 7.5   Principles of the flapper/nozzle system.

the nozzle. Figure 7.5b shows the relationship between signal pressure and gap distance. Note the great sensitivity in the central region. A nozzle/flapper is designed to operate in the central region where the slope of the line is greatest. In this region, the response will be such that a very small motion of the flapper can change the pressure by an order of magnitude. More discussion of this system is given in Chapter 10 under the discussion of pneumatic controllers.

Current-to-pressure converters

The current-to-pressure converter, or simply *I/P converter*, is a very important element in process control. Often, when we want to use the low-level electric

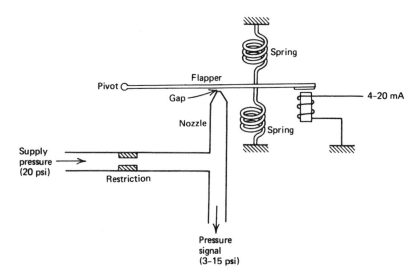

Figure 7.6   Principles of a current-to-pressure converter.

current signal to do work, it is much easier to let the work be done by a pneumatic signal. The I/P converter gives us a linear way of translating the 4–20 mA current into a 3–15 psig signal. There are many designs for these converters, but the basic principle almost always involves the use of a nozzle/flapper system. Figure 7.6 illustrates a simple way to construct such a converter. Notice that the current through a coil produces a force that will tend to pull the flapper down and close off the gap. A high current produces a high pressure so that the device is direct acting. Adjustment of the springs and perhaps the position relative to the pivot to which they are attached allows the unit to be calibrated so that 4 mA corresponds to 3 psig and 20 mA corresponds to 15 psig.

7.4 ACTUATORS

If a valve is used to control fluid flow, some mechanism must physically open or close the valve. If a heater is to warm a system, some device must turn the heater ON or OFF or vary its excitation. These are examples of the requirement of an *actuator* in the process-control loop. Notice the distinction of this device from both the input control signal and the control element itself (valve, heater, and so on, as shown in Figure 7.1). Actuators take on many diverse forms to suit the particular requirements of process-control loops. We will consider several types of electrical and pneumatic actuators.

7.4.1 Electrical Actuators

In the following paragraphs a short description of several common types of electrical actuators is given. The intention is to present only the essential features of the devices and not an in-depth study of operational principles and characteristics. In any specific application, one would be expected to consult detailed product specifications and books associated with each type of actuator.

Solenoid

A solenoid is an elementary device that converts an electrical signal into mechanical motion, usually rectilinear, that is, in a straight line. As shown in Figure 7.7, the solenoid consists of a coil and plunger. The plunger may be free-standing or spring-loaded. The coil will have some voltage or current rating and may be dc or ac. Solenoid specifications include the electrical rating and the plunger pull or push force when excited by the specified voltage. This force may be expressed in newtons or kilograms in the SI system and in pounds or ounces in the English system. Some solenoids are rated only for intermittent duty because of thermal constraints. In this case, the maximum duty cycle (percentage on to total time) will be specified. Solenoids are used when a large sudden force must be applied to perform some job. In Figure 7.8, a solenoid is used to change the gears of a two-position transmission. An SCR is used to activate the solenoid coil.

Electrical motors

Electrical motors are devices that accept electrical input and produce a continuous rotation as a result. Motor styles and sizes vary as demands for rotational speed (revolutions per minute or rpm), starting torque, rotational torque, and other specifications vary. There are numerous cases where electrical motors are employed as actuators in process control. Probably the most common control situation is

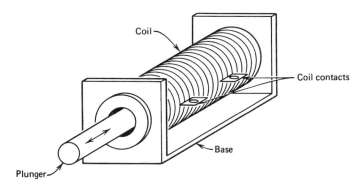

Figure 7.7   A solenoid converts an electrical signal to a physical displacement.

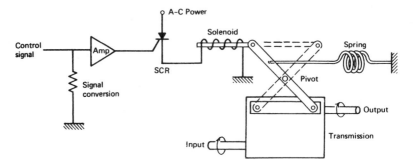

Figure 7.8 A solenoid used to change gears.

where motor speed is driving some part of a process and that speed must be controlled to control some variable in the process; the drive of a conveyor system, for example. There are many types of electrical motors, each with its special set of characteristics. We will simply discuss the three most common varieties: the dc motor, ac motor, and stepping motor.

Dc motor

In its simplest form, a dc motor uses a permanent magnet (PM) to produce a static magnetic field across two pole pieces. Between the poles is connected a coil of wire that is free to rotate (the armature) and that is connected to a source of dc current through a switch mounted on the shaft (a commutator). This system is shown schematically in Figure 7.9a. For the condition shown, the current in the coil will produce a magnetic field with a north/south orientation like that shown in Figure 7.9b. The repulsion of the PM south and the coil south (and the norths) will cause a torque that will rotate the coil as shown. If the commutator were *not split*, the coil would simply rotate until the PM and coil north and south poles were lined up and then stop, *but* because of the commutator, the coil finds that when rotated around the current direction through the coil reverses so that the condition shown in Figure 7.9c occurs. Thus, the rotational torque is again present, and the coil continues to rotate. From this simple model you can see that the coil will continue to rotate. The speed will depend on the current. Actually, the armature current is not determined by the coil resistance because of a counter emf produced by the rotating wire in a magnetic field. Thus, the effective voltage, which determines the current from the wire resistance and Ohm's law, is the difference between the applied voltage and the counter emf produced by the rotation.

Many dc motors use an electromagnet instead of a PM to provide the static field. The coil used to produce this field is called the *field coil*. The current for this field coil can be provided by placing the coil in series with the armature or in parallel (shunt). In some cases the field is composed of two windings, one of each type. This is a compound dc motor. The schematic symbols of each type of

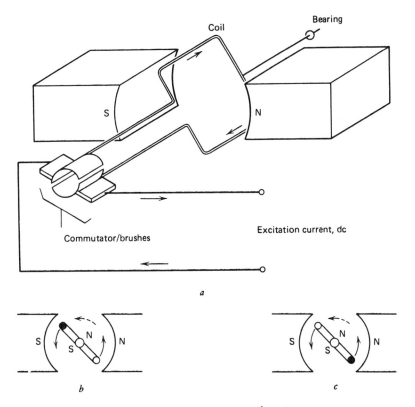

Figure 7.9 Permanent magnet dc motor.

motor are shown in Figure 7.10. Characteristics of dc motors with a field coil are as follows:

1. *Series field*   This motor has large starting torque but is difficult to speed control. Good in applications of starting heavy, nonmobile loads and where speed control is not very important, such as quick-opening valves.

2. *Shunt field*   This motor has a smaller starting torque but very good speed control characteristics by varying armature excitation current. Good in applications where speed is to be controlled, such as conveyor systems.

3. *Compound field*   This motor attempts to obtain the best features of both of the two previous types. Generally, starting torque and speed control capability fall predictably between the two pure cases.

Ac motors

There are many types of ac motors. A synchronous ac motor's speed of rotation is determined by the frequency of the ac voltage that drives it. Its primary application is in timing because of the high stability of the power-line frequency.

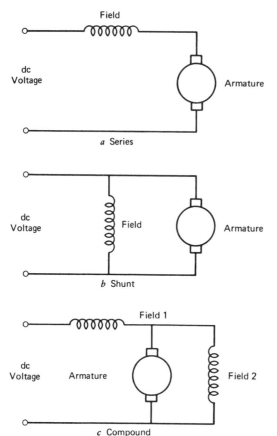

Figure 7.10 Three dc motor configurations.

Operation of this type of motor can be seen from a simple example shown in Figure 7.11. The rotor is a PM, and the field is provided by coils driven from the ac line. Because of the inertia of the PM, the starting torque is not very high, but once rotation is started the PM will rotate in-phase with the field reversals caused by the oscillations of the ac line voltage. It is clear then that the rate of rotation is determined by the ac line frequency. An induction motor replaces the PM with a very heavy wire coil, into which is induced a current from the changing field of the ac excited field coils. Figure 7.12 illustrates this motor. As before, once rotation is started the rotor will continue rotation in-phase with the line frequency-induced changes of field coil excitation. The difficulty with these motors is that they are not self-starting and special modifications are necessary to get them to begin rotation. Clearly then, the starting torque is very low. One method of providing self-starting is to drive the motor with two or more phases of ac excitation. In general, however, ac motors do not have a high starting torque or convenient methods of speed control.

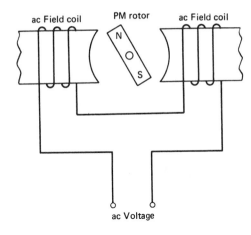

Figure 7.11 Simple ac motor with a PM rotor.

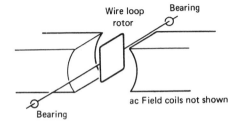

Figure 7.12 The induction motor depends on a rotor field induced by the ac field coils (not shown).

Stepping motor

The stepping motor has increased in importance in recent years because of the ease with which it can be interfaced with digital circuits. A stepping motor is a rotating machine that actually completes a full rotation by sequencing through a series of discrete rotational steps. Each step position is an equilibrium position in that, without further excitation, the rotor position will stay at the latest step. Thus, continuous rotation is achieved by the input of a train of pulses, each of which causes an advance of one step. It is not really continuous rotation, but discrete, stepwise rotation. The rotational rate is determined by the number of steps per revolution and the rate at which the pulses are applied. A driver circuit is necessary to convert the pulse train into proper driving signals for the motor.

Example 7.3

A stepper motor has 10° per step and must rotate at 250 rpm. What input pulse rate, in pulses per second, is required?

Solution   A full revolution has 360° so that with 10° per step it will take 36 steps to complete one revolution. Thus

$$\left(250 \ \frac{\text{rev}}{\text{min}}\right)\left(36 \ \frac{\text{pulses}}{\text{rev}}\right) = 9000 \ \text{pulses/min}$$

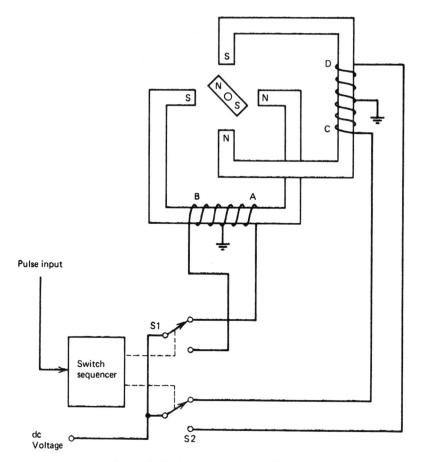

Figure 7.13   An elementary stepping motor.

Therefore,

$$(9000 \text{ pulses/min})(1 \text{ min/60 s}) = \mathbf{150 \text{ pulses/s}}$$

The operation of a stepping motor can be understood from the simple model shown in Figure 7.13, which has 90° per step. In this motor, the rotor is a PM that is driven by a particular set of electromagnets. In the position shown, the system is in equilibrium and no motion occurs. The switches are typically solid-state devices, such as transistors, SCRs, or TRIACs. The switch sequencer will direct the switches through a sequence of positions as the pulses are received. The next pulse in Figure 7.13 will change S2 from C to D, resulting in the poles of that electromagnet reversing fields. Now, because the pole north/south orientation is different, the rotor is repelled and attracted so that it moves to the new position of equilibrium shown in Figure 7.14b. With the next pulse, S1 is changed to B, causing the same kind of pole reversal and rotation of the PM to

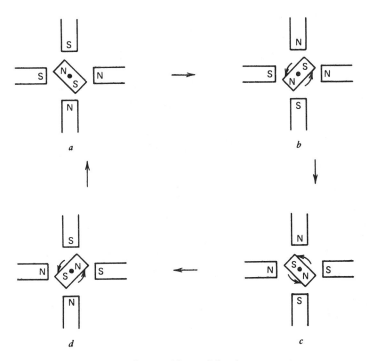

Figure 7.14 The four positions of the elementary stepper.

a new position, as shown in Figure 7.14c. Finally, the next pulse causes S2 to switch to C again and the PM rotor again steps to a new equilibrium position, as in Figure 7.14d. The next pulse will send the system back to the original state and the rotor to the original position. This sequence is then repeated as the pulse train comes in, resulting in a stepwise continuous rotation of the rotor PM. Although this example illustrates the principle of operation, the most common stepper motor does not use a PM, but uses instead a rotor of magnetic material (not a magnet) with a certain number of teeth. This rotor is driven by a phased arrangement of coils with a different number of poles so that the rotor can never be in perfect alignment with the stator. Figure 7.15 illustrates this for a rotor with 8 "teeth" and a stator with 12 "poles." One set of four teeth are aligned, but the other four are not. If excitation is placed on the next set of poles (B) and taken off the first set (A), then the rotor will step once to come into alignment with the B set of poles. The direction of rotation of stepper motors can be changed by just changing the order in which different poles are activated and deactivated.

7.4.2 Pneumatic Actuators

The actuator often translates a control signal into a large force or torque as required to manipulate some control element. The *pneumatic actuator* is most useful for such translation. The principle is based on the concept of pressure as force

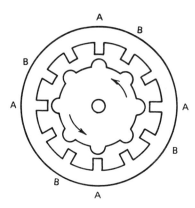

Figure 7.15 Cross section of a stepper with 8 rotor teeth and 12 stator poles. Note that the rotor lines up with the A poles. With the next step, the rotor will line up with the B poles.

per unit area. If we imagine that a net pressure difference is applied to a diaphragm of surface area A, then a net force acts on the diaphragm given by

$$F = (p_1 - p_2)A \tag{7.2}$$

where

$p_1 - p_2$ = pressure difference (Pa)
A = diaphragm area (m^2)
F = force (N)

If we need to double the available force for a given pressure, it is merely necessary to double the diaphragm area. Very large forces can be developed by standard signal pressure ranges of 3–15 psi (20–100 kPa). Many types of pneumatic actuators are available, but perhaps the most common are those associated with *control valves*. We will consider these in some detail to convey the general principles.

The action of a *direct* pneumatic actuator is shown in Figure 7.16. Figure 7.16a shows the condition in the *low* signal pressure state where the spring S maintains the diaphragm and the connected control shaft in a position as shown. The pressure on the opposite (spring) side of the diaphragm is maintained at atmospheric pressure by the open hole H. Increasing the control pressure (gauge pressure) applies a force on the diaphragm, forcing the diaphragm and connected shaft down against the spring force. Figure 7.16b shows this in the case of maximum control pressure and maximum travel of the shaft. The pressure and force are linearly related, as shown in Equation (7.2), and the compression of a spring is linearly related to forces as discussed in Chapter 5. Then we see that the shaft position is linearly related to the applied control pressure

$$\Delta x = \frac{A}{k} \Delta p \tag{7.3}$$

where

Δx = shaft travel (m)
Δp = applied gauge pressure (Pa) pascals

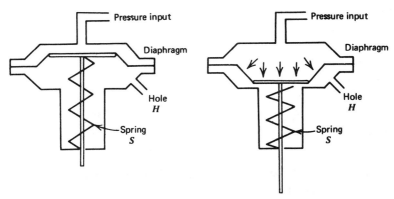

a) Direct actuator in the low–pressure state

b) Direct actuator in the high–pressure state

Figure 7.16   A direct pneumatic actuator for converting pressure signals into mechanical shaft motion.

A = diaphragm area (m^2)
k = spring constant (N/m)

A *reverse* actuator, shown in Figure 7.17, moves the shaft in the opposite sense from the direct actuator, but obeys the same operating principle. Thus, the shaft is pulled in by the application of a control pressure.

Example 7.4

Suppose a force of 400 N must be applied to open a valve. Find the diaphragm area if a control gauge pressure of 70 kPa (~10 psi) must provide this force.

Solution   We must calculate the area from

$$F = A(p_1 - p_2) \tag{7.2}$$

where our applied pressure is $p_1 - p_2$ because a gauge pressure is specified. Then

$$A = \frac{F}{p} = \frac{400 \text{ N}}{7 \times 10^4 \text{ Pa}}$$

$$A = 5.71 \times 10^{-3} \text{ m}^2$$

or about **8.5 cm** in diameter.

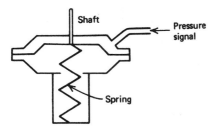

Figure 7.17   A reverse pneumatic actuator.

The inherent compressibility of gases causes an upper limit to the usefulness of gas for propagating force. Consider the pneumatic actuator of Figure 7.16. When we want motion to occur under low shaft load, we simply increase pressure in the actuator via a regulator. The regulator allows more gas from a high-pressure reservoir to enter the actuator. This increases the diaphragm force until it is able to move the shaft.

Suppose the shaft is connected to a very high load, that is, something requiring a very large force for movement. In principle, it is simply a matter of increasing the pressure of the input gas until the pressure × diaphragm area equals the required force. What we find, however, is that as we try to raise the input gas pressure, large volumes of gas must be passed into the actuator to bring about any pressure rise because the gas is compressing; that is, its density is increasing.

7.4.3 Hydraulic Actuators

We have seen that there is an upper limit to the forces that can be applied using gas as the working fluid. Yet there are many cases when large forces are required. In such cases, a hydraulic actuator may be employed. The basic principle is shown in Figure 7.18. The basic idea is the same as for pneumatic actuators except that an incompressible fluid is used to provide the pressure, which can be made very large by adjustment of the area of the forcing piston A_1. The hydraulic pressure is given by

$$p_H = F_1/A_1 \tag{7.4}$$

where

p_H = hydraulic pressure (Pa)
F_1 = applied piston force (N)
A_1 = forcing piston area (m^2)

The resulting force on the working piston is

$$F_w = p_H A_2 \tag{7.5}$$

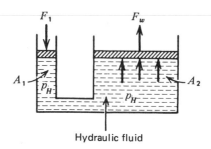

Figure 7.18 A hydraulic actuator converts a small force F_1 into an amplified force F_w.

where

    F_w = force of working piston (N)
    A_2 = working piston area (m^2)

Thus, the working force is given in terms of the applied force by

$$F_w = \frac{A_2}{A_1} F_1 \tag{7.6}$$

Example 7.5

Find the working force resulting from 200 N applied to a 1-cm radius forcing piston (a) if the working piston has a radius of 6 cm. Then (b) find the hydraulic pressure.

Solution   (a) We can find the working force from

$$F_w = \frac{A_2}{A_1} F_1 \tag{7.6}$$

or

$$F_w = \left(\frac{R_2}{R_1}\right)^2 F = \left(\frac{6 \text{ cm}}{1 \text{ cm}}\right)^2 (200 \text{ N})$$

$$F_w = \textbf{7200 N}$$

(b) Thus, the 200-N force provides 7200 N of force. The hydraulic pressure is

$$p_H = F_w/A_2 = \frac{7200 \text{ N}}{(\pi)(6 \times 10^{-2} \text{ m})^2}$$

$$p_H = \textbf{6.4} \times \textbf{10}^5 \textbf{ Pa}$$

This pressure is approximately 92 lb/in^2.

Hydraulic servos

In some cases it is desired to control the position of very large loads as part of the control system. This often can be done by using the low-energy controller output as the setpoint input to a hydraulic control system. This concept is illustrated in Figure 7.19.

In this system, high-pressure hydraulic fluid can be directed to either side of a force piston, which causes motion in either direction. The direction is determined by the position of a control valve piston in the *hydraulic servo valve*. The position of this valve piston is controlled by a linear motor driven by the output of an amplifier and error detector. The inputs to the error detector are the process controller output, which forms the setpoint of the hydraulic servo, and a feedback from the force piston shaft. Thus, the amplifier will drive the hydraulic servo until the feedback matches the setpoint input.

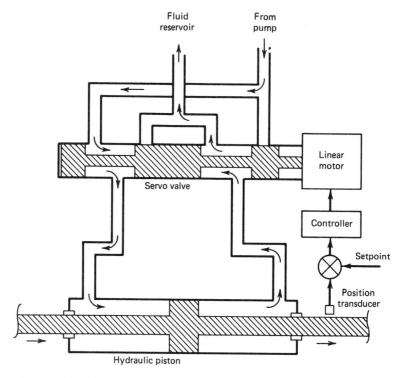

Figure 7.19   A hydraulic servo system. The process-control system provides the setpoint of the servo system.

7.5 CONTROL ELEMENTS

The actual *control element* (which is a part of the process itself) can be many different devices. It is not the intention of this text to present many of these devices, but a general survey of standard devices is valuable for a complete picture of process control. Several examples of control elements are described later in terms of different control problems.

7.5.1 Mechanical

Control elements that perform some *mechanical* operation in a process (by virtue of operations) are called *mechanical control elements*. Examples of these types are as follows:

Solid material hopper valves

Consider the grain supply bin of Figure 7.20. The control system is to maintain the flow of grain from the storage bin to provide a constant flow rate on the

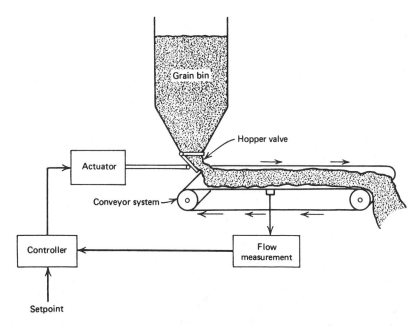

Figure 7.20   An example of a mechanical control element in the form of a hopper valve.

conveyer. This flow depends on the height of grain in the bin, and hence the *hopper valve* must open or close to compensate for the variation. In this case an actuator operates a vane-type valve to control the grain flow rate. The actuator could be a motor to adjust shaft position, a hydraulic cylinder, or others.

Paper thickness

In Figure 7.21, the essential features of a system for controlling paper thickness are shown. The paper is in a wet fiber suspension and is passed between rollers. By varying the roller separation, paper thickness is regulated. The mechanical control element shown is the movable roller. The actuator could be electrical, pneumatic, or hydraulic, and adjusts roller separation based on a thickness measurement.

7.5.2 Electrical

There are numerous cases where a direct electrical effect is impressed in some process-control situation. The following examples illustrate some typical cases of electrical control elements.

Motor speed control

The speed of large electrical motors depends on many factors, including supply voltage level, load, and others. A process-control loop regulates this speed through

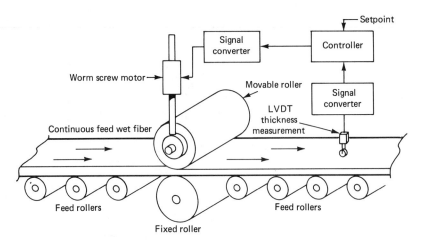

Figure 7.21   A continuous operation paper thickness controlling system using the mechanical final control elements.

direct change of operating voltage or current, as shown in Figure 7.22 for a dc motor. Voltage measurements of engine speed from a tachometer are used in a process-control loop to determine the power applied to the motor brushes. In some cases motor speed control is an intermediate operation in a process-control application. Thus, in the operation of a kiln for solid chemical reaction, the rotation (feed) rate may be varied by motor speed control based on, for example, reaction temperature, as shown in Figure 7.23.

Temperature control

Temperature often is controlled by using electrical heaters in some application of industrial control. Thus, if heat can be supplied through heaters electrically in

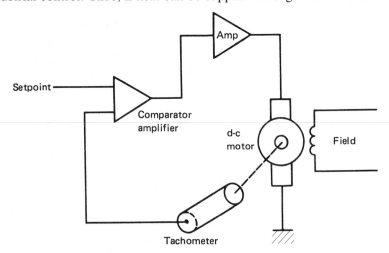

Figure 7.22   Electrical final control as found in the control of a dc motor speed.

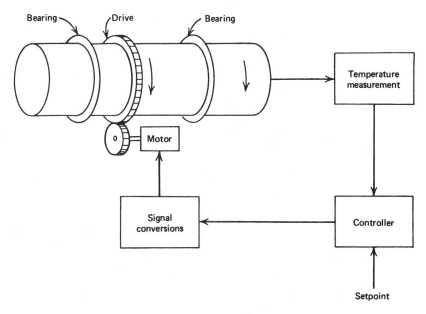

Figure 7.23   An electrical control system with an electrical final control element that varies the rotational rate of a reaction kiln.

an endothermic reaction, then the process-control signal can be used to ON/OFF cycle a heater or set the heater within a continuous span of operating voltages, as in Figure 7.24. In this example, a reaction vessel is maintained at some constant temperature using an electrical heater. The process-control loop provides this by smoothly varying excitation to the heater.

7.5.3 Fluid Valves

The chemical and petroleum industries have many applications requiring control of fluid processes. Many other industries also depend in part on operations that involve fluids and the regulation of fluid parameters. The word *fluid* here represents either gases, liquids, or vapors. Many principles of control can be equally applied to any of these states of matter with only slight corrections. Many fluid operations require regulation of such quantities as density and composition, but by far the most important control parameter is flow rate. A regulation of flow rate emerges as the regulatory parameter for reaction rate, temperature, composition, or a host of other fluid properties. We will consider in some detail that process-control element specifically associated with flow—the *control valve*.

Control valve principles

Flow rate in process control is usually expressed as volume per unit time. If a mass flow rate is desired, it can be calculated from the particular fluid density. If a given fluid is delivered through a pipe, then the volume flow rate is

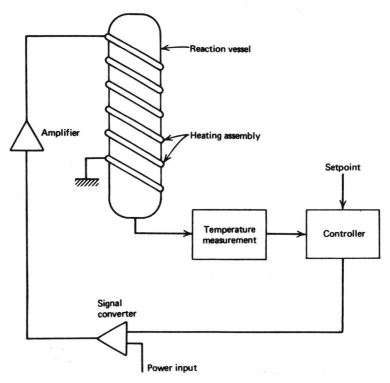

Figure 7.24 Control of heat to a reaction vessel can be provided by purely electrical means.

$$Q = Av \tag{7.7}$$

where

Q = flow rate (m³/s)
A = pipe area (m²)
v = flow velocity (m/s)

Example 7.6

Alcohol is pumped through a pipe of 10-cm diameter at 2 m/s flow velocity. Find the volume flow rate.

Solution   A pipe of 10-cm diameter has a cross-sectional area of

$$A = \frac{\pi D^2}{4} = \frac{(\pi)(10^{-1} \text{ m})^2}{4}$$

$$A = 7.85 \times 10^{-3} \text{ m}^2$$

Thus, the flow rate is

$$Q = Av = (7.85 \times 10^{-3} \text{ m}^2)(2 \text{ m/s}) \tag{7.7}$$
$$Q = \textbf{0.0157 m}^3\textbf{/s}$$

A control valve regulates the flow rate in a fluid delivery system. In general, a close relation exists between the pressure along a pipe and the flow rate so that if the pressure is changed, then the flow rate is also changed. A control valve changes flow rate by changing the pressure in a flow system because it introduces a constriction in the delivery system. In Figure 7.25, the placement of a constriction in a pipe introduces a pressure difference across the pipe. We can show that the flow rate through the constriction is given by

$$Q = K \sqrt{\Delta p} \tag{7.8}$$

where

K = proportionality constant ($\text{m}^3\text{/s/Pa}^{1/2}$)
$\Delta p = p_2 - p_1$ = pressure difference (Pa)

Example 7.7

A pressure difference of 1.1 psi occurs across a constriction in a 5-cm diameter pipe. The constriction constant is 0.009 $\text{m}^3\text{/s}$ per $\text{kPa}^{1/2}$. Find (a) the flow rate in $\text{m}^3\text{/s}$ and (b) the flow velocity in m/s.

Solution   First we note that 1.1 psi is

$$\Delta p = (1.1 \text{ psi})(6.895 \text{ kPa/psi})$$

$$\Delta p = 7.5845 \text{ kPa}$$

(a) The flow rate is

$$Q = K \sqrt{\Delta p} = (0.009)(7.5845)^{1/2} \tag{7.8}$$
$$Q = \textbf{0.025 m}^3\textbf{/s}$$

(b) The flow velocity is found from

$$Q = Av$$

$$v = \frac{Q}{A} = 4 \left[\frac{0.025 \text{ m}^3\text{/s}}{\pi(5 \times 10^{-2})^2} \right] \tag{7.7}$$

$$v = \textbf{12.7 m/s}$$

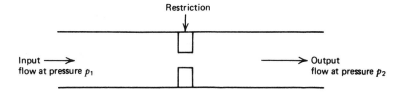

Figure 7.25   Flow rate through a restriction in a line is a function of the pressure drop across the restriction.

The constant K depends on the size of the valve, the geometrical structure of the delivery system, and, to some extent, on the material flowing through the valve. Now the actual pressure of the entire fluid delivery (and sink) system in which the valve is used (and, hence, the flow rate) is not a predictable function of the valve opening only. But because the valve opening does change flow rate, it provides a mechanism of flow control.

Control valve types

The different types of control valves are classified by a relationship between the valve stem position and the flow rate through the valve. This *control valve characteristic* is assigned with the assumptions that the stem position indicates the extent of the valve opening and that the pressure difference is determined by the valve alone. Correction factors allow one to account for pressure differences introduced by the whole system. Figure 7.26 shows a typical control valve using a pneumatic actuator attached to drive the stem and, hence, open and close the valve. There are three basic types of control valves, whose relationship between stem position (as percentage of full range) and flow rate (as a percentage of maximum) is shown in Figure 7.27.

1. *Quick opening*   This type of valve is used predominantly for full ON/full OFF control applications. The valve characteristic of Figure 7.27 shows that a relatively small motion of valve stem results in maximum possible flow rate

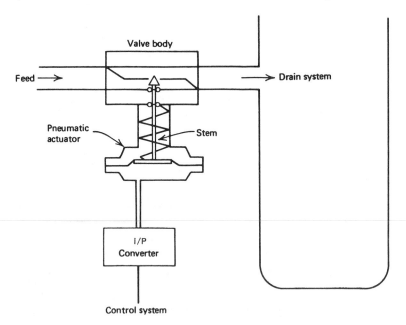

Figure 7.26   The essential features of a control valve are shown. Many variations in the construction of the valve body exist.

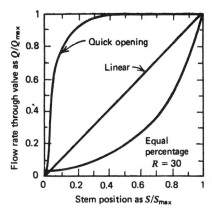

Figure 7.27 Different responses of the three main types of control valves with respect to stem position.

through the valve. Such a valve, for example, may allow 90% of maximum flow rate with only a 30% travel of the stem.

2. *Linear*   This type of valve, as shown in Figure 7.27, has a flow rate that varies linearly with the stem position. It represents the ideal situation where the valve alone determines the pressure drop. The relationship is expressed as

$$\frac{Q}{Q_{max}} = \frac{S}{S_{max}} \tag{7.9}$$

where

$$\begin{aligned}
Q &= \text{flow rate (m}^3\text{/s)} \\
Q_{max} &= \text{maximum flow rate (m}^3\text{/s)} \\
S &= \text{stem position (m)} \\
S_{max} &= \text{maximum stem position (m)}
\end{aligned}$$

3. *Equal percentage*   A very important type of valve employed in flow control has a characteristic such that a given percentage change in stem position produces an equivalent change in flow, that is, an equal percentage. Generally, this type of valve does not shut off the flow completely in its limit of stem travel. Thus, Q_{min} represents the minimum flow when the stem is at one limit of its travel. At the other extreme, the valve allows a flow Q_{max} as its maximum, open valve, flow rate. For this type, we define *rangeability R* as the ratio

$$R = \frac{Q_{max}}{Q_{min}} \tag{7.10}$$

The curve in Figure 7.27 shows a typical equal percentage curve that depends on the rangeability for its exact form. The curve shows that increase in flow rate for a given change in valve opening depends on the extent to which the valve is already open. This curve is typically exponential in form and is represented by

$$Q = Q_{min} R^{S/S_{max}} \tag{7.11}$$

where all terms have been defined previously.

Example 7.8

An equal percentage valve has a maximum flow of 50 m³/s and a minimum of 2 m³/s. If the full travel is 3 cm, find the flow at a 1-cm opening.

Solution   The rangeability is

$$R = Q_{max}/Q_{min} \tag{7.10}$$
$$R = (50 \text{ m}^3/\text{s})/(2 \text{ m}^3/\text{s}) = 25$$

Then the flow at a 1-cm opening is

$$Q = Q_{min}R^{S/S_{max}}$$
$$Q = (2 \text{ m}^3/\text{s})(25)^{1 \text{ cm}/3 \text{ cm}} \tag{7.11}$$
$$Q = 5.85 \text{ m}^3/\text{s}$$

Control valve sizing

Another important factor associated with all control valves involves corrections to Equation (7.8) because of the nonideal characteristics of the materials that flow. A standard nomenclature is used to account for these corrections depending on the liquid, gas, or steam nature of the fluid. These correction factors allow selection of the proper size of valve to accommodate the rate of flow that the system must support. The correction factor most commonly used at present is measured as the number of U.S. gallons of water per minute that flow through a fully open valve with a pressure differential of 1 pound per square inch. The correction factor is called the *valve flow coefficient* and is designated as C_v. Using this factor, a liquid flow rate in U.S. gallons per minute is

$$Q = C_v \sqrt{\frac{\Delta p}{S_G}} \tag{7.12}$$

where

Δp = pressure across the valve (psi)
S_G = specific gravity of liquid

Typical valves of C_v for different size valves are shown in Table 7.1. Similar equations are used for gases and vapors to determine the proper valve size in specific applications.

Example 7.9

Find (a) the proper C_v for a valve that must pump 150 gallons of ethyl alcohol per minute with a specific gravity of 0.8 at maximum pressure of 50 psi and (b) the required valve size.

TABLE 7.1   Control Valve Flow Coefficients

Valve Size (inches)	C_v
$\frac{1}{4}$	0.3
$\frac{1}{2}$	3
1	14
$1\frac{1}{2}$	35
2	55
3	108
4	174
6	400
8	725

Solution   (a) We find the correct sizing factor from

$$Q = C_v \sqrt{\frac{\Delta p}{S_G}} \tag{7.12}$$

Then

$$C_v = Q \sqrt{\frac{S_G}{\Delta p}}$$

$$C_v = \left(150 \, \frac{\text{gal}}{\text{min}}\right) \sqrt{\frac{0.8}{50 \, \text{lb/in}^2}}$$

$$C_v = \mathbf{18.97}$$

(b) A $1\frac{1}{2}$-in diameter valve (3.8 cm) is selected from Table 7.1.

Fluid control example

The chemical and process-control industry uses fluid control systems extensively. Examples of such applications are many and varied. Consider, for example, con-

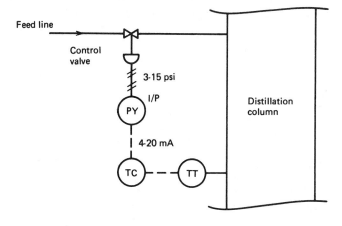

Figure 7.28 Feed control to a distillation column based on temperature.

trol of distillation column composition by regulation of a fixed-point column temperature. Such regulation is achieved by controlling the feed rate as shown in Figure 7.28. A thermocouple measures temperature that is transmitted to the controller as a 4–20 mA control signal. The controller outputs a 4–20 mA signal proportional to proper control valve position. This is converted to a 3–15 psi (20–100 kPa) pneumatic signal by an I/P converter that, in turn, operates a pneumatic actuator connected to the control valve. The valve size is determined by the characteristics of the gas or vapor that is flowing. The size of the required actuator is determined from the valve size.

SUMMARY

The operation of the final control element has three separate functions; the ultimate goal is to translate a low-level control signal into a large-scale process. The following specific details were considered:

1. The final control function can be implemented by *signal conditioning*, an *actuator*, and a *final control element*.

2. Signal conditioning involves changing a control signal into that form and power necessary to energize the actuator. Simple electronic amplification, digital-to-analog conversion, electrical-to-pneumatic conversion, and pneumatic-to-hydraulic conversion are all typical signal conditioning operations.

3. The current-to-pressure converter is frequently employed in process-control systems. This device is based on a flapper/nozzle (nozzle/baffle) system that converts linear displacement into a pressure change.

4. Actuators are an intermediate step between the converted control signal and the final control element. Common electrical actuators are solenoids, digital stepping motors, and ac and dc motors.

5. Pneumatic and hydraulic actuators are often used in process control because they allow very large forces to be produced from modest pressure systems. A pneumatic actuator converts a pressure signal to a shaft extension according to

$$\Delta x = \frac{A}{k} \Delta p \tag{7.3}$$

where the force that causes this extension is given by

$$F = A(p_1 - p_2) \tag{7.2}$$

6. Actual final control elements are as varied as the applications of process control in industry. Examples include motor-driven conveyor belts, paper thickness roller assemblies, and heating systems.

7. The most general type of final control element is a *control valve*. This device is designed for use in process-control applications involving liquid, gas, or vapor flow rate control. Three types are commonly used: *quick opening*, *linear*, and *equal percentage*.

PROBLEMS

Section 7.3

7.1 A 4–20 mA control signal is loaded by a 100-Ω resistor and must produce a 20–40 volt motor drive signal. Find an equation relating the input current to the output voltage.

7.2 Implement the equation of Problem 7.1 if a power amplifier is available that can output 0–100 volts and has a gain of 10.

7.3 A motor to be driven by a digital signal has a speed variation of 200 rev/min per volt with a minimum rpm at 5 volts and a maximum at 10 volts. Find the minimum speed word, maximum speed word, and the speed change per LSB change. Use a 5-bit, 15-volt reference DAC.

Section 7.4

7.4 A stepping motor has 130 steps per revolution. Find the digital input rate that produces 10.5 revolution per second.

7.5 A stepping motor has 7.5° per step. Find the rpm produced by a pulse rate of 2000 pps on the input.

7.6 What force is generated by 90 kPa acting on a 30 cm^2 area diaphragm?

7.7 A hydraulic system uses pistons of diameter 2 cm and 40 cm. What force on the small piston will raise a 500-kg mass?

7.8 What pneumatic pressure is required on the small piston of Problem 7.7 to produce the necessary force?

7.9 The SCR in Figure 7.8 requires a 4-volt trigger. Design a system by which the gears are shifted when a CdS photocell resistance drops below 2.5 kΩ.

7.10 Design a system by which a control signal of 4–20 mA is converted into a force of 200 to 1000 N. Use a pneumatic actuator and specify the required diaphragm area if the pressure output is to be in the range of 20 to 100 kPa. An IP converter is available that converts 0–5 volts into 20 to 100 kPa.

7.11 A feed hopper requires 30 lb of force to open. Find the pneumatic actuator area to provide this force from a 9-psi input signal.

Section 7.5

7.12 Find the proper valve size in inches and centimeters for pumping a liquid flow rate of 600 gal/min with a maximum pressure difference of 55 psi. The liquid specific gravity is 1.3.

7.13 An equal precentage control valve has a rangeability of 32. If the maximum flow rate is 100 ms^3/hr, find the flow at 2/3 and 4/5 open settings.

7.14 The level of water in a tank is to be controlled at 20 m and the output flow rate is nominally 65 m³/hr through a control valve, as shown in Figure 7.29. Under nominal conditions, determine the required valve size in inches and centimeters.

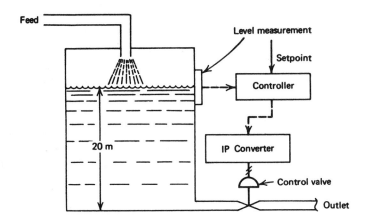

Figure 7.29   Figure for Problem 7.14.

7.15 If the valve actuator of Problem 7.14 has a rangeability of 30, a maximum stem travel of 5 cm, and is to be half open under the nominal conditions, find the minimum flow, maximum flow, and stem opening for 100 m³/hr flow.

7.16 A quick-opening valve moves from closed to maximum open with five turns of a shaft. The shaft is driven through a 10:1 reducer from a stepping motor of 3.6° per step. If the maximum input pulse rate to the motor is 250 steps per second, find the fastest time for the valve to move from closed to open.

7.17 A control valve operates from a 3–15 psi control signal. To have a 40 gal/min flow rate, express the signal input in both psi and percent of range if (a) it is a linear valve from 0 to 90 gal/min, and (b) if it is an equal percentage valve with $R = 6$ and $Q_{min} = 15$ gal/min.

CHAPTER 8

DISCRETE-STATE PROCESS CONTROL

INSTRUCTIONAL OBJECTIVES

The objectives of this chapter are to provide an understanding of process-control operations for which the process variables can take on only discrete values. After you have read this chapter and developed solutions to the problems, you should be able to

1. Define the nature of discrete-state process-control systems.
2. Give three examples of applications of discrete-state process control in industry.
3. Explain how a discrete-state process can be described in terms of the objectives and hardware of the process.
4. Construct a table of ladder diagram symbols with an explanation of the function of each symbol.
5. Develop a ladder diagram from the narrative event sequence description of a discrete-state control system.
6. Describe the nature of a programmable controller and how it is used in discrete-state process control.
7. Develop a programmable controller program from the ladder diagram of a discrete-state process-control application.

8.1 INTRODUCTION

The majority of industrial process-control installations involve more than simply regulating a controlled variable. The requirement of regulation means that some variable tends to vary in a continuous fashion because of external influences. But there are a great many processes in industry in which it is not a variable that has to be controlled but a *sequence of events*.

This sequence of events typically leads to the production of some product from a set of raw materials. For example, a process to manufacture toasters inputs various metals and plastics and outputs toasters. The expression *discrete state* is used to explain that each event in the sequence can be described by specification of the condition of all operating units of the process. Such condition description might be presented by expressions such as: valve A is open, valve B is closed, conveyor C is on, limit switch S_1 is closed, and so on. A particular set of conditions is described as a *discrete state* of the whole system.

In this chapter the nature of discrete-state process control will be studied. In addition to the nature of such control, a special technique for designing and describing the sequence of process events, called a *ladder diagram*, will be presented. The ladder diagram evolved from the early use of electromechanical relays to control the sequence of events in such processes. Relay control systems have mostly given way to computer-based methods of control, the most common of which is called a *programmable controller*. The characteristics and programming of programmable controllers will be studied in this chapter along with numerous applications.

8.2 DEFINITION OF DISCRETE-STATE PROCESS CONTROL

To better understand the material of this chapter, it is helpful to have a general definition of a discrete-state control system. Then you can see how the detailed considerations of the characteristics of such control systems fit within the overall scheme.

Discrete-state process control

Figure 8.1 is a symbolic representation of a manufacturing process and the controller for the process. Let us suppose that all measurement input variables (S_1, S_2, S_3) and all control output variables (C_1, C_2, C_3) of the process can take on or be assigned only two values. For example, valves are open/closed, motors are on/off, temperature is high/low, limit switches are closed/open, and so on.

Now we define a *discrete state* of the process at any moment to be the set of all input and output values. Each state is discrete in the sense that there are only a discrete number of possible states. If there were three input variables and three output variables, then a state consists of specification of all six values.

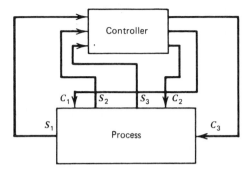

Figure 8.1   Process-control system.

Because each variable can take on two values, there are a total of 64 possible states.

An *event* in the system is defined by a particular state of the system, that is, particular assignment of all output values and a particular set of the input variables. The event lasts for as long as the input variables remain in the same state and the output variables are left in the assigned state. For a simple oven, we can have the temperature low and the heater on. This state is an event that will last until the temperature rises.

With these definitions in mind, *discrete-state process control* is a particular *sequence of events* through which the process accomplishes some objective. For a simple heater such a sequence might be

1. Temperature low, heater off
2. Temperature low, heater on
3. Temperature high, heater on
4. Temperature high, heater off

The objective of the controller of Figure 8.1 is to direct the discrete-state system through a specified event sequence. In the following sections we will consider how the event sequence is specified, how it is described, and how a controller can be developed to direct the sequence of events.

8.3 CHARACTERISTICS OF THE SYSTEM

The objective of an industrial process-control system is to manufacture some product from the input raw materials. Such a process will typically involve many operations or steps. Some of these steps must occur in series and some can occur in parallel. Some of the events may involve the discrete setting of states in the plant, that is, valves open or closed, motors on or off, and so on. Other events may involve regulation of some continuous variable over time or the duration of an event. For example, it may be necessary to maintain the temperature in some vat at a setpoint for a given length of time. In the sense of the previous statements, the discrete-state process-control system is the *master control system* for the entire plant operation.

Example 8.1

Use the definitions of this section to construct a description of the frost-free refrigerator/freezer shown in Figure 8.2 as a process with a discrete-state control system. Define the input variables, output variables, and sequence of serial/parallel events.

Solution   The discrete state input variables are

1. Door open/closed
2. Cooler temperature high/low
3. Freezer temperature high/low
4. Frost eliminator timer time out/not time out
5. Power switch on/off
6. Frost detector on/off

The discrete state output variables are

1. Light on/off
2. Compressor on/off

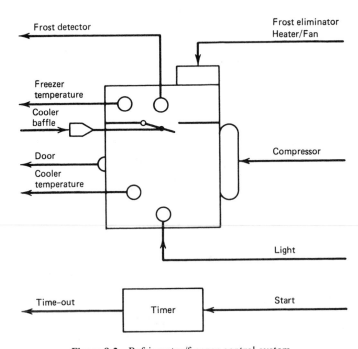

Figure 8.2   Refrigerator/freezer control system.

3. Frost eliminator timer started/not started
4. Frost eliminator heater and fan on/off
5. Cooler baffle open/closed

This is a total of 11 two-state variables. In principle, there are $2^{11} = 2048$ possible states or events. Of course, only a few of these are necessary. The event sequences are

(a) If the door is opened, the light is turned on.

(b) If the cooler temperature is high and the frost eliminator is off, the compressor is turned on and the baffle is opened until the cooler temperature is low.

(c) If the freezer temperature is high and the frost eliminator is off, the compressor is turned on until the temperature is low.

(d) If the frost detector is on, the timer is started, the compressor is turned off, and the frost eliminator heater/fan are turned on until the timer times out.

The events of (a) can occur in parallel with any of the others. The events of (b) and (c) can occur in parallel. Event (d) can only be serial with (b) or (c).

8.3.1 Discrete-State Variables

It is important to be able to distinguish between the nature of variables in a discrete-state system and those in continuous control systems. To define the difference carefully, we will consider an example contrasting a continuous variable situation with a discrete-state variable situation for the same application. Later it will be shown that continuous variable regulation can be itself a part of a discrete-state system.

Continuous control

Consider for a moment the problem of liquid level in a tank. Figure 8.3 shows a tank with a valve that controls flow of liquid into the tank and some unspecified flow out of the tank. A transducer is available to measure the level of liquid in the tank. Also shown is the block diagram of a control system whose *objective* is to maintain the level of liquid in the tank at some preset or setpoint value.

The controller will operate according to some mode of control to maintain the level against variations induced from external influences. Thus, if the outflow increases, the control system will increase the opening of the input valve to compensate by increasing the input flow rate. The level is thus *regulated*. This is a continuous variable control system because both the level and the valve setting can vary over a range. Even if the controller is operating in an ON/OFF mode, there is still variable regulation although the level will now oscillate as the input valve is opened and closed to compensate for output flow variation.

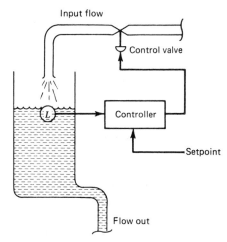

Figure 8.3 Continuous control of level.

Discrete state control

Now consider the revised problem shown in Figure 8.4. We have the same situation as in Figure 8.3, but the objectives are different and the variables, level and valve settings, are discrete because they can take on only two values. This means that the valves can only be open or closed and the level is either above or below the specified value.

Now the *objective* is to fill the tank to a certain level with no outflow. To do this, we specify an event sequence:

1. Close the output valve.

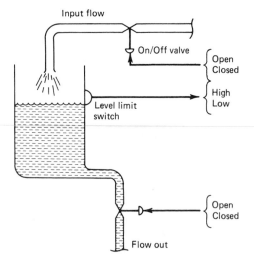

Figure 8.4 Discrete control of level.

2. Open the input valve and let the tank fill to the desired level, as indicated by a simple switch.

3. Close the input valve.

The level is certainly not going to change until, at some later time, the output valve is opened to let the liquid flow out. Notice that the variables, level measurement, input valve setting, and output valve setting, are two-state quantities. There is no continuous measurement or output over a range.

Composite discrete/continuous control

It is possible for a continuous control system to be *part* of a discrete-state process-control system. As an example, consider the problem of the tank system described in Figure 8.3. In this case we specify that the outlet valve is to be closed and the tank filled to the required level as in Figure 8.4. We now specify, however, that periodically a bottle comes into position under the outlet valve, as shown in Figure 8.5. The level must be maintained at the setpoint *while the outlet valve is opened and the bottle filled.* This requirement may be necessary to assure a constant pressure head during bottle filling.

This process will require that a continuous-level control system be used to

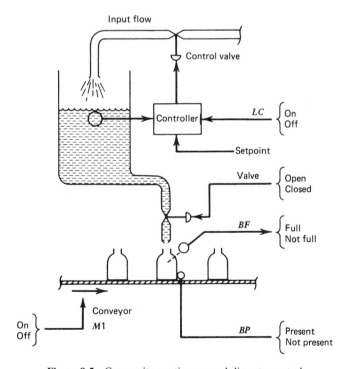

Figure 8.5 Composite continuous and discrete control.

adjust the input flow rate during bottle-fill through the output valve. The continuous control system will be "turned on or off" just as a valve or motor or other discrete device. You can see that the continuous control process is but a part of the overall discrete-state process.

8.3.2 Process Specifications

Specification of the sequence of events in some discrete-state process is directly tied to the process itself. The process is specified in two parts. The first part consists of the objectives of the process, and the second is the nature of the hardware assembled to achieve the objectives. To participate in the design and development of a control system for the process, it is essential that you understand both parts.

Process objectives

The objectives of the process are simply statements of what the process is supposed to accomplish. Objectives are usually associated with knowledge of the industry. Often a *global objective* is defined as the end result of the plant. This is then broken down into individual, mostly independent, *secondary objectives* to which the actual control is applied.

For example, in a food industry plant, a particular global objective might be to produce crackers. Clearly, this means that the plant takes in raw materials, processes them in specified ways, and outputs packaged and labeled crackers, ready for sale.

The overall objective can be broken down into many secondary objectives. Figure 8.6 suggests some of the secondary objectives that might be involved. There can be further subdivisions into simpler operations. The objectives of the process are formed by the objectives of each independent part of the whole operation. A discrete-state control system then will be applied to each independent

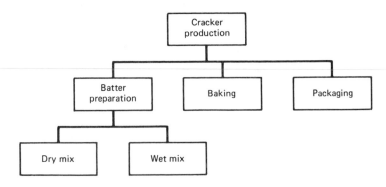

Figure 8.6   Objectives and subobjectives of a process.

part. Thus, in Figure 8.6 the operations within cracker batter preparation can probably be viewed as a stand-alone process.

A process-control specialist typically will not be responsible for the development of the objectives. That is the job of the industry experts. Thus, for crackers we need experts in food chemistry, for petrochemical industries we need chemical engineers, for steel production we need metal specialists, and so on. Nevertheless, it is important for the control system specialist to study the industry and come to an understanding of the products, the process, and the objectives of the process.

Process hardware

With determination of the objectives of the process comes the design of hardware to implement these objectives. This hardware is closely tied to the nature of the industry, and its design must come from the joint efforts of process, production, and control personnel. For the control system specialist, the essential thing is to develop a good understanding of the nature of the hardware and its characteristics. The control system developed will have to use this hardware.

Figure 8.7 shows a pictorial representation of process hardware for a conveyor system. The objective is to fill boxes moving on two conveyors from a common feed hopper and material conveyor system. A process-control system specialist may not have been involved in the development of this system. To develop the control system, he or she must study the hardware carefully and understand the characteristics of each element.

In general, the hardware is analyzed by considering how each part is related to the control system. There are really only two basic categories.

1. *Input devices*   Those hardware elements that provide inputs *to* the control system. These devices are similar to the measurement function of continuous control systems. In the case of discrete-state process control, the inputs are two-

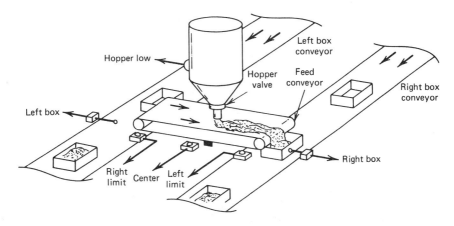

Figure 8.7 Discrete process.

state specifications, such as

Limit switches: open or closed
Comparators: high or low
Push-buttons: depressed or not depressed

2. *Output devices*   Those hardware elements that accept output commands *from* the control system. The final control element of continuous control systems is the same thing. In discrete-state process control, these output devices accept only two-state commands, such as

Light: on or off
Motor: rotating or not rotating
Solenoid: engaged or not engaged

Example 8.2

Study the pictorial process of Figure 8.7. Identify the input and output devices and the characteristics of each device.

Solution   A study of the system described in Figure 8.7 shows the following distribution of elements:

INPUT DEVICES (ALL SWITCHES)

Right box present
Left box present
Feed conveyor right travel limit
Feed conveyor left travel limit
Hopper low
Feed conveyor center

OUTPUT DEVICES

Hopper valve solenoid
Feed stock conveyor motor off
Feed stock conveyor motor right
Feed stock conveyor motor left
Right box conveyor motor
Left box conveyor motor

It is not enough to simply identify the input and output devices. In addition, it is important to note how the two states of the devices relate to the process. Thus, if a level-limit switch is open, does that mean the level is low or at the required value? If a command is to be used to turn on a cooler, does this require a high or low output command?

Finally, a full study of the hardware also should include the nature of the electrical, pneumatic, or hydraulic signals required for the element. Thus, a motor may be started by application of a 110-vac, low-current signal to a motor start relay or it may be started by a 5-vdc TTL-type signal to an electronic starter. Obviously, this type of information will be essential to the development of the control system interface to the process hardware.

8.3.3 Event Sequence Description

Now that the subobjectives of a process and the necessary hardware have been defined, the job remains to describe how this hardware will be manipulated to accomplish the objective. A *sequence of events* must be described that will direct the system through the operations to provide the desired end result.

Narrative statements

Specification of the sequence of events starts with narrative descriptions of what events must occur to achieve the objective. In many cases, this first attempt at specification reveals modifications that must be made in the hardware, such as extra limit switches. This specification describes in narrative form what must happen during the process operation. In systems that run continuously, there is typically a *start-up* or *initialization* phase and a *running* phase.

As an example, consider the system described by Figure 8.7. The start-up phase is used to position the feed conveyor in a known position. This initialization might be accomplished by the following specification:

I. Initialization Phase
   A. All motors off, feed valve solenoid off
   B. Test for right limit switch
      1. If engaged, go to C
      2. If not, set feed motor for right motion
      3. Start feed-conveyor motor
      4. Test for right limit switch
         a. If engaged, go to C
         b. If not, go to 4
   C. Set feed motor for left motion and start
   D. Test for center switch
      1. If engaged go to E
      2. If not, go to D
   E. Open feed hopper valve
   F. Test for left limit switch
      1. If engaged, go to G

    2. If not, go to F

G. All motors off, hopper feed valve closed

H. Go to running phase

Completion of this phase means that the feed conveyor is positioned at the left limit position and the right half of the conveyor has been filled from the feed hopper. The system is in a known configuration, as shown in Figure 8.8.

The running phase is described by a similar set of statements of the sequence of events. For the example of Figure 8.7, this phase might be described as follows:

II. Running

  A. Start right box conveyor

  B. Test right box present switch

    1. If set, go to C

    2. If not, go to B

  C. Start feed conveyor motor, right motion

  D. Test center switch

    1. If engaged, go to E

    2. If not, go to D

  E. Open hopper feed valve

  F. Test right limit switch

    1. If engaged, go to G

    2. If not, go to F

  G. Close hopper feed valve, stop feed conveyor

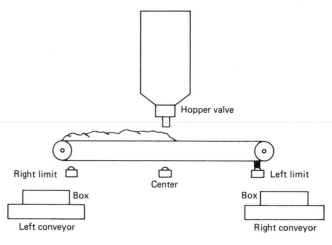

Figure 8.8 Completion of initialization phase.

   H. Start left box conveyor

   I. Test left box present switch

     1. If set, go to J

     2. If not, go to I

   J. Start feed conveyor, left motion

   K. Test center switch

     1. If engaged, go to L

     2. If not, go to K

   L. Open hopper feed valve

   M. Test left limit switch

     1. If engaged go to A

     2. If not, go to M

Note that the system cycles from step M to step A. The description is constructed by simple analysis of what events must occur and what the input and outputs must be to support these events.

Example 8.3

Construct a narrative statement outline of the event sequence for the system shown in Figure 8.5. The objective is to fill bottles moving on a conveyor.

Solution   We assume that when a command is given to stop the continuous control system, the input valve is driven to the closed position. Then the sequence would be

   I. Initialization (prefill of tank)

     A. Conveyor stopped, output valve closed

     B. Start the level control system

       1. Operate for a sufficient time to reach the setpoint, or

       2. Add another sensor so that the system knows when the setpoint has been reached

     C. When level is reached, stop the level control

     D. Go to the running phase

   II. Running

     A. Start the bottle conveyor

     B. When a bottle is in position:

       1. Stop the conveyor

       2. Open the output valve

       3. Turn on a level control system to keep the level constant during bottle-fill

     C. When the bottle is full:

       1. Close the output valve

    2. Stop the level control system

  D. Go to step A and repeat

Notice that hardware was added to the system when the event sequence was constructed. Hardware and software are often developed in conjunction, because the development of one demonstrates extra needs in the other.

Flowcharts of the event sequence

It is often easier to visualize and construct the sequence of events if a flowchart is used to pictorially present the flow of events. Although there are many sophisticated types of flowcharts, the concept can be presented quite easily by using the three symbols shown in Figure 8.9.

    The narrative statements are then simply reformatted into flowchart symbols. Often it is easier to express the sequence of events directly in terms of the flowchart symbols. Figure 8.10 shows part of the initialization phase of the conveyor system of Figure 8.7 expressed in the flowchart format.

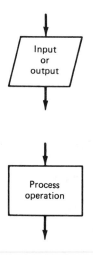

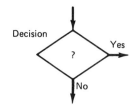

Figure 8.9 Basic flowchart symbols.

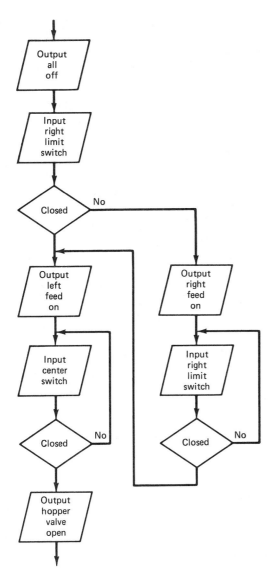

Figure 8.10 Partial initialization flowchart.

Binary-state variable descriptions

Each event that makes up the sequence of events described by the narrative scheme corresponds to a *discrete state* of the system. Thus, it is also possible to describe the sequence of events in terms of the sequence of discrete states of the system. To do this simply requires that for each event the state, including both input and output variables, be specified.

The input variables cause the state of the system to change because operations within the system cause a change of one of the state variables; for example, a limit switch becomes engaged. The output variables, on the other hand, are changes in the system state that are caused by the control system itself.

The control system works like a look-up table. The input state variables with the output become like a memory address, and the new output state variables are the contents of that memory.

Example 8.4

Define the state variable description for the process shown in Figure 8.11 and described by the following event sequence:

1. Fill the tank to level A from valve A.
2. Fill the tank to level B from valve B.
3. Start a timer, heat, and stir for 5 minutes.
4. Open output valve C until the empty switch engages

Solution   To provide the solution, we first form the state variable representation of the system by assignment of binary states. There are four input variables (LA, LB, LF, TU) and six output variables (VA, VB, VC, TM, S, H).

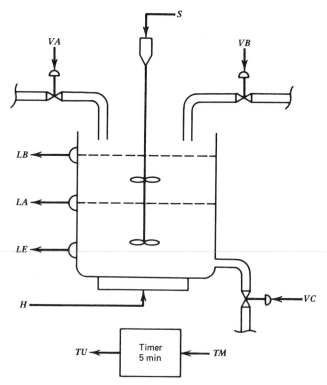

Figure 8.11   Tank process for Example 8.4.

A discrete state of the system is defined by specification of these variables. Because each variable is a two-state variable, we use a binary representation: true = 1 and false = 0. Thus, for input, if level A has not been reached, then $LA = 0$, and if it *has* been reached, then $LA = 1$. Also, for output, if valve C is to be closed, then we take $VC = 0$, and if it is commanded to be open, then $VC = 1$. Let us take the binary "word" describing the state of the system to be defined by bits in the order

$$(LA)(LB)(LE)(TU)(VA)(VB)(VC)(TM)(S)(H)$$

The sequence of events is now translated into an expression of the discrete state as a binary word per state. (An X means we do not care what that input variable is.)

Condition	Input State		Output
1. Open valve A	00XX000000	→	100000
2. Test for LA:			
a. Not true, maintain	00XX100000	→	100000
b. True, close A, open B	10XX100000	→	010000
3. Test for LB:			
a. Not true, maintain	10XX010000	→	010000
b. True, close B, start stir, heat, timer	11XX010000	→	000111
4. Test for time up:			
a. Not true, maintain	11X0000111	→	000111
b. True, heat off, stir off, open C	11X1000111	→	001000
5. Test for tank empty:			
a. Not true, maintain	XX0X001000	→	001000
b. True, close C	XX1X001000	→	000000
6. Go to 1			

Typically, this approach to specification of the event sequence is used when a computer will be used to implement the control functions.

Boolean equations

Because the discrete state of the system is described by variables that can take on only two values, it is natural to think of using binary numbers to represent these variables, as in the previous example. It is also natural to consider use of Boolean algebra techniques to deduce the output states from the input states. Although this technique is used, there are generally easier ways to view and solve the problems than with traditional Boolean techniques.

When this technique is used, it is necessary to write a Boolean equation for each *output* variable in the system. This equation will then determine when that variable is taken to its true state. The equation may depend not only on the set of input variables, but on some of the other output variables. Problems of this type are often considered in digital electronics courses; for example, the common stop-light sequence problem. The following simple example applies this description technique to a more traditional control problem.

Example 8.5

Figure 8.12 shows a pictorial view of an oven along with the associated input and output signals. All of the inputs and outputs are two-state variables, and the relation of the states and the variables is indicated. Construct Boolean equations that implement the following events:

1. The heater will be on when the on-switch is activated, the door is closed, and the temperature is below the limit.
2. The fans will be placed on when the heater is on, or when the temperature is above the limit and the door is closed.
3. The light will be turned on if the light switch is on or whenever the door is opened.

Solution The solution of problems of this type is developed by simply translating the narrative statements of the events into Boolean equations. In this case, referring to the variables defined in Figure 8.12, you can see that the solution is

Heater: $H = D \cdot \overline{T} \cdot P$
Fans: $F = H + D \cdot T$
Light: $L = \overline{D} + S$

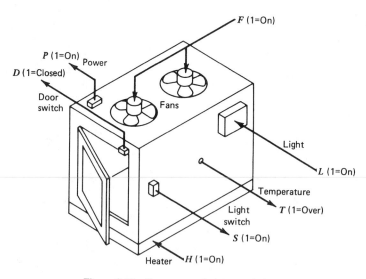

Figure 8.12 Oven control characteristics.

8.4 LADDER DIAGRAM

The previous section showed how a discrete-state control system is described in terms of the hardware of the system and the sequence of events through which that hardware is taken. These two elements are now combined to show how the hardware should be driven so that the proper sequence of events can be accomplished. In essence, this amounts to a "program" for the system written with symbols for the hardware.

A special schematic representation of the hardware and its connection has been developed that makes combination of the hardware and event sequence description clear. This schematic is called a *ladder diagram*. It is an outgrowth of early controllers that operated from ac lines and used relays as the primary switching elements.

8.4.1 Background

An industrial control system will typically involve electric motors, solenoids, heaters or coolers, and other equipment that is operated from the ac power line. Thus, when a control system specifies that a "conveyor motor be turned on," it may mean starting a 50-HP motor. This is not done by a simple toggle switch. Instead, one would logically assume that a small switch may be used to energize a relay with contact ratings that can handle the heavy load such as that shown in Figure 8.13. In this way, the relay became the primary control element of discrete-state control systems.

Control relays

Relays can be used for much more than just an energy-level translator. For example, Figure 8.14 shows a relay used as a latch where a green light is on when the relay is not latched and a red light is on when the relay is latched. In this case, when the normally open (NO) push-button switch *PB*1 is depressed, control relay *RL*1 is energized. But then its normally open (NO) contact closes, bypassing *PB*1, so that the relay stays closed. Thus, it is latched. To de-energize or unlatch

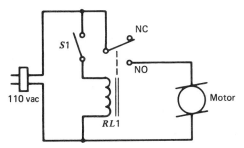

Figure 8.13 Use of relay and switch to start a motor.

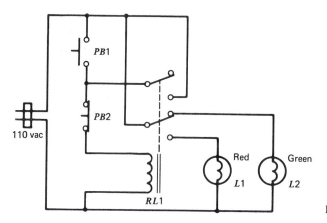

Figure 8.14　Use of a relay for a latch.

the relay, the normally closed push-button *PB2* is depressed. *PB2* opens the circuit, and the relay is released.

When an entire control system is implemented using relays, the system is called a *relay sequencer*. A relay sequencer consists of a combination of many relays, including special time-delay types, wired up to implement the specified sequence of events. Inputs are switches and push-buttons that energize relays, and outputs are closed contacts that can turn on or off lights, start motors, energize solenoids, and so on.

Schematic diagrams

The wiring of a relay control system can be described by traditional schematic diagrams, such as those shown in Figures 8.13 and 8.14. Such diagrams become very cumbersome, however, when many relays, each with many contacts, are used in a system. Simplified diagrams were gradually adopted by the industry over the years. An example of such simplification is not to require a relay's contacts to be placed directly over the coil symbol, but anywhere in the circuit diagram with a number to associate it with a particular coil. These simplifications resulted in the ladder diagram in use today.

8.4.2 Ladder Diagram Elements

The ladder diagram is a symbolic and schematic way of representing both the system hardware and the process controller. It is called a ladder diagram because the various circuit devices connected in parallel across the ac line form something that looks like a ladder, with each parallel connection a "rung" on the ladder.

In the construction of a ladder diagram, it is understood that each rung of the ladder is composed of a number of conditions or input states and a single command output. The nature of the input states determines if the output is to be energized or not energized.

Special symbols are used to represent the various circuit elements in a ladder diagram. The following sections present these symbols.

Relays

A relay coil is represented by a circle identified as *CR* for control relay and an associated identifying number. The contacts for that relay will be either normally open (NO) or normally closed (NC) and can be identified by the same number. The coil and its associated contact descriptions are presented in Figure 8.15a.

It is also possible to designate a *time-delay relay* as one for which the contacts do not activate until a specified time delay has occurred. The coil is still indicated by a circle, but with the designation of *TR* to indicate timer relay. The contacts, as shown in Figure 8.15b, have an arrow to indicate NO-to-close after delay or NC-to-open after delay. This is called an *on-delay* timer relay. When the coil is energized, the contacts are not energized until the time delay has lapsed.

There is also an *off-delay* timer relay. In this case, the contacts engage when the coil is energized. When the coil is de-energized, however, there is a time delay before the contacts go to the de-energized state.

Motors and solenoids

The symbol for a motor is a circle with a designation of *M* followed by a number, as shown in Figure 8.16a. The control system treats this circle as the actual motor, although, in fact, this may be a motor start system. The control system uses this symbol to represent the *fact* of the motor, even though other operations may be necessary in the actual hardware to start the motor.

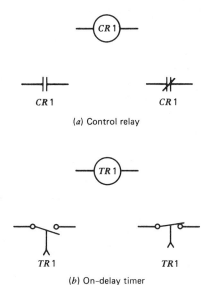

(a) Control relay

(b) On-delay timer

Figure 8.15 Symbols for control and timer relays.

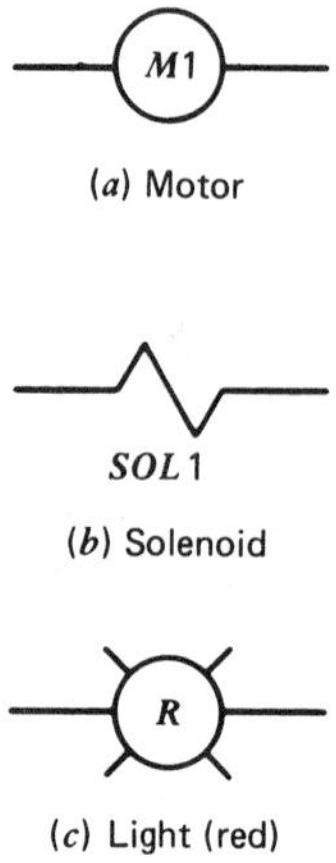

(a) Motor

(b) Solenoid

(c) Light (red)

Figure 8.16   Symbols for output devices.

The solenoid symbol is shown in Figure 8.16b. Of course, the symbol itself tells nothing of what function the solenoid plays in the process. For example, it may be a solenoid to open a flow valve or move material off of a conveyor or a host of other possibilities. The solenoid is designated by *SOL* and a number.

Lights

A light symbol, such as that shown in Figure 8.16c, is used to give operators information about the state of the system. The color of the light is indicated by a capital letter in the circle; for example, *R* stands for red, *G* for green, *A* for amber, and *B* for blue.

Switches

One of the primary input elements in a discrete-state control system is a switch. The switch may be normally open (NO) or normally closed (NC) and may be activated from many sources. In the ladder diagram, different symbols are used to distinguish between different types of switches.

Figure 8.17a shows the symbols for *push-button* switches. Both the NO and NC types are employed. These switches are typically used for operator input, such as to stop and/or to start a system.

Figure 8.17b shows the symbols for the NO and NC *limit switches*. These devices are used to detect physical motion limits within the process.

Figure 8.17c shows the symbol for pressure switches, both NO and NC. Thermally activated switches, such as for ovens or overheating protection of a motor, are indicated by the symbol shown in Figure 8.17d. Figure 8.17e shows the symbols for level switches. In all of these switches, the NO is closed by rising pressure, temperature, or level.

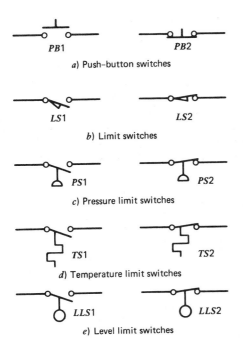

Figure 8.17 Symbols for limit switches.

8.4.3 Ladder Diagram Examples

In many cases, it is possible to prepare a ladder diagram directly from the narrative description of a control event sequence. A most elementary and common example is the relay latch illustrated by the electrical schematic of Figure 8.14. In terms of a ladder diagram, the same situation is described in Figure 8.18. This diagram has three rungs. The first is a latch involving control relay $CR1$; the second rung is for the green OFF light; the third rung is for the red ON light.

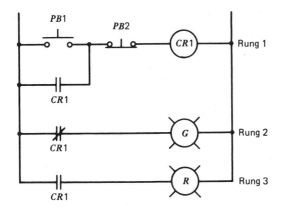

Figure 8.18 Ladder diagram for a light latch like Figure 8.14.

The following example illustrates many features of ladder diagram construction and its application to control problems.

Example 8.6

The elevator system shown in Figure 8.19 employs a platform to move objects up and down. The global objective is that when the UP button is pushed, the platform carries something to the UP position, and when the DOWN button is pushed, the platform carries something to the DOWN position.

The following hardware specifications define the equipment used in the elevator:

OUTPUT ELEMENTS

$M1$ = Motor to drive the platform UP
$M2$ = Motor to drive the platform DOWN

INPUT ELEMENTS

$LS1$ = NC limit switch to indicate UP position
$LS2$ = NC limit switch to indicate DOWN position
START = NO push-button for START
STOP = NO push-button for STOP
UP = NO push-button for UP command
DOWN = NO push-button for DOWN command

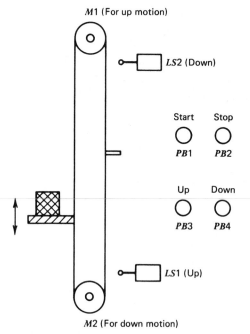

Figure 8.19 Elevator system for Example 8.6.

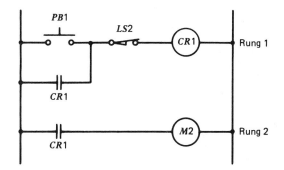

Figure 8.20   Initialization to move platform down on START.

The following narrative description indicates the required sequence of events for the elevator system.

1. When the START button is pushed, the platform is driven to the DOWN position.
2. When the STOP button is pushed, the platform is halted at whatever position it occupies at that time.
3. When the UP button is pushed, the platform, if it is not in DOWN motion, is driven to the UP position.
4. When the DOWN button is pushed, the platform, if it is not in UP motion, is driven to the DOWN position.

Prepare a ladder diagram to implement this control function.

Solution   Let us prepare a solution by breaking the requirements into individual tasks. For example, the first task is to move the platform to the DOWN position when the START button is pushed.

This task can be done by using the START button to latch a relay, whose contacts also energize $M2$ (the DOWN motor). The relay is released, stopping $M2$, when the $LS2$ limit switch opens. Figure 8.20 shows ladder rungs 1 and 2 that provide these functions. Pushing START energizes $CR1$ if $LS2$ is not open (platform not DOWN). $CR1$ is latched by the contacts across the START button. Another set of

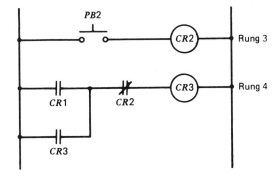

Figure 8.21   Ladder diagram for STOP sequence.

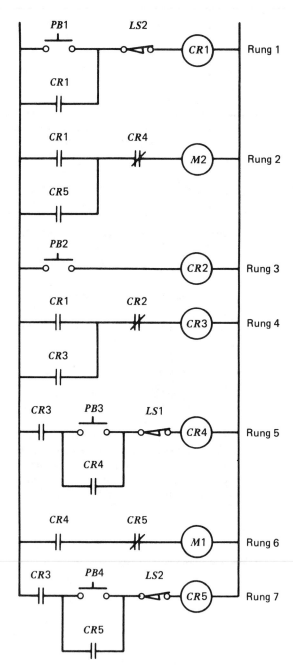

Figure 8.22 Complete ladder diagram for elevator.

$CR1$ contacts starts $M2$ to drive the platform DOWN. When $LS2$ opens, indicating the DOWN position has been reached, $CR1$ is released, unlatched, and $M2$ stops. These two rungs will only operate when the START button is pushed.

For the STOP sequence, let us assume a relay $CR3$ is the master control for the rest of the system. Because STOP is a NO switch, we cannot use it to release $CR3$ in the same sense used in previous examples. Instead, we use STOP to energize another relay $CR2$, and use the NC contacts of that relay to release $CR3$. This is shown in Figure 8.21. You can see that when START is pushed, $CR3$ in rung 4 is energized by the latching of the $CR1$ contact and the NC contact of $CR2$. When STOP is pushed, $CR2$ in rung 3 is energized, which causes the NC contact in rung 4 to open and release $CR3$.

Finally, we come to the sequences for UP and DOWN motion of the platform. In each case, a relay is latched to energize a motor if $CR3$ is energized, the appropriate button has been pushed, the limit has not been reached, and the other direction is not energized. The entire ladder diagram is shown in Figure 8.22. An NC relay connection is used to assure that the UP motor is not turned on if the DOWN motor is on, and vice versa. Also, it was necessary to add a contact to rung 2 to be sure $M2$ could not start if there was UP motion and some joker pushed the START button.

The solution to Example 8.6 can be simplified by considering the fact that $M1$ and $M2$ are actually relays used to turn on the motors via contacts. If we assume that these relays can have added contacts to drive other ladder diagram operations, then some of the control relays can be eliminated. Figure 8.23 shows a simplified

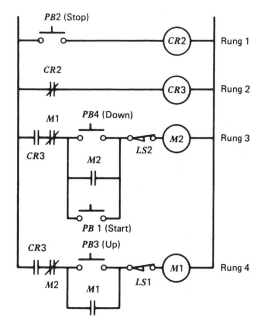

Figure 8.23 Simplified ladder diagram for the elevator.

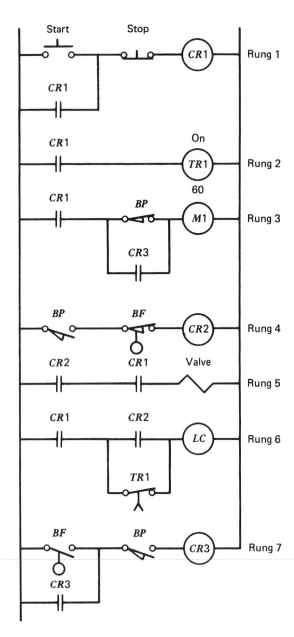

Figure 8.24   Solution to Example 8.7.

solution of Example 8.6. We use the $M1$ and $M2$ designations for contacts in other parts of the diagram, just as with control relays. You should work through this diagram to see how the sequence of events is satisfied.

Example 8.7

Construct the ladder diagram that will provide a solution to the discrete-state control problem defined by Figure 8.5 and Example 8.3. Assume that when the level-control system is commanded off, the input valve is closed and a 1-minute prefill is required for initialization.

Solution   A START/STOP latch is provided to define the initial start-up of the system. The ladder diagram is shown in Figure 8.24.

Initialization is accomplished by a 60-second timer in rung 2 that turns on the level control system for 1 minute following the start button. It is never energized again during running.

Rung 3 drives the conveyor motor until a bottle is in position, as indicated by the bottle position switch opening. Rung 4 is used to detect the bottle-full condition by energizing $CR2$. The contacts of $CR2$ turn on both the valve solenoid (rung 5) and the level-control system (rung 6). Note the timer in rung 6 for initialization. Rung 7 is necessary to detect that the bottle is full and to restart the conveyor until the bottle is moved out of position and the bottle-present switch is opened. Continuous running now occurs between rung 3 and rung 7.

8.5 PROGRAMMABLE CONTROLLERS

The previous sections of this chapter have explained *what* a discrete-state control system is and *why* such a system is needed in industrial processes. The last section shows how ladder diagrams are used to *describe* the event sequence that makes up such a control system. Finally, in this section, you will learn *how* to actually *provide* the control system; for example, using the controller shown in Figure 8.1 that inputs the state of the system and generates the required output states to make the process follow the proper event sequence.

8.5.1 Relay Sequencers

One way to provide a discrete-state controller is to use physical relays to put together a circuit that satisfies the requirements of the ladder diagram. Such a control system is called a *relay sequencer* or *relay logic panel*. In the early days of industrial control processes, this was the only way to provide control. It is still used in many applications today, although modern computer-based controllers have replaced many relay-based systems.

The ladder diagram technique of describing discrete-state control systems originated from relay logic systems, which is why the diagram contains so many relay-related terms and symbols. The ladder diagram continues to be used today because it has evolved into a very efficient method of defining the event sequence required in a discrete-state control system.

It is important to realize that with relay control *each* rung of the ladder is evaluated simultaneously and continuously, because the switches and relays are all hardwired to ac power. If any switch anywhere in the ladder diagram changes state, the consequences are immediate. This is not true for computer-based programmable controllers, to be discussed in the next section.

Special functions

To build a relay-based control system, it is necessary to provide certain kinds of special functions not normally associated with relays. These functions are often provided using analog and digital electronic techniques. Included in the special functions are such features as time-delay relays, up/down counters, and real-time clocks.

Hardwired programming

When a relay panel has been wired to implement a ladder diagram, we say that it has been *programmed* to satisfy the ladder diagram; that is, the event sequence required and described by the ladder diagram will be provided by the relay system when power is applied. Thus, the program has been wired into the relays that make up the relay logic panel.

If the event sequence is to be changed, it is necessary to rewire all or part of the panel. It may even be necessary to add more relays to the system or to use more relays than in the previous program.

Obviously, such a task is quite troublesome and time-consuming. A number of ingenious methods are used to ease some of the problems of changing the relay program. One is the use of *patch-panels* for the programming. In these systems all relay contacts and coils are brought to an array of sockets. Cords with plugs in each end are then used to patch the required coils, contacts, inputs, and outputs together in the manner required by the ladder diagram.

The patch-panel acts like a memory in which the program is placed. With the development of reliable computers, it was a very clear and easy decision to replace relay logic-based systems by computer-controlled systems.

8.5.2 Programmable Controller

The modern solution for the problem of how to provide discrete-state control is to use a computer-based device called a *programmable controller* (PC) or *programmable logic controller* (PLC).

The move from relay logic controllers to computer-based controllers was an obvious one because

1. The input and output variables of discrete-state control systems are binary in nature, just as with a computer.
2. Many of the "control relays" of the ladder diagram can be replaced by software, which means less hardware failure.
3. It is easy to make changes in a programmed sequence of events when it is only a change in software.
4. Special functions, such as time-delay actions and counters, are easy to form in software.
5. The semiconductor industry developed solid-state devices that can control high-power ac/dc in response to low-level commands from a computer, including SCRs and TRIACs.

A programmable controller can be studied by considering the basic elements shown in Figure 8.25: the processor, the input/output modules, and the software.

Processor

The processor is a computer that executes a program to perform the operations specified in a ladder diagram or a set of Boolean equations. The processor performs arithmetic and logic operations on input variable data and determines the proper state of the output variables. The processor functions under a permanent supervisory operating system that directs the overall operations from data input and output to execution of user programs.

Of course, the processor, being a computer, can only perform one operation at a time. That is, like most computers, it is a serial machine. Thus, it must sequentially sample each of the inputs, evaluate the ladder diagram program,

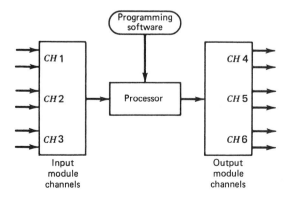

Figure 8.25   Programmable controller.

provide each output, and then repeat the whole process. The speed of the processor is important. This is discussed further under *scan and execution modes*.

Input modules

The input modules examine the state of physical switches and other input devices and put their state into a form suitable for the processor. The processor is usually able to accommodate a number of inputs, called *channels*.

In keeping with the industrial settings of most programmable controller applications and the history of relay control, the input state systems are often designed to provide 0 or 110 vac to the input module. This type of connection assumes that switches, for example, are wired to the programmable controller, as shown in Figure 8.26. If the switch is *closed*, the input will be 110 vac, and if open, the input will be 0 vac. The input module converts this into the 1 or 0 state needed by the processor. In many cases programmable controllers are now being designed to operate from dc voltages for which the switch is simply connected to the input, with no need for power to be supplied.

The input modules have a certain number of channels per module. Each channel is often equipped with an indicator light to show if the particular input is ON or OFF.

Output modules

The output modules supply ac power to external devices such as motors, lights, solenoids, and so on, just as required in the ladder diagram. The output module can supply a certain maximum power. If the required power is greater, an external relay may be used, as shown in Figure 8.27.

Internally, the output module accepts a 1 or 0 input from the processor and

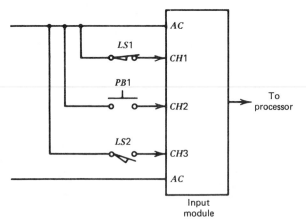

Figure 8.26 Typical wiring to an input module.

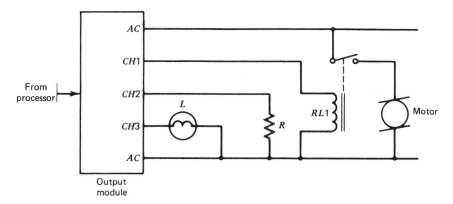

Figure 8.27 Typical wiring to an output module.

uses this to turn on or off an ac power control device such as a TRIAC. In this sense the output module is a solid-state relay.

Programmable controllers also are designed with output modules to provide other outputs, such as dc voltages or variable rate pulse outputs (such as would be required by a stepping motor).

An output module will have one to several channels per unit. Each channel is usually provided with an indicator light to show if the particular channel is being driven ON or OFF.

8.5.3 Programmable Controller Operation

Let us consider the typical operation of a programmable controller. First of all, the operation is *not* simultaneous for the entire ladder diagram and is *not* continuous as it is for relay sequencers. This is very important and can have significant impact if not taken into consideration in a design. Operation of the programmable controller can be considered in two modes, the *I/O scan mode* and the *execution mode*.

I/O scan mode

During the I/O scan mode, the processor updates all outputs and inputs the state of all inputs one channel at a time. The time required for this depends on the speed of the processor.

Execution mode

During this mode the processor evaluates each rung of the ladder diagram program that is being executed sequentially, starting from the first rung and proceeding to the last rung. As a rung is evaluated, the last known state of each switch and relay contact in the rung is considered, and if any TRUE path to the output device is detected, then that output is indicated to be energized, that is, set to ON.

At the end of the ladder diagram, the I/O mode is entered again and all output devices are provided with the ON or OFF state determined from execution of the ladder diagram program. All inputs are sampled, and the execution mode starts again.

Scan Time

An important characteristic of the programmable controller is how much time is required for one complete cycle of I/O scan and execution. Of course, this depends on how many input and output channels are involved and on the length of the ladder diagram program. A typical maximum scan/execution time is 20 milliseconds.

The speed of the controller is dependent on the clock frequency of the processor. The higher the clock frequency, the greater is the speed and the faster is the scan/execution time.

Programming unit

The programming unit is an external electronic package that is connected to the programmable controller when programming occurs. The unit will usually allow input of a program in ladder diagram symbols. The unit then transmits that program into the memory of the programmable controller.

Programming units may be small, self-contained units, such as that shown in Figure 8.28. In this unit the ladder diagram is displayed one rung at a time in a special liquid crystal display (LCD). The user can enter a program, perform diagnostic tests, run the program through the programmable controller, and perform editing of the installed program. The installed program is stored in a temporary memory that will be lost without ac power or battery backup. The program can be permanently "burned" into a ROM for final installation.

Another type of programming unit is able to display many rungs of the ladder diagram. Such a unit is shown in Figure 8.29. The same capabilities are available for entering, testing, and changing a program. The program is still in temporary memory.

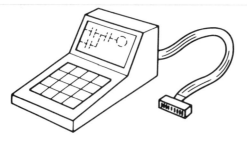

Figure 8.28 Hand-held programming unit.

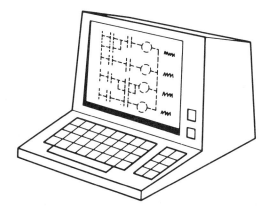

Figure 8.29   Terminal-type programming unit.

Once the program has been debugged, the programming unit can be disconnected and the programmable controller can now operate the process according to the ladder diagram program. There is the danger of loss of the program because of power failure, but this can be prevented by placing the program into the permanent memory.

RAM/ROM

The temporary memory used during ladder diagram program testing and evaluation is called RAM. This is read and write memory. Once the program is stored in RAM, it can be easily modified. As in any computer program, it is necessary to perform testing and evaluation of programmable controller programs, and corrections are usually necessary.

When a program has been debugged and is considered finished, it is "burned" into a ROM. This is a read-only memory that cannot be changed and is not affected by power failure. The ROM often can be programmed directly by the controller programming unit. When the ROM is plugged into the programmable controller, the device is ready to be placed into service in the industrial setting.

8.5.4 Programming

Although the programmable controller can be programmed directly in ladder diagram symbols through the programming unit, there are some special considerations. These considerations include the availability of special functions and the relation between external I/O devices and their programmed representations.

The programmable controller has no "real" relays or relay contacts. The only real devices are those that are actually part of the process being controlled, that is, limit switches, motors, solenoids, and so on. We continue to use symbols for relays and relay contacts, even though they are software symbols.

Addressing

When the ladder diagram for some event sequence was developed in a previous section, each switch device, output device, and relay was referred to by a label. For example, $CR1$ referred to control relay 1, and the contacts for that relay were referred to by the same label. Other designations included $LS1$ for a limit switch, $M1$ for a motor relay, and so on.

The programmable controller uses a similar method of identifying devices, but it is referred to as the device *address*. The addresses are used to identify both the physical and software devices according to the following categories:

1. Physical input devices—ON or OFF
2. Physical output devices—energized (ON) or de-energized (OFF)
3. Programmed control relay coils and contacts
4. Programmed time-delay relay coils and contacts
5. Programmed counters and contacts

The address designation depends on the type of programmable controller. Some controllers may reserve certain addresses for physical I/O devices, other addresses for software control relays, and yet others for special functions.

For the purpose of examples and problems to be considered in this chapter, we make the following definitions of addresses:

Function	Address
Input channels:	00 to 07
Output channels:	08 to 15
Internal relays:	16 to 31
Timers:	32 to 39

Programmed diagram interpretation

There is an important difference between the interpretation of a physical ladder diagram and a programmed ladder diagram. This difference arises from the fact that the programmed diagram bases the state of a rung on a logical interpretation of the symbol rather than its physical state.

In a programmed diagram rung, the ON or OFF state of the output of the rung is determined by testing the elements of the rung for a TRUE or FALSE condition. If a complete TRUE element path to the output exists in the rung, then the output will be ON.

In a physical diagram, the symbol for a NO contact indicates a normally open contact through which current cannot flow unless the contact has been closed. If it is a push-button switch, then someone must close the contacts by pushing. If it is the contact of a relay, then the relay coil must be energized.

For the NC contact, the idea is that current will flow until the contact has been opened. If it is a push-button switch, then someone must open the contact and stop the current flow by pushing. If it is the contact of a relay, then the relay must be energized to open the contact and stop current flow through the contacts.

In a programmed diagram, the symbol for a NO contact indicates that the device should be interpreted as FALSE if the contact is tested and found to be open and TRUE if it is found to be closed. We often say it is to be "examined ON," and if ON it is TRUE. This is much the same as for the physical diagram.

Consider the programmed NC symbol. This means if it is tested and found to be closed, then it is FALSE, and if tested to be open it is ON. This is not like the physical. We often say this is an "examine OFF," and if it is OFF it is TRUE.

The diagrams of Figure 8.30 illustrate this concept. Suppose we have a physical NC push-button switch and that we want to turn on a red light when the

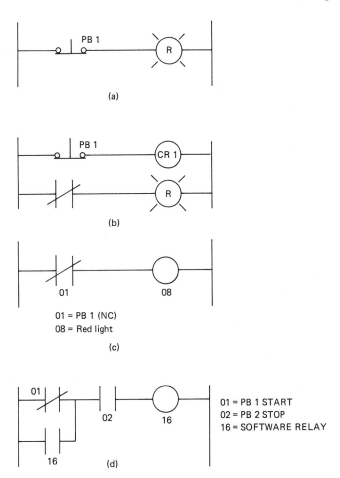

Figure 8.30   Software latch.

switch is pushed. First let us look at the physical interpretation. Figure 8.30a shows that we cannot simply wire the light to the switch. In this case the light will go out (OFF) when the switch is pushed.

Figure 8.30b shows how to provide the answer in a physical system with a control relay, $CR1$. Now, $CR1$ is normally energized so its NC contacts are open and the light will be OFF. When the push-button is pushed, $CR1$ is de-energized and its NC contact, which is open because $CR1$ has been energized, closes and the red light comes ON. So this works.

Now, in the programmed system we do not need either physical or programmed control relay to do this. Figure 8.30c shows that we simply refer to the push-button with an "examine OFF" symbol connected directly to the light. So, if the switch has not been pushed, a test of the symbol shows it to be closed; therefore, it interprets logically as FALSE and the light is not energized by the program. When pushed it is tested to be open, and therefore there is a logic TRUE and the light is energized.

Figure 8.30d shows how a latch can be programmed where both push-button switches are NO. Since switch contact 02 is tested as "examine OFF," the symbol will be TRUE until the button is pushed.

Of course, it is very important to carefully document a programmable controller ladder diagram program. Such documentation shows what physical devices are actually connected to the various input and output addresses.

Time-delay and counters

The operations of time-delay relays and counters also are provided in software. The symbol is the same as a control relay with an appropriate address to tie contacts to the relay. The contacts are simply shown as NO or NC designations. Only the address to the relay coil shows that it is a timer contact. The timer is activated by a condition in a rung that leads to the timer being energized (TRUE). Then the indicated time delay (in seconds, for example) is inserted before the associated contacts change state.

The counter works in a similar fashion, except that it is the counting of events that determines when the associated contacts change state. Thus, if the counter is loaded with the number 20, after being energized 20 times the associated contacts will be changed. Counters can be configured to count up, count down, count from preset values, and be reset before the count is finished.

Example 8.8

Prepare the physical and programmed ladder diagram for the control problem shown in Figure 8.31. The global objective is to heat a liquid to a specified temperature and keep it there for 30 minutes.

The hardware has the following characteristics:

    1. START push-button is NO, STOP is NC.

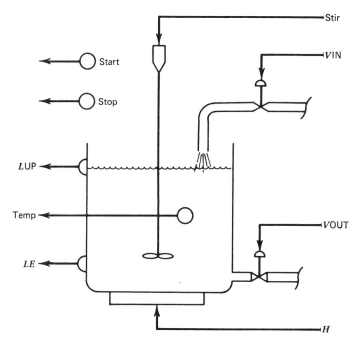

Figure 8.31 Tank system for Example 8.8.

2. NO and NC are available for the limit switches.

The event sequence is

1. Fill the tank.
2. Heat and stir the liquid to the temperature setpoint and hold for 30 minutes.
3. Empty the tank.
4. Repeat from step 1.

Solution   The solution is provided by first constructing the physical ladder diagram. Once this is done, addresses are assigned to all the elements and the ladder diagram program is prepared.

Figure 8.32 shows the six-rung physical ladder diagram. Rung 2 opens the input valve, provided the output valve is not open, until the full level is reached. When the full level is reached, rung 3 turns on the stir provided the output valve is not open. Rung 4 starts a 30-minute timer. The heater is controlled by rung 5. The rung is energized and de-energized as the temperature goes below and above the limit. When the timer times out, the rung is de-energized and rung 6 is energized to open the output valve. The output valve remains open until the empty limit switch opens. The output valve cannot be opened as long as the input valve is open.

The programmed ladder diagram is shown in Figure 8.33. Addresses for the input, output, and internal devices have been assigned. The external switches are not designated as normally open or normally closed but simply as a contact.

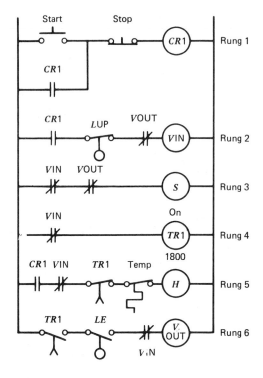

Figure 8.32 Physical ladder diagram for Example 8.8.

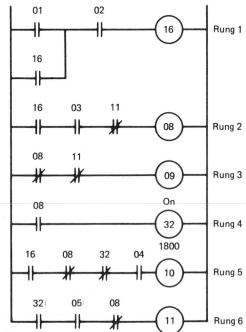

Figure 8.33 Programmed ladder diagram for Example 8.8.

8.5.5 Advanced Features

As the technology of computers and programmable controllers advances, many new features are included, such as computer control of continuous processes, DDC, as a built-in feature of the programmable controller. The ladder diagram includes rungs for such a control system with specification of the appropriate proportional, integral, and derivative gains.

SUMMARY

This chapter has presented the concepts of discrete-state control systems. The topics covered include the following:

1. A discrete-state process is one for which the process variables can take on only two states.

2. A discrete-state process-control system is one that causes the process to pass through a sequence of events. Each event is described by a unique specification of the process variables.

3. The hardware of the process must be carefully defined in terms of the nature of its two states and its relation to the process.

4. The sequence of events can be described in narrative fashion, as a flow-chart, or in terms of Boolean equations.

5. A ladder diagram is a schematic way of describing the sequence of events of a discrete-state control system.

6. A programmable controller is a computer-based device that implements the required sequence of events of a discrete-state process.

PROBLEMS

Section 8.3

8.1 Describe a microwave oven in the framework of a discrete-state process. Define input variables, output variables, and the sequence of serial/parallel events.

8.2 Develop the control system of an automatic coffee-vending machine. Insertion of a coin and pushing of buttons provides a paper cup with coffee that can be black, with sugar, with cream, or with both. Describe the features of the machine as a discrete-state system.

8.3 Develop a flowchart to describe the event sequence for the running phase of the bottle-filling system of Example 8.3 and Figure 8.5.

8.4 Prepare a flowchart of the operations required to support the coffee-vending machine of Problem 8.2.

8.5 Design a state variable solution to the system of Figure 8.5 such as in Example 8.4. Assume another binary input R that is **1** when the system is running.

8.6 Develop Boolean equations to satisfy the requirement of the process of Example 8.4 and Figure 8.11. *Note:* You may need to specify additional hardware.

Section 8.4

8.7 Develop the physical ladder diagram for a motor with the following: NO start button, NC stop button, thermal overload limit switch opens on high temperature, green light when running, red light for thermal overload.

8.8 When turned ON, the tank system of Figure 8.34 alternately fills to level L and then empties to level E. The level switches are activated on a rising level. Both NO and NC connections are available for the level switches and the ON/OFF push-buttons. Prepare a physical ladder diagram for this system.

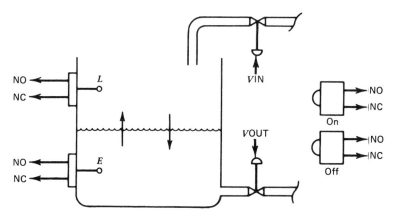

Figure 8.34 System for problems.

8.9 Develop a ladder diagram that provides for the running phase of the process described in Example 8.3 and Figure 8.5.

8.10 Prepare a ladder diagram for the process described in Example 8.4 and Figure 8.11.

8.11 The system of Figure 8.35 has the objective of drilling a hole in an object moved on a conveyor belt. Develop the ladder diagram that accomplishes this objective.

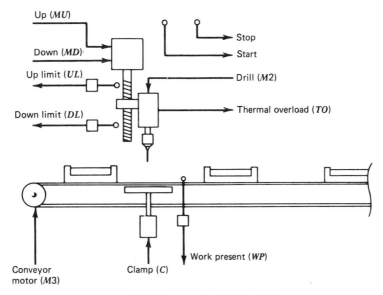

Figure 8.35 Process for Problem 8.11.

Section 8.5

Note: For the following problems, use the programmable controller address given in Section 8.5.4.

8.12 Design a programmable controller program to solve the control problem for the process described in Example 8.4 and Figure 8.11.

8.13 Prepare a programmable controller ladder diagram for Problem 8.7.

8.14 Prepare a programmable controller ladder diagram for Problem 8.8.

8.15 When the system of Figure 8.36 is turned ON, the motor is to alternate rotation CW

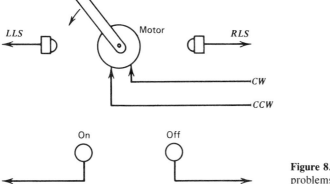

Figure 8.36 Motor process for problems.

and then CCW, cycling, as the shaft extension contacts the two limit switches, RLS and LLS. All four switches have only normally closed positions. Prepare physical and programmed ladder diagrams to solve the problem.

8.16 Prepare a programmed ladder diagram for Problem 8.11.

CHAPTER 9

CONTROLLER PRINCIPLES

INSTRUCTIONAL OBJECTIVES

This chapter presents the *operational* modes of a *process-control loop*. A mode is determined by the nature of controller responses to a controlled variable measurement and setpoint comparison. After you have read this chapter, you should be able to

1. Define process load, process lag, and self-regulation.
2. Describe two-position and floating-control mode.
3. Define the proportional controller mode.
4. Give an example and description of an integral-control mode.
5. Describe the derivative-control mode.
6. Contrast proportional-integral and proportional-derivative control modes.
7. Describe three mode controllers.
8. Provide a description of the controller output for a fixed error input of any of the controller modes.

9.1 INTRODUCTION

In the previous chapter we studied the nature and implementation of controllers for discrete-state control processes. In those processes the operations and variables were in only one of two states: ON or OFF. In the present chapter we consider the nature of controller action for systems with operations and variables that range over continuous values. The controller inputs the result of a measurement of the controlled variable and determines an appropriate output to the final control element. Essentially, the controller is some form of computer, either analog or digital, pneumatic or electronic, that, using input measurements, solves certain equations to calculate the proper output. The equations necessary to obtain control exist in only a few forms, independent both of the process itself and whether the controller function is provided by an analog or digital computer. These equations describe the *modes* or *action* of controller operation. The nature of the process itself and the particular variable controlled determine which mode or modes of control are to be used and the value of certain constants in the mode equations. In this chapter we will study the various modes of controller operation. Later chapters will examine how the modes are implemented by analog or digital means and by pneumatic and electronic means.

9.2 PROCESS CHARACTERISTICS

The selection of what controller modes to use in a process is a function of the characteristics of the process. It is not our intention to discuss how the modes are selected but to define the meaning of each mode. At the same time, it is helpful in understanding the modes if certain pertinent characteristics of the process are considered. In this section we will define a few properties of processes that are important for selecting the proper modes.

9.2.1 Process Equation

A process-control loop regulates some *dynamic variable* in a process. This *controlled* variable, a process parameter, may depend on many other parameters (in the process) and thus suffer changes from many different sources. We have selected one of these other parameters to be our *controlling parameter*. If a measurement of the controlled variable shows a deviation from the setpoint, then the controlling parameter is changed, which in turn changes the controlled variable.

As an example, consider the control of liquid temperature in a tank, as shown in Figure 9.1. The *controlled variable* is the liquid temperature T_L. This temperature depends on many parameters in the process, for example, the input flow rate via pipe A, the output flow rate via pipe B, the ambient temperature T_A, the steam temperature T_S, inlet temperature T_0, and the steam flow rate Q_S. In this case, the steam flow rate is the *controlling parameter* chosen to provide control over the variable (liquid temperature). If one of the other parameters changes, a change in temperature results. To bring the temperature back to the setpoint value, we only change the steam flow rate, that is, heat input to the process. This process could be described by a *process equation* where liquid temperature T_L is a function

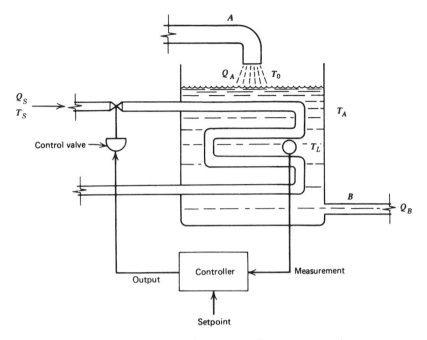

Figure 9.1   Control of temperature by process control.

as

$$T_L = F(Q_A, Q_B, Q_S, T_A, T_S, T_0) \tag{9.1}$$

where

$$Q_A, Q_B = \text{flow rates in pipes } A \text{ and } B$$
$$Q_S = \text{steam flow rate}$$
$$T_A = \text{ambient temperature}$$
$$T_0 = \text{inlet fluid temperature}$$
$$T_S = \text{steam temperature}$$

To provide control via Q_S, we do not need to know the functional relationship exactly, nor do we require linearity of the function. The control loop adjusts Q_S and thereby regulates T_L, regardless of how the other parameters in Equation (9.1) vary with each other. In many cases the relationship of Equation (9.1) is not even analytically known.

9.2.2 Process Load

From the process equation, or knowledge of and experience with the process, it is possible to identify a set of values for the process parameters that results in the controlled variable having the setpoint value. This set of parameters is called the *nominal set*. *Process load* refers to this set of all parameters, *excluding* the controlled variable. When all parameters have their nominal values, we speak of the *nominal load* on the system. The required controlling variable value under these conditions is the *nominal value* of that parameter. If the setpoint is changed, the control parameter is altered to cause the variable to adopt this new operating point. The load is still nominal, however, because the other parameters are assumed unchanged. Suppose one of the parameters changes from nominal, causing a corresponding shift in the controlled variable. We then say that a *process load* change has occurred. The controlling variable is adjusted to compensate for this load change and its effect on the dynamic variable to bring it back to the setpoint. In the example of Figure 9.1, a process load change is due to any change in any of the five parameters affecting liquid temperature. The extent of the load change on the controlled variable is formally determined by process equations such as Equation (9.1). In practice, we are only concerned that variation in the *controlling parameter* brings the controlled variable back to the setpoint. We are not necessarily concerned with the cause, nature, or extent of the load change.

Transient

Another type of change involves a temporary variation of one of the load parameters. After the excursion, the parameter returns to its nominal value. This variation is called a *transient*. A transient causes variations of the controlled variable, and the control system must make equally transient changes of the controlling variable to keep error to a minimum. A transient is not a load change because it is not permanent.

9.2.3 Process Lag

As previously noted, process-control operations are essentially a time-variation problem. At some point in time, a process load change or transient causes a change in the controlled variable. The process-control loop responds to assure, some finite time later, that the variable returns to the setpoint value. Part of this time is consumed by the process itself and is called the *process lag*. Thus, referring to Figure 9.1, assume the inlet flow is suddenly doubled. Such a large process load change radically changes (reduces) the liquid temperature. The control loop responds by opening the steam inlet valve to allow more steam and heat input to bring the liquid temperature back to the setpoint. The loop itself reacts faster than the process. In fact, the physical opening of the control valve is the slowest part of the loop. Once steam is flowing at the new rate, however, the body of liquid must be heated by the steam before the setpoint value is reached again. This time delay or *process lag* in heating is a function of the process and *not* the control system. Clearly, there is no advantage in designing control systems *many* times faster than the process lag.

9.2.4 Self-Regulation

A significant characteristic of some processes is the tendency to adopt a specific value of the controlled variable for nominal load with no control operations. The control operations may be significantly affected by such *self-regulation*. The process of Figure 9.1 has self-regulation, as shown by the following argument.

(1) Suppose we fix the steam valve at 50% and open the control loop so that no changes in valve position are possible. (2) The liquid heats up until the energy carried away by the liquid equals that input energy from the steam flow. (3) If the load changes, a new temperature is adopted (because the system temperature is *not* controlled). (4) The process is *self-regulating*, however, because the temperature will not "run away," but stabilizes at some value under given conditions.

An example of a process *without* self-regulation is tank from which liquid is pumped out at a fixed rate. Assume that the influx just matches the outlet rate. Then the liquid in the tank is fixed at some nominal level. If the influx increases slightly, however, the level rises until the tank overflows. No self-regulation of the level is provided.

9.3 CONTROL SYSTEM PARAMETERS

We have just described the basic characteristics of the process that are related to control. Let us now examine the general properties of the controller shown in Figure 9.2.

To review: (1) inputs to the controller are a measured indication of both the *controlled variable* and a *setpoint* representing the desired value of the variable, expressed in the same fashion as the measurement; (2) the controller output is a signal representing action to be taken when the measured value of the controlled variable deviates from the setpoint.

The measured indication of a variable is denoted by b, while the actual variable is denoted by c. Thus if a sensor measures temperature by conversion to resistance, the actual variable is temperature in degrees Celsius but the measured indication is resistance in ohms.

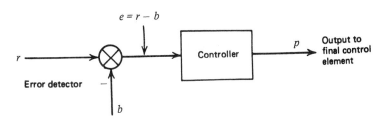

Figure 9.2   The error detector and controller block diagram.

Further conversion may be performed by transducers or transmitters to provide a current in mA, for example. In such a case, the current becomes the measured indication of the variable.

9.3.1 Error

The deviation or error of the controlled variable from the setpoint is given by

$$e = r - b \tag{9.2}$$

where

e = error
b = measured indication of variable
r = setpoint of variable (reference)

Equation (9.2) expresses error in an absolute sense, or in units of the measured analog of the control signal. Thus, if the setpoint in a 4–20 mA range corresponds to 9.9 mA and the measured value is 10.7 mA, we have an error of -0.8 mA. Obviously, this current error has little direct meaning unless related to the controlled variable. We could work back through the loop and prove that it corresponds to flow rate of 1.1 m^3/hr, for example. This would show the significance of the error relative to the actual process-control loop.

To describe controller operation in a general way, it is better to express the error as percent of the measured variable range (i.e., the span). The measured value of a variable can be expressed as percent of span over a range of measurement by the equation

$$c_p = \frac{c - c_{min}}{c_{max} - c_{min}} \cdot 100$$

where

c_p = measured value as percent of measurement range
c = actual measured value
c_{max} = maximum of measured value
c_{min} = minimum of measured value

The previous equation is in terms of the actual measured variable c, but the same equation can be expressed in terms of the measured indication b. It is only necessary to translate the measured minimum and maximum to b_{max} and b_{min}.

To express error as percent of span, it is only necessary to write both the setpoint and measurement in terms of percent of span and take the difference as per Equation (9.2). The result is

$$e_p = \frac{r - b}{b_{max} - b_{min}} \cdot 100 \tag{9.3}$$

where

e_p = error expressed as percent of span

You can see the convenience of using a standard measured indication range like 4–20 mA because the span is always 16 mA. Suppose we have a setpoint of 10.5 mA and a measurement of 13.7 mA. Then, without even knowing what is being measured, we know the error is

$$e_p = \frac{10.5 \text{ mA} - 13.7 \text{ mA}}{20 \text{ mA} - 4 \text{ mA}} \cdot 100$$

$$e_p = -20\%$$

Example 9.1

The temperature in Figure 9.1 has a range of 300 to 440 K and a setpoint of 384 K. Find the percent of span error when the temperature is 379 K.

Solution   The percent error is

$$e_p = \frac{r - b}{b_{\max} - b_{\min}} \cdot 100$$

$$e_p = \frac{384 - 379}{440 - 300} \cdot 100 \tag{9.3}$$

$$e_p = \pm 3.6\%$$

A *positive* error indicates a measurement *below* the setpoint, and a *negative* error indicates a measurement *above* the setpoint.

9.3.2 Variable Range

Generally, the variable under control has a range of values within which control is to be maintained. This range can be expressed as the minimum and maximum value of the variable or the nominal value plus and minus the spread about this nominal. If a standard 4–20 mA signal transmission is employed, then 4 mA represents the minimum value of the variable and 20 mA the maximum.

When a computer-based control system is used, the dynamic variable is converted to an *n*-bit digital signal. Often, the transformation is made so that all 0's are the minimum value of the variable and all 1's are the maximum value.

9.3.3 Control Parameter Range

Another range is associated with the *controller output*. Here we assume the final control element has some minimum and maximum effect on the process. The controller output range is the translation of output to the range of possible values of the final control element. This range also is expressed as the 4–20 mA standard signal again with the minimum and maximum effects in terms of the minimum and maximum current.

Similarly, in computer-based control, the output will range over all states of the *n*-bit output. Generally, all 0's are the minimum output and all 1's the maximum. These numbers do not necessarily represent the minimum and maximum of the final control element, however. We may never wish a valve to be fully closed, for example; therefore, all 0's might represent some percentage of full open.

Often, the output is expressed as a percentage where 0% is the minimum controller output and 100% the maximum (obviously). Thus, in the example of Figure 9.1, the valve in the fully open position corresponds to a 100% controller signal output. Very often the minimum does *not* correspond to zero effect. For example, it may be that the steam flow should never be less than that flow with the valve half open. In this case, a 0% minimum controller corresponds to the flow rate with a half-open valve.

The controller output as a percent of full scale when the output varies between specified limits is given by

$$p = \frac{u - u_{min}}{u_{max} - u_{min}} \cdot 100 \tag{9.4}$$

where

p = controller output as percent of full scale
u = value of the output
u_{max} = maximum value of controlling parameter
u_{min} = minimum value of controlling parameter

Example 9.2

A controller outputs a 4–20 mA signal to control motor speed from 140–600 rpm with a linear dependence. Calculate (a) current corresponding to 310 rpm and (b) the value of (a) expressed as the percent of control output.

Solution
(a) We find the slope m and intersect S_0 of the linear relation between current I and speed S, where

$$S_p = mI + S_0$$

by knowing S_p and I at the two given positions, we write two equations:

$$140 = 4m + S_0$$

$$600 = 20m + S_0$$

Solving these simultaneous equations, we get m = 28.75 rpm/mA and S_0 = 25 rpm. Thus, at 310 rpm we have 310 = 28.75I + 25, which gives I = 9.91 mA.

(b) Expressed as a percentage of the 4–20 mA range, this controller output is

$$p = \frac{u - u_{min}}{u_{max} - u_{min}} \cdot 100$$

$$= \left[\frac{9.91 - 4}{20 - 4} \right] \cdot 100 \qquad (9.4)$$

$$p = \mathbf{36.9\%}$$

9.3.4 Control Lag

The control system also has a lag associated with its operation that must be compared to the process lag (Section 9.2.3). When a controlled variable experiences a sudden change, the process-control loop reacts by outputting a command to the final control element to adopt a new value to compensate for the detected change. *Control lag* refers to the time for the process-control loop to make necessary adjustments to the final control element. Thus, in Figure 9.1, if a sudden change in liquid temperature occurs, it requires some finite time for the control system to physically actuate the steam control valve.

9.3.5 Dead Time

Another time variable associated with process control is both a function of the process-control system and the process. This is the elapsed time between the instant a deviation (error) occurs and the corrective action first occurs. An example of *dead time* occurs in the control of a chemical reaction by varying reactant flow rate through a very long pipe. When a deviation is detected, a control system quickly changes a valve setting to adjust flow rate. But if the pipe is quite long, there is a period of time during which no effect is felt in the reaction vessel. This is the time required for the new flow rate to move down the length of the pipe. Such *dead times* can have a very profound effect on the performance of control operations on a process.

9.3.6 Cycling

We frequently refer to the behavior of the dynamic variable error under various modes of control. One of the most important modes is an *oscillation* of the error about zero. This means the variable is *cycling* above and below the setpoint value. Such cycling may continue indefinitely, in which case we have *steady-state cycling*. Here we are interested in both the *peak amplitude* of the *error* and the *period* of the *oscillation*.

    If the cycling amplitude decays to zero, however, we have a *cyclic transient error*. Here we are interested in the *initial error*, the *period* of the *cyclic oscillation*, and *decay time* for the error to reach zero.

9.3.7 Controller Modes

The controller was defined (Section 9.3) by the statement that a controller generates a control signal to the final element, based on a measured deviation of the controlled variable from the setpoint. It is natural to ask how the controller responds to the deviation. In a thermostatically controlled temperature system used in the home, the controller response is simple. If the temperature drops below the thermostat setpoint, a bimetallic relay turns on a heater.

But consider the case of the system shown in Figure 9.1. No simple ON-OFF decision can be made because the setting of the steam valve can be smoothly varied from one extreme to another. Thus, if a deviation from liquid temperature setpoint occurs, what should the controller do? Should it open the valve a little or a lot? Should it open the valve fast or slow?

These questions are answered by specifying the *mode* of the controller operation. One distinction is clear from the earlier examples. The domestic thermostat involves a mode that is *discontinuous*, where the controller command initiates a discontinuous change in the control parameter. The process of Figure 9.1 is *continuous* because smooth variation of the control parameter is possible. Section 9.4 covers various controller modes in detail.

The choice of operating mode for any given process-control system is a complicated decision. It involves not only process characteristics but cost analysis, product rate, and other industrial factors. At the outset, the process-control technologist should have a good understanding of the operational mechanism of each mode and its advantages and disadvantages. The operation of each mode is defined later, and examples are given with some general statements of application details. In each case, the output of the controller is described by a factor p. This is the *percent of controller output* relative to its total range as defined in Equation (9.4).

For example, if a controller outputs a 4–20 mA current signal to the final control element and has a $p = 25\%$, then the corresponding current is

$$I = I_{min} + p(I_{max} - I_{min})$$

$$I = 4 \text{ mA} + (0.25)(20 - 4) \text{ mA}$$

$$I = 8 \text{ mA}$$

If this current is used to drive a value actuator for which 4 mA is closed and 20 mA full open, then the valve is 25% open. If the valve is an equal percentage type with a rangeability of 30, then the flow rate from Section 7.5.3 is

$$Q = Q_{min}R^{S/S_{max}}$$

$$Q = Q_{min}(30)^{0.25} \tag{7.11}$$

$$Q = 2.34Q_{min}$$

The example shows that if the percentage output of the controller is known, then the actual value of the controlled variable can be determined.

The input of the controller is described by the error e_p, defined in Equation (9.3) as the percentage error of measured variable from the setpoint relative to range. In general, the controller operation is expressed as a relation:

$$p = F(e_p) \tag{9.5}$$

where $F(e_p)$ represents the relation by which the appropriate controller output is determined. In some cases, a graph of p versus e_p is also employed to aid in a definition of the control mode.

Reverse and direct action

The error that results from measurement of the controlled variable may be positive or negative, because the value may be greater or less than the setpoint. How this polarity of the error changes the controller output can be selected according to the nature of the process.

A controller operates with *direct action* when an increasing value of the controlled variable causes an increasing value of the controller output. An example would be a level-control system which outputs a signal to an output valve. Clearly, if the level rises (increases), the valve should be opened (i.e., its drive signal should be increased).

Reverse action is the opposite case, where an increase in controlled variable causes a decrease in controller output. An example of this would be a simple temperature control from a heater. If the temperature increases, the drive to the heater should be decreased.

9.4 DISCONTINUOUS CONTROLLER MODES

This section discusses the various controller modes that show discontinuous changes in controller output as controlled variable error occurs. It is important that you understand these modes, both because of their frequent use in process control, and because they form the basis of the continuous modes to be discussed in the next section.

9.4.1 Two-Position Mode

The most elementary controller mode is the ON-OFF or two-position mode. This is an example of a *discontinuous* mode. It is the simplest and the cheapest, and often suffices when its disadvantages are tolerable. Although an analytic equation cannot be written, we can, in general, write

$$p = \begin{cases} 0\% & e_p < 0 \\ 100\% & e_p > 0 \end{cases} \tag{9.6}$$

This relation shows that when the measured value is less than the setpoint, *full* controller output results. When it is more than the setpoint, the controller output is zero. A space heater is a common example. If the temperature drops below a setpoint, the heater is turned ON. If above the setpoint, it turns OFF.

Neutral zone

In virtually any practical implementation of the two-position controller, there is an overlap as e_p increases through zero or decreases through zero. In this span, *no change* in controller output occurs. This is best shown in Figure 9.3, which plots p versus e_p for a two-position controller. We see that until an increasing error changes by Δe_p *above zero*, the controller output will not change state. In decreasing, it must fall Δe_p *below zero* before the controller changes to the 0% rating. The range $2\Delta e_p$, which is referred to as the *neutral zone* or *differential gap*, is often purposely designed above a certain minimum quantity to prevent excessive cycling. The existence of such a neutral zone is an example of desirable hysteresis in a system.

Example 9.3

A liquid-level control system linearly converts a displacement of 2–3 meters into a 4–20 mA control signal. A relay serves as the two-position controller to open or close an inlet valve. The relay closes at 12 mA and opens at 10 mA. Find (a) the relation between displacement level and current, and (b) the neutral zone or displacement gap in meters.

Solution

(a) The relation between level and a current is a linear equation such as

$$H = KI + H_0$$

We find K and H_0 by writing two equations

$$2 \text{ m} = K(4 \text{ mA}) + H_0$$

$$3 \text{ m} = K(20 \text{ mA}) + H_0$$

Solving these simultaneous equations yields $K = 0.0625$ m/mA and $H_0 = 1.75$ m, at the intersection of the linear relations.

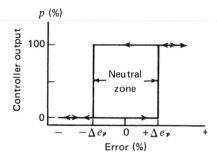

Figure 9.3 Two-position controller action with neutral zone.

(b) The relay closes at 12 mA, which is a high level H_H of

$$H_H = (0.0625 \text{ m/mA})(12 \text{ mA}) + 1.75 \text{ m}$$

$$H_H = 2.5 \text{ m}$$

The low level H_L occurs at 10 mA, which is

$$H_L = (0.0625 \text{ m/mA})(10 \text{ mA}) + 1.75 \text{ m}$$

$$H_L = 2.375 \text{ m}$$

Thus, the neutral zone is $H_H - H_L = (2.5 - 2.375)$ m or **0.125 m**.

Applications

Generally, the two-position control mode is best adapted to large-scale systems with relatively *slow* process rates. Thus, in the example of either a room heating or air-conditioning system, the capacity of the system is very large in terms of air volume, and the overall effect of the heater or cooler is relatively slow. Sudden large-scale changes are not common to such systems. Other examples of two-position control applications are liquid bath temperature control and level control in large volume tanks. The process under two-position control must allow continued oscillation in the controlled variable because, by its very nature, this mode of control always produces such oscillation. For large systems, these oscillations are of long duration, which is partly a function of the neutral-zone size. To illustrate this, consider the following example.

Example 9.4

As a water tank loses heat, the temperature drops by 2 K per minute. When a heater is on, the system gains temperature at 4 K per minute. A two-position controller has a 0.5-minute control lag and a neutral zone of $\pm 4\%$ of the setpoint about a setpoint of 323 K. Plot the heater temperature versus time. Find the oscillation period.

Solution   Let us assume we start at the setpoint value; then the temperature will drop linearly at

$$T_1(t) = T(t_s) - 2(t - t_s) \tag{9.7}$$

where t_s = time at which we start observation. The heater will start at a temperature of 310 K (4% below setpoint), after which the temperature will rise according to

$$T_2(t) = T(t_h) + 4(t - t_h) \tag{9.8}$$

where t_h = time at which heater goes on. When the temperature reaches 336 K, the heater goes off and the system temperature drops by 2 K/min until 310 K is reached. The system response is then plotted as in Figure 9.4, using Equations (9.7) and (9.8). Notice the period is 21.5 minutes. There is also a 1-K undershoot and a $+2$-K overshoot because of the lag.

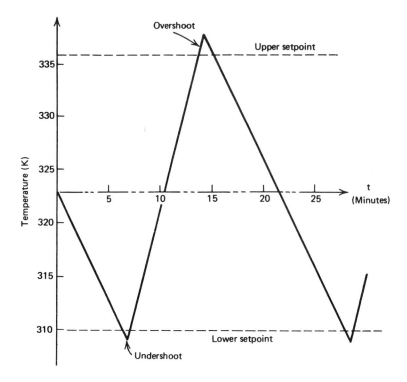

Figure 9.4   Figure for Example 9.4.

In general, some overshoot and undershoot of the controlled variable will occur as in Example 9.4. This is due to the finite time required for the control element to impress its full effect on the process. Thus, the finite warmup time and cooloff time of the heater (included in Example 9.4) caused some overshoot and undershoot of temperature. In some cases, if the final control element lag is large, substantial errors can result, and the neutral zone must be reduced to reduce these errors. In general, the cycling, as noted in the previous example, is a function of the neutral zone. If the neutral zone of this example is reduced to ±2%, the reader can verify that, although the control is now maintained tighter, the cycling period is reduced to 11.91 min. The solution of these values is left as a problem for the reader.

9.4.2 Multiposition Mode

A logical extension of the previous two-position control mode is to provide *several intermediate* rather than only two settings of the controller output. This discontinuous control mode is used in an attempt to reduce the cycling behavior and overshoot and undershoot inherent in the two-position mode. In fact, however, it is usually more expedient to use some other mode when the two-position is not satisfactory. This mode is represented by

$$p = p_i \qquad e_p > |e_i| \quad i = 1, 2, \cdots n \qquad (9.9)$$

As the error exceeds certain set limits $\pm e_i$, the controller output is adjusted to preset values p_i. The most common example is the three-position controller where

$$p = \begin{cases} 100 & e_p > e_2 \\ 50 & -e_1 < e_p < e_2 \\ 0 & e_p < -e_1 \end{cases} \qquad (9.10)$$

As long as the error is between e_2 and e_1 of the setpoint, the controller stays at some nominal setting indicated by a controller output 50%. If the error exceeds the setpoint by e_1 or more, then the output is increased by 100%. If it is less than the setpoint by $-e_1$ or more, the controller output is reduced to zero.

Figure 9.5 illustrates this mode graphically. Some small neutral zone usually exists about the change points, but not by design and thus they are not shown. This type of control mode usually requires a more complicated final control element because it must have more than two settings. Figure 9.6 shows a graph of dynamic variable and final control element setting versus time for a hypothetical case of three-position control. Note the change in control element setting as the variable changes about the two trip points. On this graph the finite time required for the final control element to change from one position to another also is shown. Notice the overshoot and undershoot of the error around the upper and lower setpoints. This is due to both the process lag time and the controller lag time indicated by the finite time required for the control element to reach a new setting.

9.4.3 Floating-Control Mode

In the two previous modes of controller action, the output was uniquely determined by the magnitude of the error input. If the error exceeded some preset limit, the output was changed to a new setting as quickly as possible. In floating

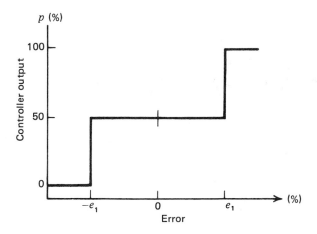

Figure 9.5 Three-position controller action.

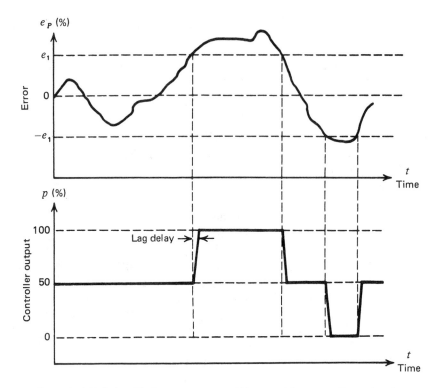

Figure 9.6 Relationship between error and three-position controller action, including the effects of lag.

control, the specific output of the controller is *not* uniquely determined by the error. If the error is zero, the output will not change but remains (floats) at whatever setting it was when the error went to zero. When the error moves off zero, the controller output again begins to change. Actually, as with the two-position mode, there is typically a neutral zone about zero error where no change in controller position occurs.

Single speed

In the single-speed floating-control mode, the output of the control element changes at a fixed rate when the error exceeds the neutral zone. An equation for this action is

$$\frac{dp}{dt} = \pm K_F \qquad |e_p| > \Delta e_p \qquad (9.11)$$

where

$\dfrac{dp}{dt}$ = rate of change of controller output with time

K_F = rate constant (%/s)

Δe_p = half the neutral zone

If Equation (9.11) is integrated for the actual controller output, we get

$$p = \pm K_F t + p(0) \qquad | e_p | > \Delta e_p \qquad (9.12)$$

where

$p(0)$ = controller output at $t = 0$

which shows that the present output depends on the time history of errors that have previously occurred. Because such a history usually is not known, the actual value of p floats at an *undetermined* value. If the deviation persists, then Equation (9.11) shows that the controller saturates at either 100% or 0% and remains there until an error drives it toward the opposite extreme. A graph of single-speed floating control is shown in Figure 9.7a.

Example 9.5

Suppose a process error lies within the neutral zone with $p = 25\%$. At $t = 0$, the error falls *below* the neutral zone. If $K = +2\%$ per second, find the time when the output saturates.

Solution   The relation between controller output and time is

$$p = K_F t + p(0) \qquad (9.12)$$

when $p = 100$

$$100\% = (2\%/s)(t) + 25\%$$

that, when solved for t, yields

$$t = \textbf{37.5 s}$$

In Figure 9.7b a graph shows controller output versus time and error versus time for a hypothetical case illustrating typical operation. In this example, we assume the controller is reverse acting, which means the controller output decreases when the error exceeds the neutral zone. This corresponds to a negative K_F in Equation (9.11). Most controllers can be adjusted to act in either the reverse or direct mode. Here the controller starts at some output $p(0)$. At time t_1, the error exceeds the neutral zone. The controller output decreases at a constant rate until t_2 when the error *again* falls below the neutral zone limit. At t_3, the error falls below the lower limit of the neutral zone, causing controller output to change until the error again moves within the allowable band.

Multiple speed

In the floating multiple-speed control mode, not one but several possible speeds (rates) are changed by controller output. Usually, the rate increases as the de-

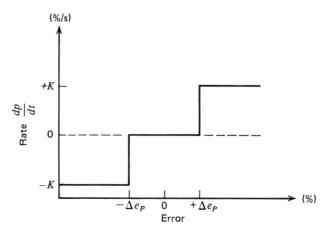

a) Single-speed floating controller action. The ordinate is the rate of change of controller output with time

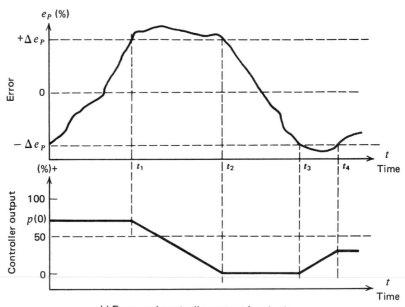

b) Error and controller output for single-speed floating action.

Figure 9.7　Single-speed floating controller.

viation exceeds certain limits. Thus, if we have certain speed change points e_{pi} depending on the error, then each has its corresponding output rate change K_i. We can then say

$$\frac{dp}{dt} = \pm K_{Fi} \qquad |e_p| > e_{pi} \tag{9.13}$$

If the error exceeds e_{pi}, then the speed is K_{Fi}. If the error rises to exceed e_{p2}, the speed is increased to K_{F2}, and so on. Actually, this mode is a discontinuous attempt to realize an integral mode (Section 9.4.4). A graph of this mode is shown in Figure 9.8.

Applications

Primary applications of the floating-control mode are for the single-speed controllers with a neutral zone. This mode has an inherent cycle nature much like the two-position, although this cycling can be minimized depending on the application. Generally, the method is well suited to self-regulation processes with very small lag or dead time, which implies small-capacity processes. When used with large-capacity systems, the inevitable cycling must be considered.

An example of single-speed floating control is a liquid flow rate through a control valve. Such a system is shown in Figure 9.9. The load is determined by the inlet and outlet pressures P_{in} and P_{out}, and the flow is determined in part by the pressure P within the DP cell and control valve. This is an example of a system with self-regulation. We assume some valve opening has been found commensurate with the desired flow rate. If the load changes (either P_{in} or P_{out}), then an

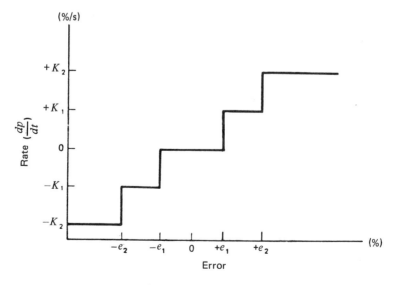

Figure 9.8 Multiple-speed floating-control mode action.

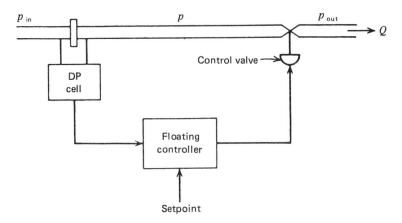

Figure 9.9   Single-speed floating-control action applied to a flow-control system.

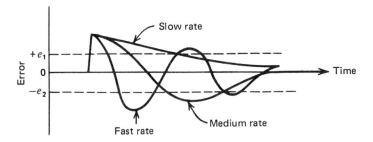

Figure 9.10   The rate of controller output change has a strong effect on error recovery in a floating controller.

error occurs. If larger than the neutral zone, the valve begins to open or close at a constant rate until an opening is found that supports the proper flow rate at the new load conditions. Clearly, the rate is very important because very fast process lags cause the valve to continue opening (or closing) beyond that optimum self-regulated position. This is shown in Figure 9.10, where the response to a sudden deviation is shown for various floating rates.

9.5 CONTINUOUS CONTROLLER MODES

The most common controller action used in process control is one or a combination of continuous controller modes. In these modes, the output of the controller changes smoothly in response to the error or rate of change of error. These modes are an extension of the discontinuous types discussed in the previous section.

9.5.1 Proportional Control Mode

The two-position mode had the controller output of either 100% or 0% depending on the error being greater or less than the neutral zone. In multiple-step modes, more divisions of controller outputs versus error are developed. The natural extension of this concept is the *proportional mode*, where a smooth, linear relationship exists between the controller output and the error. Thus, over some range of errors about the setpoint, each value of error has a unique value of controller output in one-to-one correspondence. The range of error to cover the 0% to 100% controller output is called the *proportional band* because the one-to-one correspondence exists only for errors in this range. This mode can be expressed by

$$p = K_P e_p + p_0 \qquad (9.14)$$

where

K_P = proportional gain between error and controller output (% per %)
p_0 = controller output with no error (%)

Direct and reverse action

Recall that the error in Equation (9.14) is expressed using the difference between setpoint and the measurement, $r - b$. This means that as the measured value increases above the setpoint, the error will be negative and the output will decrease. That is, the term $K_P e_p$ will subtract from p_0. Thus, Equation (9.14) represents reverse action. Direct action would be provided by putting a negative sign in front of the correction term.

A plot of the proportional mode output versus error for Equation (9.14) is shown in Figure 9.11. In this case, p_0 has been set to 50% and two different gains have been used. Note that the proportional band is dependent on the gain. A high gain means large response to an error but also a narrow error band within which the output is not saturated.

In general, the proportional band is defined by the equation

$$PB = \frac{100}{K_P} \qquad (9.15)$$

Let us summarize the characteristics of the proportional mode and Equation (9.14).

1. If the error is zero, the output is a constant equal to p_0.
2. If there is error, for every 1% of error a correction of K_P percent is added to or subtracted from p_0, depending on the reverse or direct action of the controller.
3. There is a band of error about zero of magnitude PB within which the output is not saturated at 0% or 100%.

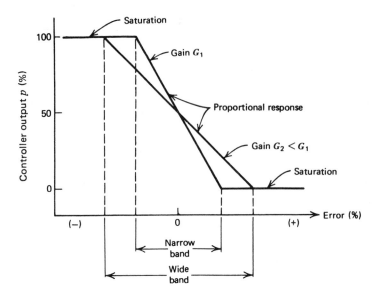

Figure 9.11 The proportional band of a proportional controller depends on the gain, in an inverse fashion.

Offset

An important characteristic of the proportional control mode is that it produces a permanent *residual error* in the operating point of the controlled variable when a change in load occurs. This error is referred to as *offset*. It can be minimized by a larger constant K_P, which also reduces the proportional band. To see how offset occurs, consider a system under nominal load with the controller at 50% and the error zero, as shown in Figure 9.12. If a transient error occurs, the system responds by changing controller output in correspondence with the transient to effect a return to zero error. Suppose, however, a load change occurs that requires a permanent change in controller output to produce the zero error state. Because a one-to-one correspondence exists between controller output and error, it is clear that a new, zero error controller output can *never* be achieved. Instead, the system produces a small permanent offset in reaching a compromise position of controller output under new loads.

Example 9.6

Consider the proportional mode level-control system of Figure 9.13. Valve A is linear with a flow scale factor of 10 m³/hr per percent controller output. The controller output is nominally 50% with a constant of $K_P = 10\%$ per %. A load change occurs when flow through valve B changes from 500 m³/hr to 600 m³/hr. Calculate the new controller output and offset error.

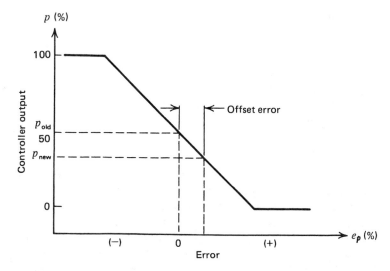

Figure 9.12   An offset error must occur if a proportional controller requires a new nominal controller output following a load change.

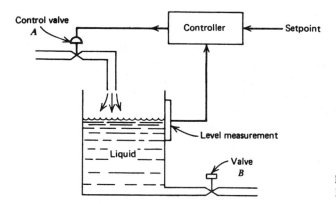

Figure 9.13   Level-control system for Example 9.6.

Solution   Certainly, valve A must move to a new position of 600 m³/hr flow or the tank will empty. This can be accomplished by a 60% new controller output because

$$Q_A = \left(10 \, \frac{\text{m}^3/\text{hr}}{\%} \right) (60\%) = 600 \, \text{m}^3/\text{hr}$$

as required. Because this is a proportional controller, we have

$$p = K_P e_p + p_0 \tag{9.14}$$

with the nominal condition $p_0 = 50\%$. Thus

$$e_p = \frac{p - p_0}{K_P} = \frac{60 - 50}{10} \%$$

$$e_p = 1\%$$

so that a 1% offset error occurred because of the load change.

Application

The offset error limits use of the proportional mode to only a few cases, particularly those where a manual reset of the operating point is possible to eliminate offset. Proportional control generally is used in processes where large load changes are unlikely or with moderate to small process lag times. Thus, if the process lag time is small, the proportional band can be made very small (large K_P), which reduces offset error. Figure 9.11 shows that if K_P is made very large, the *PB* becomes very small and the proportional mode acts just like an ON/OFF mode. Remember that the ON/OFF mode exhibited oscillations about the setpoint. From these statements it is clear that for high gain the proportional mode causes oscillations of the error.

9.5.2 Integral Control Mode

This mode represents a natural extension of the principle of floating control in the limit of infinitesimal changes in the rate of controller output with infinitesimal changes in error. Instead of single speed or even multiple speeds, we have a continuous change in speeds depending on error. This mode is often referred to as *reset action*. Analytically, we can write

$$\frac{dp}{dt} = K_I e_p \tag{9.16}$$

where

$\dfrac{dp}{dt}$ = rate of controller output change (%/s)

K_I = constant relating the rate to the error ((%/s)/%)

In some cases the inverse of K_I, called the integral time $T_I = 1/K_I$, expressed in seconds or minutes, is used to describe the integral mode.

If we integrate Equation (9.16), we can find the actual controller output at any time as

$$p(t) = K_I \int_0^t e_p(t)dt + p(0) \tag{9.17}$$

where

$$p(0) = \text{the controller output at } t = 0$$

This equation shows that the present controller output $p(t)$ depends on the history of errors from when observation started at $t = 0$. We see from Equation (9.16), for example, that if the error doubles, the rate of controller output change also doubles. The constant K_I expresses the scaling between error and controller output. Thus, a large value of K_I means that a *small* error produces a *large* rate of change of p and vice versa. Figure 9.14a graphically illustrates the relationship between the p rate of change and error for two different values of K_I. Figure 9.14b shows how, for a fixed error, the different K_I values produce different values

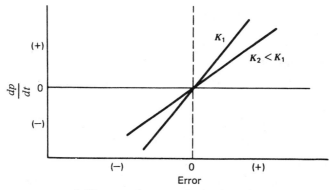

a) The rate of output change depends on gain and error

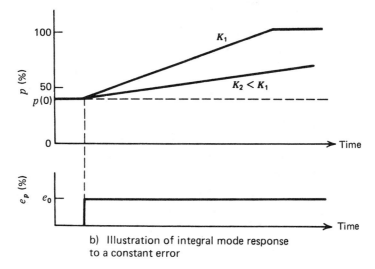

b) Illustration of integral mode response to a constant error

Figure 9.14 Integral controller mode action.

of p as a function of time as predicted by Equation (9.17). Thus, we see that the *faster* rate provided by K_I causes a *much greater* controller output at a particular time after the error is generated.

Let us summarize the characteristics of the integral mode and Equation (9.17).

1. If the error is zero, the output stays fixed at a value equal to what it was when the error went to zero.
2. If the error is not zero, the output will begin to increase or decrease at a rate of K_I percent/second for every 1% of error.

Area accumulation

From calculus we learn that an integral determines the area of the function being integrated. Thus, Equation (9.17) can be interpreted as providing a controller output equal to the net area under the *error-time* curve multiplied by K_I. We often say that the integral term *accumulates* error as a function of time. Thus, for every $1\% - s$ of accumulated error-time area, the output will be K_I percent.

Example 9.7

An integral controller is used for speed control with a setpoint of 12 rpm within a range of 10–15 rpm. The controller output is 22% initially. The constant $K_I = -0.15\%$ controller output per second per percentage error. If the speed jumps to 13.5 rpm, calculate the controller output after 2 seconds for a constant e_p.

Solution   We find e_p from

$$e_p = \frac{r - b}{b_{\max} - b_{\min}} \times 100$$

$$e_p = \frac{12 - 13.5}{15 - 10} \times 100 \tag{9.3}$$

$$e_p = -30\%$$

The rate of controller output change is then

$$\frac{dp}{dt} = K_I e_p = (-0.15 \text{ s}^{-1})(-30\%)$$

$$\frac{dp}{dt} = 4.5\%/s \tag{9.16}$$

The controller output for constant error will be

$$p = K_I \int_0^t e_p dt + p(0) \tag{9.17}$$

but because e_p is constant

$$p = K_I e_p t + p(0)$$

After 2 seconds we have

$$P = (0.15)(30\%)(2) + 22$$

$$P = \mathbf{31\%}$$

The integral controller constant K_I may be expressed in percentage change per *minute* per percentage error, whenever a typical process-control loop has characteristic response time in minutes rather than seconds. Thus, an integral mode controller with reset action at 5.7 minutes means that K_I for our equations would be

$$K_I = \frac{1}{(5.7 \text{ min})(60 \text{ s/min})}$$

$$K_I = 2.92 \times 10^{-3} \text{ s}^{-1}$$

Applications

Use of the integral mode is shown by the flow control system in Figure 9.9, except that we now assume that the controller operates in the *integral* mode. Operation can be understood using Figure 9.15. A load change induced error occurs at some time t_1. The proper valve position under the new load to maintain the constant flow rate is shown as a dashed line in the p graph of Figure 9.15. In the integral mode, the value initially begins to change very rapidly, as predicted by Equation (9.16). As the valve opens, the error decreases and slows the valve opening rate as shown. The ultimate effect is that the system drives the error to zero at a slowing controller rate. The effect of process and control system lag is shown as simple delays in the controller output change and in the error reduction when the controller action occurs. If the process lags are too large, the error can oscillate about zero or even be cyclic. Typically, the integral mode is not used alone but can be for systems with small process lags and correspondingly small capacities.

9.5.3 Derivative Control Mode

The last *pure* mode of controller operation provides that the controller output depends on the rate of change of *error*. This mode also is known as *rate* or *anticipatory* control. The mode *cannot* be used alone because when the error is *zero or constant*, the controller has no output. The analytic expression is

$$p = K_D \frac{de_p}{dt} \tag{9.18}$$

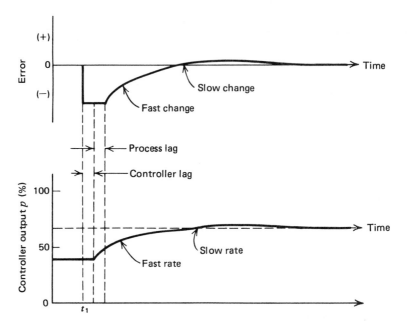

Figure 9.15 Illustration of integral mode controller output and error showing the effect of process and control lag.

where

K_D = derivative gain constant $(\% - \text{s}/\%)$

$\dfrac{de_p}{dt}$ = rate of change of error $(\%/\text{s})$

The derivative gain constant also is called the rate or derivative time and is commonly expressed in minutes. The characteristics of this device can be noted from the graph of Figure 9.16, which shows controller output for the rate of change of error. This shows that, for a given rate of change of error, there is a unique value of controller output. The time plot of error and controller response further shows the behavior of this mode, as shown in Figure 9.17. The extent of controller output depends on the rate at which this error is changed and *not* on the value of the error.

Let us summarize the characteristics of the derivative mode and Equation (9.18).

1. If the error is zero, the mode provides no output.
2. If the error is constant in time, the mode provides no output.
3. If the error is changing in time, the mode contributes an output of K_D percent for every 1% per second rate of change of error.

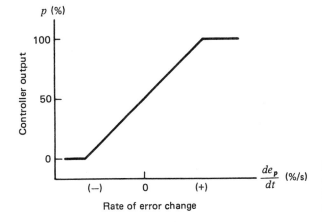

Figure 9.16 Derivative mode of controller action where an output of 50% has been assumed for the zero derivative state.

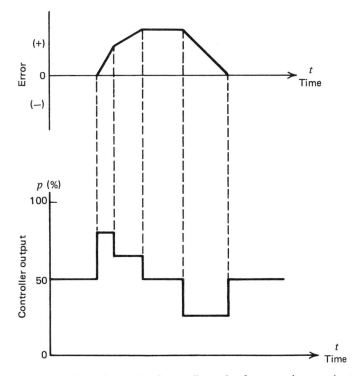

Figure 9.17 Derivative mode of controller action for a sample error signal.

4. For direct action, a positive rate of change of error produces a positive derivative mode output.

9.6 COMPOSITE CONTROL MODES

It is very common in the complex of industrial processes to find control requirements that do *not* fit the application norms of any of the previously considered controller modes. It is both possible and expedient to *combine* several basic modes, thereby gaining the advantages of each mode. In some cases, an added advantage is that the modes tend to eliminate some limitations they individually possess. We will consider only those combinations that are commonly used and discuss the merits of each mode.

9.6.1 Proportional-Integral Control (PI)

This is a control mode that results from a combination of the proportional mode and the integral mode. The analytic expression for this control process is found from a series combination of Equations (9.14) and (9.17)

$$p = K_P e_p + K_P K_I \int_0^t e_p dt + p_I(0) \tag{9.19}$$

where $p_I(0)$ = integral term value at $t = 0$ (initial value).

The main advantage of this composite control mode is that the one-to-one correspondence of the proportional mode is available and the integral mode eliminates the inherent offset. Notice that the proportional gain, by design, also changes the net integration mode gain, but that the integration gain, through K_I, can be independently adjusted. Recall that the proportional mode offset occurred when a load change required a new nominal controller output and could not be provided except by a fixed error from the setpoint. In the present mode, the integral function provides the required new controller output, thereby allowing the error to be zero after a load change. The integral feature effectively provides a *reset* of the zero error output after a load change occurs. This can be seen by the graphs of Figure 9.18. At time t_1 a load change occurs that produces the error shown. Accommodation of the new load condition requires a new controller output. We see that the controller output is provided through a sum of proportional plus integral action that finally leaves the error at zero. The proportional part is obviously just an image of the error.

Let us summarize the characteristics of the PI mode and Equation (9.19).

1. When the error is zero, the controller output is fixed at the value that the integral term had when the error went to zero. This output is given by $p_I(0)$ in Equation (9.19) simply because we chose to define the time at which observation starts as $t = 0$.

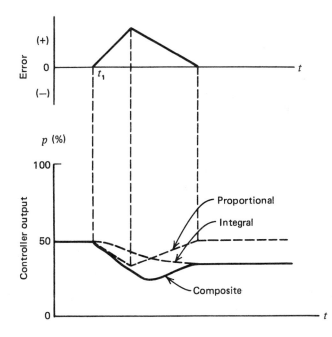

Figure 9.18 Proportional-integral (PI) action showing the reset action of the integral contribution (reverse action).

2. If the error is not zero, the proportional term contributes a correction and the integral term begins to increase or decrease the accumulated value [initially $p_I(0)$], depending on the sign of the error and the direct or reverse action.

The integral term cannot become negative. Thus, it will saturate at zero if the error and action try to drive the area to a net negative value.

Application

As noted, this composite proportional-integral mode eliminates the offset problem of proportional controllers. It follows that the mode can be used in systems with frequent or large load changes. Because of the integration time, however, the process must have relatively slow changes in load to prevent oscillations induced by the integral overshoot. Another disadvantage of this system is that during startup of a batch process, the integral action causes a considerable overshoot of the error and output before settling to the operation point. This is shown in Figure 9.19, where we see the proportional band as a dashed band. The effect of the integral action can be viewed as a *shifting* of the whole proportional band. The proportional band is defined as that positive and negative error for which the output will be driven to 0% and 100%. Therefore, the presence of an integral accumulation changes the amount of error that will bring about such saturation by the proportional term. In Figure 9.19, the output saturates whenever the error

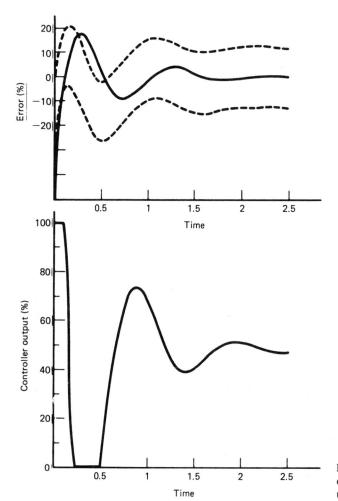

Figure 9.19 Overshoot and cycling often result when PI mode control is used in startup of batch processes.

exceeds the *PB* limits. The *PB* is constant, but its location is shifted as the integral term changes.

Example 9.8

Given the error of Figure 9.20a, plot a graph of a proportional-integral controller output as a function of time. $K_P = 5$, $K_I = 1.0$ s^{-1}, and $p_I(0) = 20\%$.

Solution   We find the solution by an application of

$$p = K_P e_p + K_P K_I \int_0^t e_p dt + p_I(0) \qquad (9.19)$$

To find the controller output, we solve Equation (9.19) in time. The error can be expressed in three time regions.

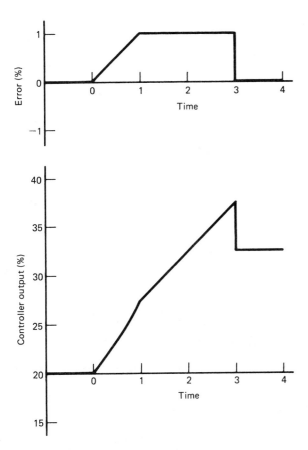

Figure 9.20   Solution for Example 9.8.

$0 \leqslant t \leqslant 1$ (t between 0 and 1 second) The error rises from 0% to 1% in 1 second. Thus, it is given by $e_p = t$.

$1 \leqslant t \leqslant 3$ For this time span the error is constant and equal to 1%; therefore it is given by $e_p = 1$.

$t \geqslant 3$ For this time the error is zero, $e_p = 0$.

We now write out and solve Equation (9.19) for each of these time spans.

$$0 \leqslant t \leqslant 1 \quad e_p = t$$

$$p_1 = 5t + 5 \int_0^t t\, dt + 20$$

$$p_1 = 5t + 5\left[\frac{t^2}{2}\right]\bigg|_0^t + 20$$

$$p_1 = 5t + 2.5t^2 + 20$$

This is plotted in Figure 9.20b from 0 to 1 second. Notice the curvature because of the squared term. At the end of 1 second, the integral term has accumulated a value of $p_1(1) = 22.5\%$.

$$1 \leq t \leq 3 \quad e_p = 1$$

$$p_2 = 5 + 5 \int_1^t 1 \, dt + 22.5$$

The integral term accumulation from 0 to 1 second forms the initial condition for this new equation.

$$p_2 = 5 + 5[t]_0^t + 22.5$$

$$p_2 = 5 + 5(t - 1) + 22.5$$

This function is plotted in Figure 9.20b from 1 to 3 seconds. At the end of this period, the integral term has accumulated a value of $p_2(3) = 32.5\%$.

$$t \geq 3 \quad e_p = 0$$

$$p_3 = 5[0] + 5 \int_3^t 0 \, dt + 32.5$$

$$p_3 = 32.5$$

Figure 9.20b shows that the output will stay constant at 32.5% from 3 seconds. The sudden drop of 5% is due to the sudden change of error from 1% to 0% at $t = 3$ seconds.

9.6.2 Proportional-Derivative Control Mode (PD)

A second combination of control modes has many industrial applications. It involves the serial (cascaded) use of the proportional and derivative modes. The analytic expression for this mode is found from a combination of Equations (9.14) and (9.18)

$$p = K_P e_p + K_P K_D \frac{de_p}{dt} + p_0 \tag{9.20}$$

where the terms are all defined in terms given by previous equations.

It is clear that this system cannot eliminate the offset of proportional controllers. It can, however, handle fast process load changes as long as the *load change offset error* is acceptable. An example of the operation of this mode for a hypothetical load change is shown in Figure 9.21. Note the effect of derivative action in moving the controller output in relation to the error rate change.

Example 9.9

Suppose the error, Figure 9.22a, is applied to a proportional-derivative controller with $K_P = 5$, $K_D = 0.5$ s, and $p_0 = 20\%$. Draw a graph of the resulting controller output.

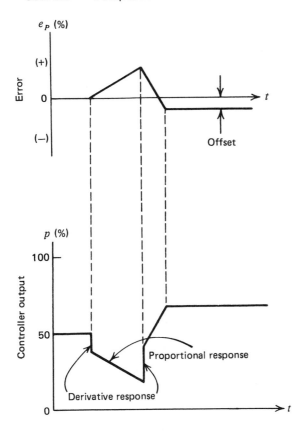

Figure 9.21 Proportional-derivative (PD) action showing the offset error from the proportional mode (reverse action).

Solution   In this case we evaluate

$$p = K_P e_p + K_D K_P \frac{de_p}{dt} + p_0 \qquad (9.20)$$

over the two spans of the error. In the time of $0 - 1$ s where $e_p = at$, we have

$$p_1 = K_P at + K_D K_P a + p_0$$

or because $a = 1\%/s$

$$p_1 = 5t + 2.5 + 20$$

Note the instantaneous change of 2.5% produced by this error. In the span from 1–3 s we have

$$p_2 = 5 + 20 = 25$$

The span from 3–5 s has an error of $e_p = -0.5t + 2.5$ so that we get for 3–5 s

$$p_3 = -2.5t + 12.5 - 1.25 + 20$$

or

$$p_3 = -2.5t + 31.25$$

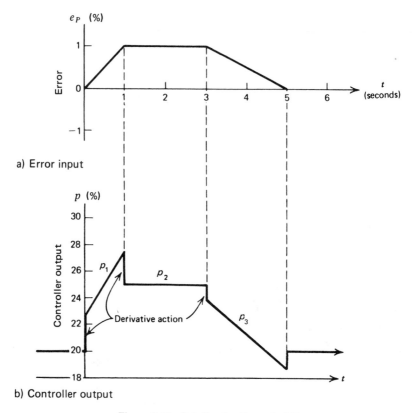

Figure 9.22   Solution for Example 9.9.

This controlled output is plotted in Figure 9.22b.

9.6.3 Three-Mode Controller (PID)

One of the most powerful but complex controller mode operations combines the proportional, integral, and derivative modes. This system can be used for virtually *any* process condition. The analytic expression is

$$p = K_P e_p + K_P K_I \int_0^t e_p dt + K_P K_D \frac{de_p}{dt} + p_I(0) \qquad (9.21)$$

where all terms have been defined earlier.

This mode eliminates the offset of the proportional mode and still provides fast response. In Figure 9.23, the response of the three-mode system to an error is shown.

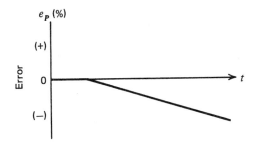

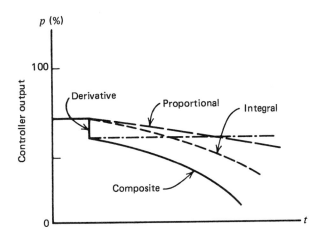

Figure 9.23 The three-mode controller action exhibits proportional, integral, and derivative action.

Example 9.10

Let us combine everything and see how the error of Figure 9.22a produces an output in the three-mode controller with $K_P = 5$, $K_I = 0.7$ s^{-1}, $K_D = 0.5$ s, and $p_I(0) = 20\%$. Draw a plot of the controller output.

Solution From Figure 9.22a, the error can be expressed as follows:

$$0\text{–}1 \text{ s} \quad e_p = t\%$$

$$1\text{–}3 \text{ s} \quad e_p = 1\%$$

$$3\text{–}5 \text{ s} \quad e_p = -\frac{1}{2}t + 2.5\%$$

We must apply each of these spans to the three-mode equation for controller output

$$p = K_P e_p + K_P K_I \int_0^t e_p \, dt + K_P K_D \frac{de_p}{dt} + p_I(0) \tag{9.21}$$

or

$$p = 5e_p + 3.5 \int_0^t e_p \, dt + 2.5 \frac{de_p}{dt} + 20$$

From 0–1 s, we have

$$p_1 = 5t + 3.5 \int_0^t t\,dt + 2.5 + 20$$

or

$$p_1 = 5t + 1.75t^2 + 22.5$$

This is plotted in Figure 9.24 in the span of 0–1 s. At the end of 1 s, the integral term has accumulated to $p_I(1) = 21.75\%$. Now from 1–3 s, we have

$$p_2 = 5 + 3.5 \int_0^t (1)\,dt + 21.75$$

or

$$p_2 = 3.5(t - 1) + 26.75$$

This controller variation is shown in Figure 9.24 from 1–3 s. At the end of 3 s, the integral term has accumulated to a value of $p_I(3) = 28.75\%$. Finally, from 3–5 s, we have

$$p_3 = 5\left(-\frac{1}{2}t + 2.5\right) + 3.5 \int_3^t \left(-\frac{1}{2}t + 2.5\right) dt - \frac{2.5}{2} + 28.75$$

or

$$p_3 = -0.875t^2 + 6.25t + 21.625$$

This is plotted in Figure 9.24 from 3–5 s. After 5 seconds, the error is zero. Therefore,

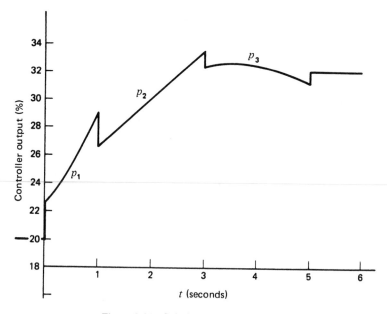

Figure 9.24 Solution for Example 9.10.

the output will simply be the accumulated integral response providing a constant output of $p_I = 32.25\%$.

The examples used in this chapter are idealized in terms of the sudden way that errors change. In the real world, changes are not instantaneous, and therefore the sharp breaks in output, such as those shown in Figure 9.24, do not occur.

9.6.4 Special Terminology

There are a number of special terms used in process control for discussing the controller modes. The following summary defines some of these terms and shows how they relate to the equations presented in this chapter.

1. *Proportional band (PB)*   Although this term was defined earlier, let us note again that this is the percentage error that results in a 100% change in controller output.

2. *Repeats per minute*   This term is another expression of the integral gain for PI and PID controller modes. The term derives from the observation that the integral gain K_I has the effect of causing the controller output to change every unit time by the proportional mode amount. You also can see this by taking the derivative of the integral term in the controller equation. This gives a change in controller output Δp of

$$\Delta p = K_I K_P e_p \Delta t$$

Because $K_P e_p$ is just the proportional contribution, in a unit time interval $\Delta t = 1$, K_I just repeats the proportional term. For example, if $e_p = 0.5\%$ and $K_P = 10\%$, then $K_P e_p = 5\%$. If $K_I = 10\%/(\%\text{-min})$, then every minute the output would increase by 5% times $10\%/(\%\text{-min})$ or 50% or "10 repeats per minute." It repeats the proportional amount 10 times per minute.

3. *Rate gain*   This is just another way of saying the derivative gain, K_D. Because K_D has the units of $\%\text{-s}/\%$ (or $\%\text{-min}/\%$), one often expresses the gain as time directly. Thus, a *rate gain* of 0.05 min or a *derivative time* of 0.05 min both mean $K_D = 0.05\%\text{-min}/\%$.

4. *Direct/reverse action*   This specifies whether the controller output should increase (direct) or decrease (reverse) for an increasing controlled variable. The action is specified by the sign of the proportional gain; $K_P > 0$ is direct and $K_P < 0$ is reverse.

SUMMARY

This chapter covers the general characteristics of controller operating modes without considering implementation of these functions. Numerous terms that are important to an understanding of controller operations are defined. The highlighted items are as follows:

1. In considering controller operating modes, it is important to know the *process load*, which is the nominal value of all process parameters, and the *process lag*, which represents a delay in reaction of the controller variable to a change of load variable.

2. Some processes exhibit *self-regulation*; that is, the characteristic that a dynamic variable adopts some nominal value commensurate with the load with no control action.

3. The controller operation is defined through a relationship between percentage *error* or *deviation* relative to full scale

$$e_p = \frac{r - b}{b_{\text{max}} - b_{\text{min}}} \cdot 100 \tag{9.3}$$

and the controller output as a percentage of the controlling parameter

$$p = \frac{u - u_{\text{min}}}{u_{\text{max}} - u_{\text{min}}} \cdot 100 \tag{9.4}$$

4. Control lag and dead time, respectively, refer to a delay in controller response when a deviation occurs and a period of no response of the process to a change in the controlling variable.

5. Discontinuous controller modes refer to instances where the controller output does not change smoothly for input error. Examples are two-position, multiposition, and floating.

6. Continuous controller modes are modes where the controller output is a smooth function of the error input or rate of change. Examples are proportional, integral, and derivative modes.

7. Composite controller modes combine the continuous modes. Examples are the proportional-integral (PI), proportional-derivative (PD), and the proportional-integral-derivative (PID) (or three-mode).

PROBLEMS

Section 9.2

9.1 Define the variables in the system of Figure 9.1 which constitute the process load.

9.2 Analyze each of the following control systems and determine if they have self-regulation, what the process load would be, what would constitute a transient, and if a process lag would be expected:
   a. A home air-conditioning system
   b. The cracker-baking system of Figure 7.2
   c. The level-control system of Figure 8.3

Section 9.3

9.3 A velocity control system has a range of 220 to 460 mm/s. If the setpoint is 327 mm/s and the measured value is 294 mm/s, calculate the error as percentage of span.

9.4 A controlling variable is a motor speed that varies from 800 to 1750 rpm. If the speed is controlled by a 25- to 50-volt dc signal, calculate (a) the speed produced by an input of 38 volts, and (b) the speed calculated as a percent of span.

Section 9.4

9.5 A 5-m diameter cylindrical tank is emptied by a constant outflow of 1.0 m³/min. A two-position controller is used to open and close a fill valve with an open flow of 2.0 m³/min. For level control, the neutral zone is 1 m and the setpoint is 12 m. (a) Calculate the cycling period, and (b) plot the level versus time.

9.6 For Example 9.4, verify that a ±2% neutral zone produces the results of limits of oscillation and period given in the text.

9.7 A floating controller with a rate gain of 6%/min and $p(0) = 50\%$ has a ±5 gal/min deadband. Plot the controller output for an input given by Figure 9.25. The setpoint is 60 gal/min.

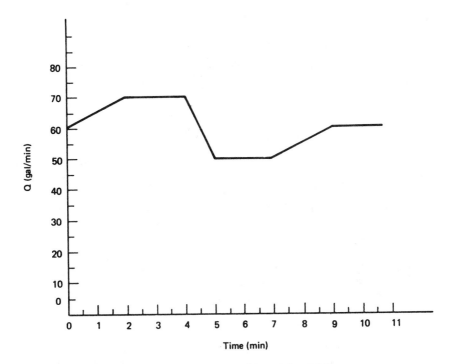

Figure 9.25 Figure for Problems 9.7 and 9.18.

Section 9.5

9.8 For a proportional controller, the controlled variable is a process temperature with a range of 50 to 130°C and a setpoint of 73.5°C. Under nominal conditions, the setpoint is maintained with an output of 50%. Find the proportional offset that results from a load change which requires a 55% output if the proportional gain is (a) 0.1, (b) 0.7, (c) 2.0, and (d) 5.0.

9.9 For the applications of Problem 9.8, find the percentage controller output with the 73.5°C setpoint and a proportional gain of 2.0 if the temperature is (a) 61°C, (b) 122°C, and (c) a ramping temperature of $(82 + 5t)$°C.

9.10 An integral controller has a reset action of 2.2 minutes. Express the integral controller constant in s^{-1}. Find the output of this controller to a constant error of 2.2%.

9.11 How would a derivative controller with $K_D = 4$ s respond to an error which varies as $e_p = 2.2 \sin(0.04t)$?

9.12 A proportional controller has a gain of $K_P = 2.0$. Plot the controller output for the error given by Figure 9.26 if $p_0 = 50\%$.

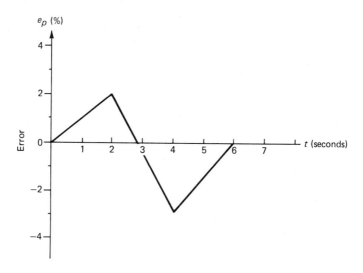

Figure 9.26 Figure for Problems 9.12, 9.13, 9.14 and 9.15.

Section 9.6

9.13 A PI controller has $K_P = 2.0$, $K_I = 2.2$ s^{-1}, and $p_I(0) = 40\%$. Plot the output for an error given by Figure 9.26.

9.14 A PD controller has $K_P = 2.0$, $K_D = 2$ s, and $p_0 = 40\%$. Plot the controller output for the error input of Figure 9.26.

9.15 A PID controller has $K_P = 2.0$, $K_I = 2.2$ s^{-1}, $K_D = 2$ s, and $p_I(0) = 40\%$. Plot the controller output for the error of Figure 9.26.

9.16 A PI controller is reverse acting, PB = 20, 12 repeats per minute. Find (a) the proportional gain, (b) the integral gain, and (c) the time that the controller output will reach 0% after a constant error of 1.5% starts. The controller output when the error occurred was 72%.

9.17 Suppose rate action was added to the controller of Problem 9.16 with a rate gain of 0.2 minutes. Specify the derivative gain and determine the time at which the controller output reaches 0% with this added mode if the input error is $e_p = 0.9t^2$.

9.18 A PI controller is used to control flow within a range of 20 to 100 gal/min. The setpoint is 60 gal/min and the controller output drives a valve with a 3 to 15 psi signal. The controller settings are direct action, $K_P = 0.9$, $K_I = 0.4$ min^{-1}. Plot the pneumatic pressure for the flow of Figure 9.25. Assume an initial pressure output of 10.8 psi.

9.19 A PI controller has $K_P = 4.5$ and $K_I = 7$ s^{-1}. Find the controller output for an error given by $e_p = 3 \sin(\pi t)$. What is the phase shift between error and controller output?

CHAPTER 10

ANALOG CONTROLLERS

INSTRUCTIONAL OBJECTIVES

The objectives of this chapter provide a more comprehensive understanding of the principles of mode implementation. The objectives have been chosen from this viewpoint. After a comprehensive study of this chapter, you should be able to

1. Recognize the essential elements of an analog controller.
2. Diagram and describe the implementation of two-position, proportional, and integral control modes, using op amps.
3. Diagram and describe the implementation of proportional-integral, proportional-derivative, and three-mode controllers using op amps; diagram and describe the operation of a three-mode pneumatic controller.
4. Design the basic elements of a process-control loop using electronic analog techniques.

10.1 INTRODUCTION

The previous chapter presented the defining principles of various controller modes. Selection of the mode to use and appropriate gains depends on many factors involved in the process operation. This decision is made by engineers who are familiar with the process itself and who are aided by process-control technique experts who understand the characteristics of each mode. This chapter will study how modes of controller action are realized using *analog* techniques. The emphasis is on *electronic* techniques, using op amps as the active element because of their widespread use. *Pneumatic* techniques also are discussed because there are many operations where a complete implementation of a process-control loop uses pneumatic methods.

Digital and computer methods of providing controller modes are considered in Chapter 11. The study of analog electronic implementation of controller modes can be considered an application of analog computer techniques because we are, in effect, looking for a solution to controller mode equations.

Specific methods of controller mode realization, using either electronics or pneumatics, are as varied as the manufacturers of this equipment. Thus the material in this chapter is presented in a general fashion, using op amps in electronics and general principles in pneumatics. If a particular manufacturer chooses to use special discrete circuits, for example, then a one-to-one correspondence with op amp methods will help in understanding the circuit. Because many controllers are now using integrated circuit (IC) op amps, there is a distinct advantage to a specific study of this method.

10.2 GENERAL FEATURES

An analog controller is a device that implements the controller modes described in Chapter 9, using analog signals to represent the loop parameters. The analog signal may be in the form of an electric current or a pneumatic air pressure. The controller accepts a measurement expressed in terms of one of these signals, calculates an output for the mode being used, and outputs an analog signal of the same type. Because the controller does solve equations, we think of it as an analog computer. The controller must be able to add, subtract, multiply, integrate, and find derivatives. It does this working with analog voltages or pressures. In this section we will examine the general physical layout of typical analog controllers.

Typical physical layout

Analog controllers are usually designed to fit into a panel assembly as a slide in/out module, as shown in Figure 10.1. The front displays all necessary information and provides adjustment capability for the operator. When the unit is pulled out part way but still connected, other less frequently required adjustments are available. When the controller is pulled further out, an extension cable can be disconnected and the entire unit removed from the panel for replacement, if necessary.

Front panel

The front panel of an analog controller displays information for operators and allows adjustment of the setpoint. Figure 10.1 shows a typical front panel. The

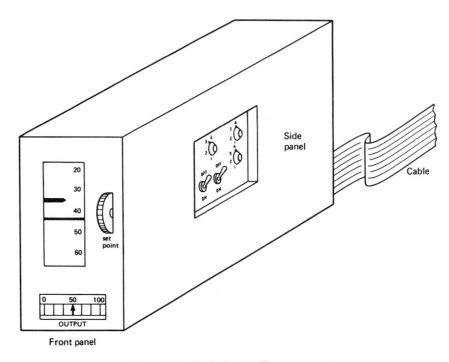

Figure 10.1 Typical controller appearance.

setpoint knob moves a sliding scale under the fixed setpoint indicator. Thus, a fixed span of measurement above and below the setpoint is visible as indicated by the measurement meter indicator. The error is the difference between the setpoint indicator and the measurement meter. The display is typically expressed in percentage of span (4–20 mA or 3–15 psi). The lower meter shows the controller output, again expressed in percentage of span. Of course, the output is actually 4–20 mA or 3–15 psi so that 0% would mean 4 mA, for example. There often is a switch on the front panel by which the controller can be placed in a *manual* control, which means that the output can be adjusted independently of the input using the output adjust knob. In *automatic* mode, this knob has no effect on the output. Connections to the controller are made through electrical or pneumatic cables connected to the rear of the unit.

Side panel

On the side of the controller, when partially pulled out, knobs are available to adjust operation of the controller modes. On this panel, as shown in Figure 10.1, the proportional, integral (reset), and derivative (rate) gains can be adjusted. In addition, filtering action and reverse/direct operation can often be selected.

10.3 ELECTRONIC CONTROLLERS

In the following treatment of electronic methods of realizing controller modes, emphasis is on the use of op amps as the circuit element. Discrete electronic components also are used to implement this function, but the basic principles are best illustrated using op amp circuits. Op amp circuits other than the ones described also can be developed.

10.3.1 Error Detector

The detection of an error signal is accomplished in electronic controllers by taking the difference between voltages. One voltage is generated by the process signal current passed through a resistor. The second voltage represents the setpoint. This is usually generated by a voltage divider using a constant voltage as a source. An example is shown in Figure 10.2. We assume a two-wire system is in use so that the current drawn from the floating power supply is the 4–20 mA signal current. The signal current is used to generate a voltage IR across the resistor R. This is placed in series opposition to a voltage V_{sp} tapped from a variable resistor R_{sp} connected to a constant negative-source V_0. The result is an error voltage V_E $= V_{sp} - IR$. This is then used in the process controller to calculate controller output.

An error detector also can be made from a differential amplifier. Such a system only can be used if the current from the transducer is referenced to ground. Figure 10.3 shows one typical configuration. The sensor signal current passes to ground through R_L providing a signal voltage, $V_m = IR_L$. The differential amplifier then subtracts this from the setpoint voltage.

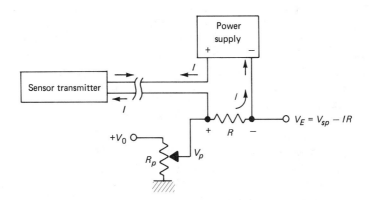

Figure 10.2   Typical divider error detector.

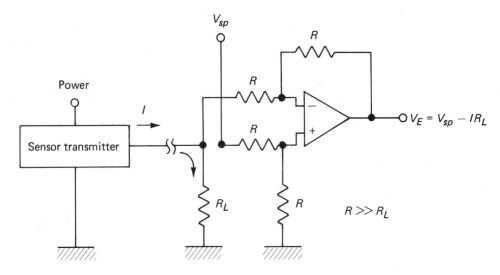

Figure 10.3 An error detector using a differential amplifier.

10.3.2 Single Mode

The following op amp circuits illustrate methods of implementing the pure modes of controller action with op amp circuits.

Two-position

A two-position controller can be implemented by a great variety of electronic and electromechanical designs. Many household air-conditioning and heating systems employ a two-position controller constructed from a bimetal strip and mercury switch, as shown in Figure 10.4. We see that as the bimetal strip bends because of a temperature decrease, it reaches a point where the mercury slides down to close an electrical contact. The inertia of the mercury tends to keep the system in that position until the temperature increases to a value above the setpoint temperature. This provides the required neutral zone to prevent excessive cycling of the system.

A method using op amp implementation of ON-OFF control with adjustable neutral zone is given in Figure 10.5. Here the controller input signal is assumed to be a voltage level with an "ON" voltage of V_H and an "OFF" voltage V_L, and the output is the comparator output of zero, or V_0. The comparator output

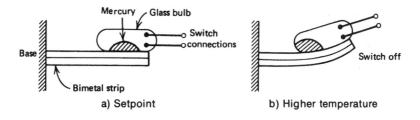

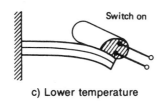

a) Setpoint   **b) Higher temperature**

c) Lower temperature

Figure 10.4 A mercury switch on a bimetal strip is often used as a two-position temperature controller.

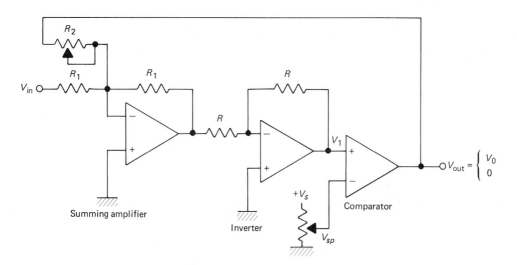

Figure 10.5 A two-position controller with neutral zone constructed from op amps and a comparator.

switches states when the voltage on its input V_1 is equal to the setpoint value V_{sp}. Analysis of this circuit shows that the high (ON) switch voltage is

$$V_H = V_{sp} \tag{10.1}$$

and the low (OFF) switching voltage is

$$V_L = V_{sp} - \frac{R_1}{R_3} V_0 \tag{10.2}$$

Figure 10.6 shows the typical two-position relationship between input and output voltage for this circuit. The width of the neutral zone between V_L and V_H can be adjusted by variation of R_2. The relative location of the neutral zone is calculated from the difference between Equations (10.1) and (10.2).

The inverter resistance in Figure 10.5 can be chosen as any convenient value. Typically, it is in the 1 to 100 kΩ range.

Example 10.1

Level measurement in a sump tank is provided by a transducer scaled as 0.2 V/m. A pump is to be turned on by application of +5 V when the sump level exceeds 2.0 m. The pump is to be turned back off when the sump level drops to 1.5 m. Develop a two-position controller.

Solution   Let us use the circuit of Figure 10.5. The high and low trip voltages will be determined by the conditions of the problem. From these the values of the resistances can be determined.

$$V_H = (0.2 \text{ V/m})(2.0 \text{ m}) = 0.4 \text{ V}$$

and

$$V_L = (0.2 \text{ V/m})(1.5 \text{ m}) = 0.3 \text{ V}$$

This gives the following relations for the resistances and V_{sp}.

$$0.4 \text{ V} = V_{sp}$$

$$0.3 \text{ V} = V_{sp} - \frac{R_1}{R_2} V_0$$

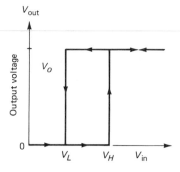

Figure 10.6   The circuit of Figure 10.5 shows the characteristic two-position response.

Therefore, $V_{sp} = 0.4$ volts and, from the second equation,

$$0.3 = 0.4 - (R_1/R_2)(5)$$

$$(R_1/R_2) = 0.02$$

Since there are two unknowns and only one condition, one unknown can be selected. Picking $R_1 = 5$ kΩ, for example, means that $R_2 = 250$ kΩ.

Reverse action

The two-position controller of Figure 10.5 can be made reverse acting by placing an inverter in the feedback and reversing the comparator. It is also possible to devise other circuits that automatically incorporate the reverse action. Multiposition controllers can be devised by a similar process of op amp circuit development. An example for a three-position controller is given in the problems at the end of this chapter.

Floating

The floating-type controller can be generated by connecting the output of a three-position controller into an integrator. Such a circuit is shown in Figure 10.7. We assume the three-position controller was designed to provide outputs of V_1, zero, or $-V_1$, depending on input. As an input to the integrator, this produces possible outputs of

$$V_{\text{out}} = \begin{cases} -\dfrac{V_1}{RC}(t - t_a) + V_a & V_m < V_{s1} \\[2mm] V_b & V_{s1} < V_{\text{in}} < V_{s2} \\[2mm] \dfrac{V_1}{RC}(t - t_c) + V_c & V_{\text{in}} > V_{s2} \end{cases} \qquad (10.3)$$

where

$$V_{s1} = \text{lower setpoint voltage}$$

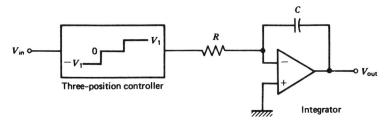

Figure 10.7 Construction of a floating controller from a three-position controller and integrator.

$$V_{s2} = \text{upper setpoint voltage}$$
$$V_a, V_b, V_c = \text{values of output when the input condition occurs}$$
$$t_a, t_b, t_c = \text{times at which input reaches the setpoints}$$

The *trip* voltages can be set to provide the desired band of inputs producing no output, a positive rate, or negative rate. The actual rate of output change depends on the values of resistor and capacitor in the integrator and the output level of the three-position circuit preceding the integrator. Remember that the output floats at whatever the latest value of output is when the input falls within the neutral zone.

Example 10.2

A control signal varies from 0–5 volts. A floating controller, such as that in Figure 10.7, has trip voltages of 2 volts and 4 volts, and a three-position controller has outputs of 0 and ± 2 volts. The integrator consists of a 1-MΩ resistor and a 1-μF capacitor. Plot the controller output in response to the input of Figure 10.8a.

Solution   We can find the output by applying the relations of Equation (10.3) to the input voltage. The result is shown in Figure 10.8b. This result is arrived at as follows:

   1. From 0–1 second, the output is zero (an assumed starting point) and remains so because the input is within the neutral zone.
   2. From 1–3 seconds, the lower setpoint has been reached and the output is given by

$$V_{\text{out}} = -\frac{V_1}{RC}(t - 1) = -2(t - 1) \text{ V}$$

At $t = 3$ s, the output is -4 V.
   3. From 3–4 seconds, the output remains at -4 because the input is in the neutral zone.
   4. From 4–7 seconds, the input reaches the upper setpoint and the output becomes

$$V_{\text{out}} = +\frac{V_1}{RC}(t - 4) - 4 \text{ V}$$

$$V_{\text{out}} = +2(t - 4) - 4 \text{ V}$$

At $t = 7$ seconds, the input again falls within the neutral zone and the output becomes

$$V_{\text{out}} = +2$$

   5. The output will remain at $+2$ until the input again hits a setpoint value.

Proportional mode

Implementation of this mode requires a circuit that has a response given by

$$p = K_P e_p + p_0 \tag{9.14}$$

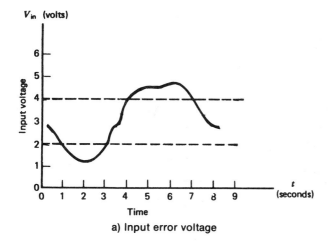

a) Input error voltage

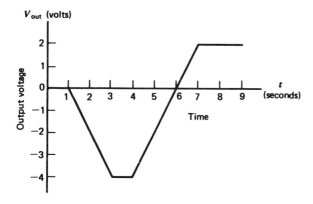

b) Output voltage

Figure 10.8 Input and output voltage for the floating controller of Example 10.2.

where

p = controller output 0–100%
K_P = proportional gain
e_p = error in percent of variable range
p_0 = controller output with no error

If we consider both the controller output and error to be expressed in terms of voltage, we see that Equation (9.14) is simply a *summing amplifier*. The op amp circuit in Figure 10.9 shows such an electronic proportional controller. In this case, the analog electronic equation for the output voltage is

$$V_{out} = G_P V_e + V_0 \qquad (10.4)$$

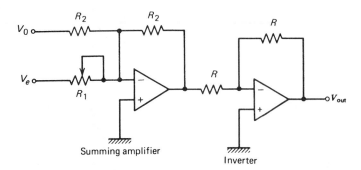

Figure 10.9 An electronic proportional controller.

where

V_{out} = output voltage
$G_P = R_2/R_1$ = gain
V_e = error voltage
V_0 = output with zero error

To use the circuit of Figure 10.9 for proportional control, a relationship must be established with the characteristics of the mode as defined in Chapter 9. In Chapter 9, the error was expressed as the percent of measurement range and the output was simply 0% to 100%. Yet Figure 10.9 deals with voltage on both the input and output.

We first can simply identify that the output voltage range of the circuit, whatever it is, represents a swing of 0% to 100%. Thus, if a final control element needs 0 to 5 V, then a zener is added, as shown in Figure 10.10, so that the op amp output can swing only between 0 and 5 volts. An output of 50% is (0.5)(5 V) = 2.5 V.

For the input it will be necessary to determine the range of measurement from which the error voltage was produced. If temperature is to be measured and controlled from 100°C to 200°C and this is converted to 2.0 to 8.0 volts, then the measurement range is (8.0 − 2.0) = 6.0 V. Error voltage can be expressed as a percent of this range. This will be correlated with the error in the equations of Chapter 9.

Finally, the gain of Figure 10.9 is *not* simply equal to K_P. The actual value

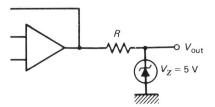

Figure 10.10 A zener diode can be used to fix the output swing.

of G_P must be determined so that its effect in voltage is the same as that required by K_P in terms of percent.

Suppose $K_P = 4\%$ per %, the output voltage range is 0 to 5 V, and the input voltage spans 2 to 8 V (a range of 6 V). Let us determine the proper value for G_P. K_P says that if the error constitutes 1% of the input range, then the output must be changed by 4% of its range. This is easy because 1% of the input range is just $(0.01)(6 \text{ V}) = 0.06 \text{ V}$ and 4% of the output is simply $(0.04)(5 \text{ V}) = 0.2 \text{ V}$. Now because G_P is the gain in terms of voltage, we must have $G_P = (0.2 \text{ V}/0.06 \text{ V}) = 3.33$. Therefore, the values of R_1 and R_2 can be selected so that $(R_2/R_1) = 3.33$.

Another way of looking at this is using the concept of proportional band $PB = 100/K_P$. The PB is that percent of error that will cause a 100% change of output. For the example of the last paragraph, $PB = 100/4 = 25\%$. Therefore, a 25% change of error must produce a 100% change of output. Twenty-five percent of the error is simply $(0.25)(6 \text{ V}) = 1.5 \text{ V}$ and 100% of the output is 5 V. Thus, $G_P = (5 \text{ V}/1.5 \text{ V}) = 3.33$, as before.

Example 10.3

A controller is shown in Figure 10.9 with scaling so that 0–10 volts corresponds to a 0–100% output. If $R_2 = 10 \text{ k}\Omega$ and full-scale error range is 10 volts, find the values of V_0 and R_1 to support a 20% proportional band about a 50% zero error controller output.

Solution   The value of V_0 is simply 50% of 10 volts or 5 volts, to provide the zero error controller output. To design for a 20% proportional band means that a change of error of 20% must cause the controller output to vary 100%. Thus, from

$$V_{\text{out}} = G_P V_e + V_0$$

we note that when the error has changed 20% of 10 volts or 2 volts, we must have full controller output change. Thus

$$G_P = \frac{\Delta V_{\text{out}}}{\Delta V_e} = \frac{10}{2}$$

$$G_P = 5$$

so that if $R_2 = 10 \text{ k}\Omega$, then

$$R_1 = R_2/K = \textbf{2 k}\boldsymbol{\Omega}$$

Example 10.4

If the load in the previous example changes such that a new controller output of 40% is required, find the corresponding offset error.

Solution   In this case we need a negative error so that the output is 40% of 10 volts = 4 volts.

$$V_{\text{out}} = G_P V_e + V_0$$

$$4 = 5V_e + 5$$

$\therefore V_e = -\dfrac{1}{5}$ volts and because the full-scale error signal is 10 volts, we have an error of

$$\frac{-0.2}{10} \times 100 = -2\%$$

Generally, a voltage-to-current converter is used on the output to convert the output voltages to a 4–20 mA range of current signals to drive the final control element.

Integral mode

In the previous chapter we saw that the integral mode was characterized by an equation of the form

$$p(t) = K_I \int_0^t e_p(t)dt + p_I(0) \tag{9.17}$$

where

$p(t)$ = controller output in percent of full scale
K_I = integration gain (s^{-1})
$e_p(t)$ = deviations in percent of full-scale variable value
$p_I(0)$ = controller output at $t = 0$

This function is easy to implement when op amps are used as the building blocks. A diagram of an integral controller is shown in Figure 10.11. The corresponding equation relating input to output is

$$V_{\text{out}} = G_I \int_0^t V_e dt + V_{\text{out}}(0) \tag{10.5}$$

where

V_{out} = output voltage

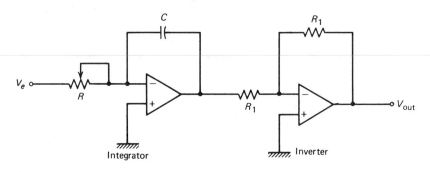

Figure 10.11 Electronic integral-mode controller.

$$G_I = 1/RC = \text{integration gain}$$
$$V_e = \text{error voltage}$$
$$V_{out}(0) = \text{initial output voltage}$$

The value of R and C can be adjusted to obtain the desired integration time. The initial controller output is the integrator output at $t = 0$. As we noted earlier, the integration time constant determines the *rate* at which controller output increases when the error is constant. If K_I is made too large, the output rises so fast that overshoots of the optimum setting occur and cycling is produced.

The actual value of G_I, and therefore R and C, is determined from K_I and the input and output voltage ranges. One way to do this is to recognize that the integral gain says that an input error of 1% must produce an output that changes as K_I percent per second. Another way is to know that if an error of 1% lasts for 1 second, the output must change by K_I percent.

Suppose we have an input range of 6 V, an output range of 5 V, and $K_I = 3.0\%/(\% - \text{min})$. Integral gain is often given in minutes because industrial processes are slow compared to a time of seconds. This gain also is often expressed as integration time T_I, which is just the inverse of the gain, so $T_I = 3.33$ min.

We must first convert the time units to seconds. So

$$[3\%/(\% - \text{min})][1 \text{ min}/60 \text{ s}] = 0.05\%/(\% - \text{s})$$

An error of 1% for 1 second is found from

$$(0.01)(6 \text{ V})(1 \text{ s}) = 0.06 \text{ V} - \text{s}$$

Furthermore, K_I percent of the output (using the seconds expression for gain) is

$$(0.0005)(5 \text{ V}) = 0.0025 \text{ V}$$

Therefore, the integral gain in terms of voltage must be

$$G_I = (0.0025 \text{ V})/(0.06 \text{ V} - \text{s}) = 0.0417 \text{ s}^{-1}$$

The values of R and C can be selected from this.

Example 10.5

An integral control system will have a measurement range of 0.4 to 2.0 V and an output range of 0 to 6.8 V. Design an op amp integral controller to implement a gain of $K_I = 4\%/(\% - \text{min})$. Specify the values of G_I, R, and C.

Solution   The input range is $2.0 - 0.4 = 1.6$ V and the output range is 6.8 V. We must convert K_I to units of seconds

$$[4\%/(\% - \text{min})][1 \text{ min}/60 \text{ s}] = 0.0667\%/(\% - \text{s})$$

$$1\% \text{ of the input for 1 s} = (0.01)(1.6 \text{ V})(1 \text{ s}) = 0.016 \text{ V} - \text{s}$$

$$0.0667\% \text{ of the output} = (0.000667)(6.8 \text{ V}) = 0.00454 \text{ V}$$

Thus, the gain is

$$G_I = (0.00454 \text{ V}/0.016 \text{ V} - \text{s}) = 0.283 \text{ s}^{-1}$$

Because $G_I = 1/(RC)$, we have

$$RC = 3.53 \text{ s}$$

If we pick $C = 100 \text{ }\mu\text{F}$, then $R = $ **35.3 kΩ.**

Derivative mode

The derivative mode is never used alone because it cannot provide a controller output when the error is zero (see Section 9.3.6). Nevertheless, we show here how it is implemented with op amps for combination with other modes in the next section. The control mode equation was given earlier as

$$p = K_D \frac{de_p}{dt} \tag{9.18}$$

where

p = controller output in percent of full output
K_D = derivative time constant (s)
e_p = error in percent of full-scale range

This function is implemented by op amps in the configuration shown in Figure 10.12. Resistance R_1 is added for stability of the circuit against rapidly changing signals. The response of this circuit for slowly varying inputs is

$$V_{\text{out}} = G_D \frac{dV_e}{dt} \tag{10.6}$$

where

V_{out} = output voltage

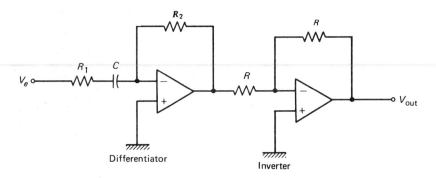

Figure 10.12 Electronic derivative-mode controller circuit.

$G_D = R_2C$ = derivative time in seconds
V_e = error voltage

Without R_1 the circuit of Figure 10.12 has a gain that increases with increasing frequency. Therefore, the circuit will be unstable and tend toward spontaneous oscillation when energized. Adding R_1 causes the circuit gain to revert to that of an inverting amplifier with a gain of R_2/R_1 at higher frequencies.

The value of $G_D = R_2C$ is determined from K_D and knowledge of the measurement and output voltage ranges. For this mode the interpretation of K_D is that for an error change of 1% in 1 second, the output should change by K_D percent. Thus, G_D is found from the quotient of K_D percent of the output voltage and 1% of the input voltage. Of course, K_D must be expressed in seconds.

The value of R_1 is determined by a requirement that derivative action stops or "breaks" at some frequency f_c that is higher than any expected signal frequency. Thus, spontaneous oscillation is prevented, but no measurement information is lost. Generally, we make f_c 10 to 100 times the maximum expected signal frequency. If f_s is the maximum signal frequency, then the requirement can be written

$$R_1C \ll 1/(2\pi f_s) \tag{10.7}$$

Of course, the derivative mode is never used alone because no output is provided for zero or constant error.

The circuits of this section show that the pure modes of controller operation are easily constructed from op amps. As stated in Chapter 9, a pure mode is seldom used in process control because of the advantages of composite modes in providing good control. In the next section, implementation of composite modes using op amps is considered.

10.3.3 Composite Controller Modes

In Chapter 9, the combination of several controller modes was found to combine the advantages of each mode and, in some cases, eliminate disadvantages. Composite modes are implemented easily using op amp techniques. Basically, this consists of simply combining the mode circuits introduced in the previous section.

Proportional-integral

A simple combination of the proportional and integral circuits provides the proportional-integral mode of controller action. The resulting circuit is shown in Figure 10.13. For this case the relation between input and output is most easily found by applying op amp circuit analysis. We get (including the inverter)

$$V_{\text{out}} = \left(\frac{R_2}{R_1}\right)V_e + \frac{1}{R_1C}\int_0^t V_e dt$$

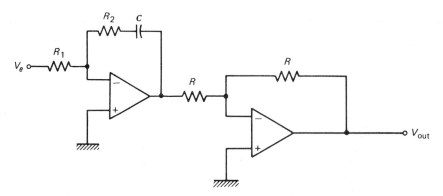

Figure 10.13   Electronic proportional-integral (PI) controller.

The definition of the proportional-integral controller includes the proportional gain in the integral term, so we write

$$V_{\text{out}} = \left(\frac{R_2}{R_1}\right)V_e + \left(\frac{R_2}{R_1}\right)\frac{1}{R_2C}\int_0^t V_e\,dt + V_{\text{out}}(0) \qquad (10.8)$$

Equation (10.8) has the same form as Equation (9.19) for this mode. The adjustments of this controller are the *proportional band* through $G_P = R_2/R_1$, and the *integration gain* through $G_I = 1/R_2C$.

Example 10.6

Design a proportional-integral controller with a proportional band of 30% and an integration gain of 0.1%/(% − s). The 4–20 mA input converts to a 0.4–2 volt signal, and the output is to be 0–10 volts. Calculate values of G_P, G_I, R_2, R_1, and C, respectively.

Solution   A proportional band of 30% means that when the input changes by 30% of range or 0.48 volts, the output must change by 100% or 10 volts. This gives a gain of

$$G_P = \frac{R_2}{R_1} = \frac{10\text{ V}}{0.48} = \mathbf{20.83}$$

A K_I of 0.1%/(% − s) says that a 1% error for 1 second should produce an output change of 0.1%. One percent of 1.6 V is 0.016 V and 0.1% of 10 V is 0.01 V, so

$$G_I = \frac{1}{R_2C} = \frac{0.01}{0.016} = 0.625\text{ s}^{-1}$$

or

$$R_2C = 1.6\text{ s}$$

As an example of values to do this, we could pick

$$C = 10 \ \mu F, \text{ which requires}$$

$$R_2 = \frac{1.6 \text{ s}}{10^{-5} \text{ F}} = 160 \text{ k}\Omega$$

Then, to get the proportional gain, we use

$$R_1 = \frac{160 \text{ k}}{20.83} = 7.68 \text{ k}\Omega$$

Proportional-derivative

A powerful combination of controller modes is the proportional and derivative modes (Section 9.6.2). This combination is implemented using a circuit similar to that shown in Figure 10.14. Analysis shows that this circuit responds according to the equation

$$V_{\text{out}} + \left(\frac{R_1}{R_1 + R_3}\right)R_3 C \frac{dV_{\text{out}}}{dt} = \left(\frac{R_2}{R_1 + R_3}\right)V_e + \left(\frac{R_2}{R_1 + R_3}\right)R_3 C \frac{dV_e}{dt} + V_0$$

where the quantities are defined in the figure and the output inverter has been included. We make the derivative coefficient on the left small to eliminate instability. One choice is

$$\frac{R_1}{R_1 + R_3} R_3 C = \frac{0.1}{2\pi} T$$

where T is the fastest variable time change to be expected in the process. Then, the equation for the proportional derivative response becomes

$$V_{\text{out}} = \left(\frac{R_2}{R_1 + R_3}\right)V_e + \left(\frac{R_2}{R_1 + R_3}\right)R_3 C \frac{dV_e}{dt} + V_0 \qquad (10.9)$$

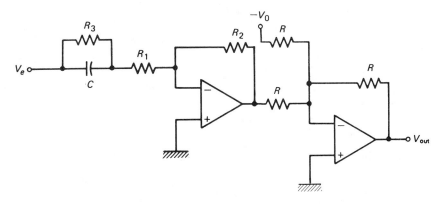

Figure 10.14   An electronic proportional-derivative (PD) controller.

where the proportional gain is $G_P = R_2/(R_1 + R_3)$ and the derivative gain is $G_D = R_3C$. This equation now corresponds to the form given by Equation (9.20) for the proportional-derivative controller. Of course, this mode still has the offset error of a proportional controller because the derivative term cannot provide reset action.

Example 10.7

A proportional-derivative controller has a 0.4- to 2.0-V input measurement range, a 0- to 5-V output, $K_P = 5\%/\%$, and $K_D = 0.08\%$ per ($\%$/min). The period of the fastest expected signal change is 1.5 seconds. Implement this controller with an op amp circuit.

Solution   To use the circuit of Figure 10.14, we first find the appropriate circuit gains, G_P and G_D.

A $K_P = 5\%/\%$ means a 20% *PB*. So we can write

$$G_P = \frac{(100\%)(5 \text{ V})}{(20\%)(1.6 \text{ V})} = \frac{5 \text{ V}}{0.32 \text{ V}} = 15.625$$

To find G_D we must first change K_D to seconds.

$$K_D = [0.08\%/(\%\text{min})]60 \text{ s/min}] = 4.8\%/(\%/s)$$

Now we can write

$$G_D = \frac{(4.8\%)(5 \text{ V})}{(1\%)(1.6 \text{ V})} = 15 \text{ s}$$

The period limitation allows us to write

$$\frac{R_1}{R_1 + R_3} R_3C = \frac{0.1}{2\pi}(1.5 \text{ s}) = 0.024 \text{ s}$$

Now we have three relations

$$\frac{R_2}{R_1 + R_3} = 15.625, \qquad R_3C = 15, \qquad \frac{R_1}{R_1 + R_3} = 0.0016$$

The last relation comes from combining the limitation equation with $R_3C = 15$. With these three relations, we have four unknowns. One can be selected. Let us try $C = 100 \text{ }\mu\text{F}$. Then the equations give

$$R_1 = 240 \text{ }\Omega, \qquad R_2 = 2.35 \text{ M}\Omega, \qquad R_3 = 150 \text{ k}\Omega$$

Three mode

The ultimate process controller is the one that exhibits proportional, integral, and derivative response to the process error input. In Chapter 9 we saw that this mode was characterized by the equation

$$p = K_Pe_p + K_PK_I \int_0^t e_p dt + K_PK_D \frac{de_p}{dt} + p_I(0) \qquad (9.21)$$

where

$$p = \text{controller output in percent of full scale}$$
$$e_p = \text{process error in percent of the maximum}$$
$$K_P = \text{proportional gain}$$
$$K_I = \text{integral gain}$$
$$K_D = \text{derivative gain}$$
$$p_I(0) = \text{initial controller integral output}$$

The zero error term of the proportional mode is not necessary because the integral automatically accommodates for offset and nominal setting. This mode can be provided by a straight application of op amp circuits resulting in the circuit of Figure 10.15. It must be noted, however, that it is possible to reduce the complexity of the circuitry of Figure 10.15 and still realize the three-mode action, but in these cases an interaction results between derivative and integral gains. We will use the circuit of Figure 10.15 because it is easy to follow in illustrating the principles of implementing this mode. Analysis of the circuit shows that the output is

$$-V_{\text{out}} = \left(\frac{R_2}{R_1}\right)V_e + \left(\frac{R_2}{R_1}\right)\frac{1}{R_IC_I}\int V_e dt + \left(\frac{R_2}{R_1}\right)R_DC_D\frac{dV_e}{dt} + V_{\text{out}}(0)$$

$$(10.10)$$

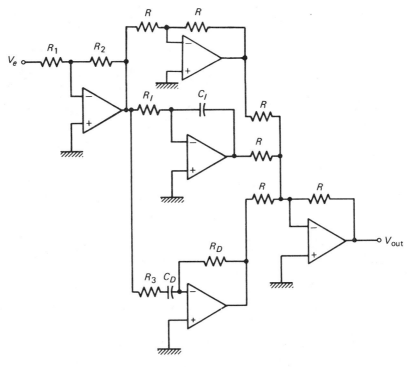

Figure 10.15 An electronic three-mode (PID) controller. It is possible to implement this mode by other circuits, some with only one op amp.

where R_3 has been chosen from $2\pi R_3 C_D \ll T$ for stability. Comparison with Equation (9.21) shows that this implements the three-mode controller if

$$G_P = \frac{R_2}{R_1}, \ G_D = R_D C_D, \ G_I = \frac{1}{R_I C_I}$$

Example 10.8

A temperature-control system inputs the controlled variable as a range from 0 to 4 V. The output is a heater requiring 0 to 8 V. A PID is to be used with $K_P = 2.4\%/\%$, $K_I = 9\%/(\%$-min$)$, $K_D = 0.7\%/(\%/$min$)$. The period of the fastest expected change is estimated to be 8 seconds. Develop the PID circuit.

Solution   The input range is 4 V and the output range is 8 V. Let us figure the circuit gains.

For the proportional mode, a 1% error means a voltage change of $(0.01)(4 \text{ V})$ = 0.04 V. This should cause an output change of 2.4% or $(0.024)(8 \text{ V})$ = 0.192 V. Thus

$$G_P = (0.192 \text{ V}/0.04 \text{ V}) = 4.8$$

For the integral term, an error of 1% should cause the output to change by 9%/min, which is $(9/60) = 0.15\%/$s. Thus

$$G_I = (0.0015 \text{ s}^{-1})(8 \text{ V})/(0.04 \text{ V})$$

$$G_I = 0.3 \text{ s}^{-1}$$

For the derivative term, an error change of 1% per min or $(0.04 \text{ V}/60) = 6.67 \times 10^{-4}$ V/s should cause an output change of 0.7% or $(0.007)(8 \text{ V}) = 0.056$ V. Thus

$$G_D = (0.056 \text{ V}/6.67 \times 10^{-4} \text{ V/s})$$

$$G_D = 84 \text{ s}$$

These results provide the following relations

$$(R_2/R_1) = 4.8$$

$$1/(R_I C_I) = 0.3 \text{ s}^{-1}$$

$$R_D C_D = 84 \text{ s}$$

From the fastest period specification, using a factor of 100 for the inequality, we form the relationship

$$2\pi R_3 C_D = (0.01)(8 \text{ s}) = 0.08 \text{ s}$$

which gives seven unknowns and four equations. We can pick three quantities. Let us try $R_1 = 10 \text{ k}\Omega$, $C_I = C_D = 10 \text{ }\mu\text{F}$. This gives

$$R_2 = 4.8 R_1 = 48 \text{ k}\Omega$$

$$R_I = 1/(0.3 C_I) = 333 \text{ k}\Omega$$

$$R_D = 84/C_D = 8.4 \text{ M}\Omega$$

$$R_3 = 0.08/(2\pi C_D) = 1.27 \text{ k}\Omega$$

8.4 MΩ seems too large for practical considerations. Let us change C_D to 100 μF. Now we get

$$R_D = 840 \text{ k}\Omega$$

$$R_3 = 127 \text{ }\Omega$$

which seems more reasonable.

These circuits have shown that the direct implementation of controller modes can be provided by standard op amp circuits. It is necessary, of course, to scale the measurement as a voltage within the range of operation selected by the circuit. Furthermore, the outputs of the circuits shown have been voltages that may be converted to currents for use in an actual process-control loop.

These circuits are only examples of basic circuits that implement the controller modes. Many modifications are employed to provide the controller action with different sets of components.

10.4 PNEUMATIC CONTROLLERS

Historically, the reason for using pneumatics in process control was probably that electronic methods were not yet competitive in cost or reliability. Safety was and still is a factor where the danger of explosion from electrical malfunctions exists. It also is true that the final control element is often pneumatically or hydraulically operated, which suggests that an all-pneumatic process-control loop might be advantageous. It appears that analog or digital electronic methods will eventually replace most pneumatic installations. But we will still have pneumatic equipment for many years until these are depreciated in industry. A good understanding of process-control principles can be applied to either electronic or pneumatic techniques, but it is necessary to consider some special features of pneumatic technology. This section provides a brief description of operations by which controller modes are pneumatically implemented.

10.4.1 General Features

The outward appearance of a pneumatic controller is typically the same as that for the electronic controller shown in Figure 10.1. The same readout of setpoint, error, and controller output appears and adjustments of gain, rate, and reset are available. The working signal is most typically the 3–15 psi standard pneumatic process-control signal, usually derived from a regulated air supply of 20–30 psi. As usual, we use the English system unit of pressure because its use is so widespread in the process-control industry. Eventual conversion to the SI unit of N/m^2 or pascals will require some alteration in scale (of measurement) to a range of 20 to 100 kPa.

The pneumatic controller is based on the nozzle/flapper described in Section 7.3.3 as the basic mechanism of operation, much the same as the op amp is used in electronics. The schematic drawings of controller mode implementation are intended to convey the operating principles. The reader is advised that specific designs may vary considerably from the systems shown.

10.4.2 Mode Implementation

In the following discussions, the essential features of controller-mode implementation using pneumatic techniques are presented. The equations are stated in general form with units in SI, but the reader should be prepared to work with English units when necessary.

Proportional

A proportional mode of operation can be achieved with the system shown in Figure 10.16. Operation is understood by noting that if the input pressure increases, then the input bellows forces the flapper to rotate to close off the nozzle. When this happens, the output pressure increases so that the feedback bellows exerts a force to balance that of the input bellows. A balance condition then occurs when torques exerted by each about the pivot are equal, or

$$(p_{out} - p_0)A_2x_2 = (p_{in} - p_{sp})A_1x_1$$

This equation is solved to find the output pressure

$$p_{out} = \frac{x_1}{x_2}\frac{A_1}{A_2}(p_{in} - p_{sp}) + p_0 \tag{10.11}$$

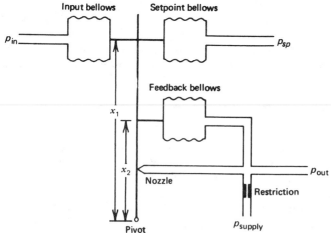

Figure 10.16 Pneumatic proportional controller.

where

p_0 = pressure with no error
p_{in} = input pressure (Pa)
A_1 = input and setpoint bellows effective area (m^2)
x_1 = level arm of input (m)
p_{out} = output pressure (Pa)
A_2 = feedback bellow effective area (m^2)
x_2 = feedback lever arm (m)
p_{sp} = setpoint pressure

This relation is based on the notion of torque equaling force time lever arm, and that a pressure in a bellows produces a force that is effectively the pressure times bellows area, much like a diaphragm. Equation (10.11) displays the standard response of a proportional mode in that output is directly proportional to input. The gain in this case is given by

$$K_P = \left(\frac{x_1}{x_2}\right)\left(\frac{A_1}{A_2}\right) \tag{10.12}$$

Because the bellows are usually of fixed geometry, the gain is varied by changing the lever arm length. In this simple representation, the gain is established by the distance between the bellows. If this separation is changed, the forces are no longer balanced and for the same pressure a new controller output will be formed corresponding to the new gain.

Example 10.9

Suppose a proportional pneumatic controller has $A_1 = A_2 = 5$ cm^2, $x_1 = 8$ cm, and $x_2 = 5$ cm. The input and output pressure ranges are 3–15 psi. Find the input pressures that will drive the output from 3–15 psi. The setpoint pressure is 8 psi and p_0 = 10 psi. Find the proportional band.

Solution   First we find the gain from

$$K_P = \left(\frac{x_1}{x_2}\right)\left(\frac{A_1}{A_2}\right) = \left(\frac{8 \text{ cm}}{5 \text{ cm}}\right)\left(\frac{5 \text{ cm}^2}{5 \text{ cm}^2}\right) \tag{10.12}$$

$$K_P = 1.6$$

Now we have

$$p_{out} = K_P(p_{in} - p_{sp}) + p_0 \tag{10.11}$$

$$p_{out} = 1.6(p_{in} - 8) + 10$$

The low input occurs when $p_{out} = 3$ psi so that

$$3 = 1.6(p_L - 8) + 10$$

which gives

$$p_L = \textbf{3.625 psi}$$

The high is found from

$$15 = 1.6(p_H - 8) + 10$$

which gives

$$p_H = \mathbf{11.125 \ psi}$$

The proportional band (PB) is

$$PB = \left(\frac{11.125 - 3.625}{15 - 3} \right) 100$$

$$PB = \mathbf{62.5\%}$$

Note that this checks with

$$PB = \frac{100}{K_P} = \frac{100}{1.6} = 62.5\%$$

which could be used because the input and output ranges are the same.

Proportional-integral

This control mode is also implemented using pneumatics by the system shown in Figure 10.17. In this case, an extra bellows with a variable restriction is added to the proportional system. Suppose the input pressure shows a sudden increase. This drives the flapper toward the nozzle, increasing output pressure until the proportional bellows balances the input as in the previous case. The integral bellows is still at the original output pressure because the restriction prevents pressure changes from being transmitted immediately. As the increased pressure on

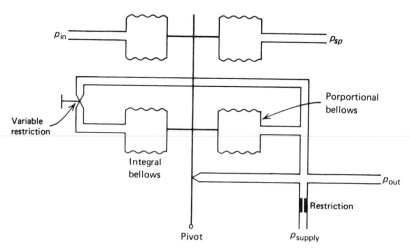

Figure 10.17 Pneumatic proportional-integral controller.

the output bleeds through the restriction, the integral bellows slowly moves the flapper closer to the nozzle, thereby causing a steady increase in output pressure (as dictated by the integral mode). The variable restriction allows for variation of the *leakage rate* and hence the *integration time*.

Proportional-derivative

This controller action can be accomplished pneumatically by the method shown in Figure 10.18. A variable restriction is placed on the line leading to the balance bellows. Thus, as the input pressure increases, the flapper is moved toward the

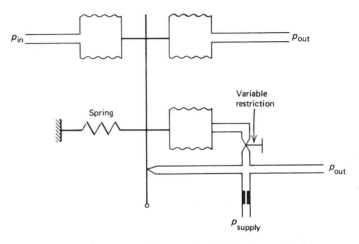

Figure 10.18 Pneumatic proportional-derivative controller.

nozzle with no impedance because the restrictions prevent an immediate response of the balance bellows. Thus, the output pressure rises very fast and then, as the increased pressure leaks into the balance bellows, decreases as the balance bellows moves the flapper back away from the nozzle. Adjustment of the variable restriction allows for changing the derivative time constraint.

Three-mode

The three-mode controller is actually the most common type produced because it can be used to accomplish any of the previous modes by setting of restrictions. This device is shown in Figure 10.19, and, as can be seen, is simply a combination of the three systems presented.

By opening or closing restrictions the three-mode controller can be used to implement the other composite modes. Proportional gain, reset time, and rate are set by adjustment of bellows separation and restriction size.

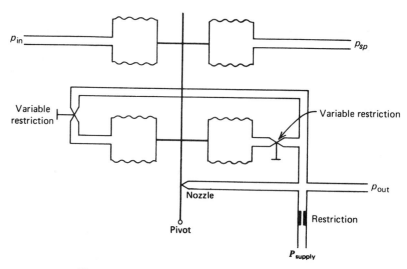

Figure 10.19   Pneumatic three-mode (PID) controller.

10.5 DESIGN CONSIDERATIONS

To illustrate some of the facets involved in setting up a process-control loop, it would be of value to follow through some hypothetical examples. The following examples assume that a process-control loop is required, and that the controller operation must be provided by electronic analog circuits.

Example 10.10

Design a process-control system that regulates light level by outputting a 0–10 volt signal to a lighting system that provides 30–180 lux. The sensor has a transfer function of $-120\ \Omega/\text{lux}$ with a 10-kΩ resistance at 100 lux. The setpoint is to be 75 lux, and proportional control with a 75% proportional band has been selected.

Solution   We solve such problems by first establishing the characteristics of each part of the system.

   1. The illumination varies from 30–180 lux. We find the resistance changes according to

$$R = 10\ \text{k}\Omega - 0.12\ \text{k}\Omega(I - 100)$$

where I is the illumination in lux.

   2. This allows us to find the resistance at 30 lux as

$$R = 10\ \text{k}\Omega - 0.12(30 - 100)$$

$$R = \mathbf{18.4\ k\Omega}$$

and at 180 lux we get **0.4 kΩ**. The setpoint (75 lux) has a resistance of **13 kΩ**.

   3. We can convert this resistance variation to voltage using the photocell in

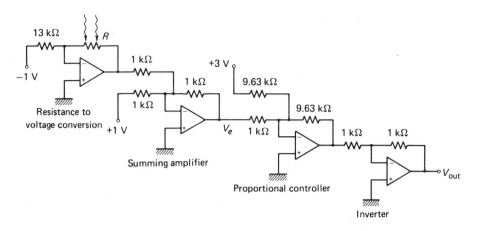

Figure 10.20   Circuit for Example 10.10.

an op amp circuit. In Figure 10.20 we use an inverting amplifier with a gain of 1 at the setpoint and a constant -1-volt input. The resistance to voltage conversion gives

$$V = -\frac{R}{13 \text{ k}\Omega}(-1 \text{ V}) = \frac{R}{13 \text{ k}}$$

Using this equation, we find the output voltage at 18.4 kΩ to be

$$V = -\frac{18.4 \text{ k}\Omega}{13 \text{ k}\Omega}(-1 \text{ V}) = +\text{ 1.42 volts}$$

and at 0.5 kΩ we get

$$V = -\frac{0.5 \text{ k}\Omega}{13 \text{ k}\Omega}(-1 \text{ V}) = \text{ 0.038 volts}$$

so the input voltage range is $1.42 - .038 = 1.382$ V.

4. Now we use a summing amplifier to find the error in Figure 10.20.

$$V_e = \frac{R}{13 \text{ k}\Omega} - 1$$

A 75% proportional band controller with a 75-lux setpoint requires a zero error output of

$$V_0 = \frac{75 - 30}{180 - 30} 10 \text{ V} = \text{ 3 V}$$

5. The 75% band means that when the illumination changes by 75% of (180 $- 30) = 112.5$ lux, the output should swing by 10 volts. Thus, in terms of resistance, this corresponds to 13.5 kΩ and in terms of error voltage it is 13.5 kΩ/13 k or 1.038 volts.

6. Finally, the gain must be

$$G_P = \frac{10 \text{ V}}{1.038} = \text{ 9.63}$$

The overall response is

$$V_{out} = 9.63V_e + 3$$

or

$$V_{out} = 9.63\left(\frac{R}{13\ k\Omega} - 1\right) + 3$$

The rest of the circuit in Figure 10.20 accomplishes this function. When $V_{out} = 0$, $R = 8.9\ k\Omega$ or 90.83 lux and for $V_{out} = 10$ V, $R = 22.4\ k\Omega$ or 203.33 lux so that the output swings 100% as the input swings

$$\frac{203.33 - 90.83}{180 - 30} = 0.75 \text{ or } 75\%$$

as required.

The proportional band could not be used to find the gain directly in Example 10.10 because the input and output were not expressed to the same scale, that is, as 0–100% or 0–10 volts, and so on.

Example 10.11

A type-J thermocouple (TC) with a 0°C reference is used to control temperature between 100°C and 200°C. Design a proportional-integral controller with a 40% band and a 0.08-minute reset (integral) time. The final control element requires a 0–10 volt range.

Solution

(a) In this problem we must perform the following steps:

    1. Amplify the low TC voltage to a more convenient value than the TC mV output.

    2. Use this amplifier output as input to the proportional-integral controller and pick a proportional gain that swings the output 0–10 V as the input swings 40% of full scale.

    3. Select values to provide a 0.08-min (4.8-s) integral time.

(b) The solution is shown in Figure 10.21.

    1. We note that a type-J TC produces a voltage of 5.27 mV at 100°C and 10.78 mV at 200°C. An amplifier with a gain of 100 will convert these to 0.528 and 1.078 volts, respectively.

    2. Now we sum this output to a properly scaled setpoint voltage to get an error signal. The setpoint value is obtained from a voltage divider. To get the proper controller values, we note that 40% of the input swing is

$$0.4(1.077 - 0.527) = 0.2204 \text{ V}$$

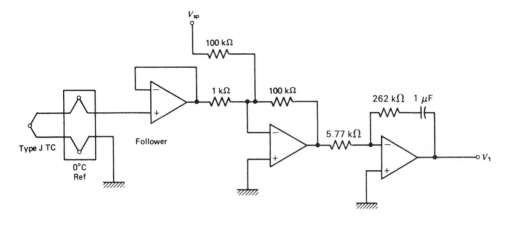

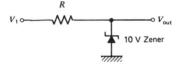

Figure 10.21 Circuit for Example 10.11.

Thus, the proportional gain is

$$G_P = \frac{10 \text{ V}}{0.2204 \text{ V}} = 45.37$$

So values are to be chosen to provide this gain.

3. For the integral term, an 0.08 minute reset means $K_I = 12.5\%/(\% -$ min) or $(12.5/60) = 0.21\%/(\% - \text{s})$. Thus, an error of 1% for 1 second must produce a change in output of 0.21%.

$$G_I = \frac{(0.0021)(10 \text{ V})}{(0.01)(0.551 \text{ V})} = 3.81 \text{ s}^{-1}$$

so

$$\frac{R_2}{R_1} = 45.37, \qquad R_2C = 0.262 \text{ s}$$

Let us try $C = 1 \text{ }\mu\text{F}$. Then

$$R_2 = 262 \text{ k}\Omega \quad \text{and} \quad R_1 = 5.77 \text{ k}\Omega$$

The overall transfer function for the final circuit shown in Figure 10.21 is found to be

$$V_{\text{out}} = 45.37V_e + 173.2 \int V_e dt$$

where

$$V_e = 100V_{TC} - V_{sp}$$

The output diode and zener limit the swing from 0–10 volts.

Example 10.12

A differential pressure gauge is used to measure flow that varies as the square root of the pressure difference (Equation 7.8). The pressure signal is a 0–2 volt range for minimum to maximum flow. A *square root extractor* circuit is available that accepts from 0–10 volts and outputs the square root of the input. Design a proportional controller with a 15% proportional band having a 0–10 volt output and a nominal (zero error) output of 5 volts.

Solution   The circuit of Figure 10.22 implements this function. The controller input is a 0–3.162 volt signal. A 15% proportional band means that if the input changes by (0.15)(3.162 V) = 0.474 V, the output must change by 10 volts. Thus, the gain is

$$G_P = \frac{10}{0.474} = \mathbf{21.1}$$

This is provided by the 1 kΩ and 21.1 kΩ resistors.

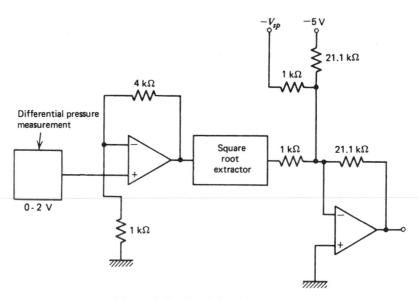

Figure 10.22   Circuit for Example 10.12.

SUMMARY

This chapter presented numerous methods of implementing the controller function of a process-control loop. It shows typical methods of obtaining controller modes from analog, electronic, and pneumatic approaches. If the reader understands *these* typical methods, then other specific methods of implementation can be analyzed and understood by analogy.

The topics covered are summarized by the following:

1. Realization of controller modes with op amps is obtained by a straight application of amplifier, integrator, and differentiator circuits using standard op amp techniques. The gains are found by the external resistors and capacitors used with the op amps.

2. Pneumatic controller mode implementation is made possible by a combination of a flapper/nozzle system, appropriate bellows, and variable flow restrictions. In general, given a three-mode controller, any of the other composite modes is obtained by opening the restrictions.

PROBLEMS

Section 10.3

10.1 A sensor converts position from 0 to 2.0 m into a 4–20 mA current. An error detector such as Figure 10.2 is used with $R = 100 \ \Omega$, $V_0 = 5.0$ V, and $R_{sp} = 1$ kΩ pot. **(a)** If the setpoint is 0.85 m, what is V_{sp}? **(b)** If $V_{sp} = 1.5$ V, what is the range of error voltage as position varies from 0 to 2.0 m?

10.2 Show how the circuit of Figure 10.3 can be applied to find the error voltage for Problem 10.1.

10.3 Using the system of Figure 10.5, design a two-position controller with a 0- to 10-volt input and a 0- or 10-volt output. The setpoint is 4.3 V and the neutral zone is to be ±1.1 volts about this setpoint.

10.4 Design a two-position controller that turns a 5-volt light relay ON when a silicon photocell output drops to 0.22 volts and OFF when the cell voltage reaches 0.78 volts.

10.5 Design a two-position controller which provides an output of 5 volts when a Type J TC junction reaches 250°C and drops to a low of 0 volts when the temperature has fallen to 240°C. Assume a 0°C reference.

10.6 A three-position controller is shown in Figure 10.23. Show how this circuit implements three-position behavior. Derive equations for the input setpoint voltages at which transition to output states occurs.

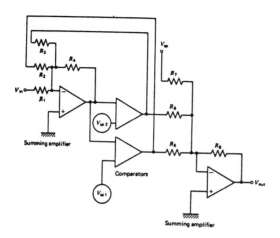

Figure 10.23　An electronic three-position controller.

10.7 **(a)** Design a 45% PB controller for motor speed control. The motor speed varies from 100 to 150 rpm for an input control voltage of 0 to 5 volts. A speed sensor linearly changes from 2.0 to 5.0 kΩ over the speed range. A setpoint of 125 rpm is desired for which the motor control circuit input is 2.5 volts. **(b)** Suppose the setpoint is changed to 120 rpm with no other adjustments. What offset error will occur?

10.8 A type J TC with a 0°C reference is used in a proportional mode temperature control system with a 140°C setpoint and a range of 100–180°C. The zero error output should be 45% and the PB = 35%. The output is 0–10 volts, and the full-scale input range is 0 to 1 volt. Design a controller according to the circuit of Figure 10.9.

10.9 An integral controller has an input range of 1 to 8 volts and an output range of 0 to 12 volts. If $K_I = 12\%/(\%\text{-min})$, find G_I, R, and C.

10.10 Design a proportional-integral controller with an 80% PB and a 0.03-min reset time. Use a 0–5 V input and a 0–12 V output. See Figure 10.13 for the circuit.

10.11 Derive the PI response of Figure 10.13 and show it is equal to Equation (10.8).

10.12 Design a PD controller with a 140% PB and a 0.2-min derivative time. The fastest signal speed is 1 minute. Measurement range is 0.4 to 2 volts, and the output is 0 to 10 V.

10.13 A liquid-level system converts a 4–10 m level into a 4–20 mA current. Design a three-mode controller that outputs 0–5 V with a 50% PB, 0.03-min reset time, and 0.05-min derivative time. Fastest expected change time is 0.8 minutes.

Section 10.4

10.14 A proportional pneumatic controller has equal area bellows. If 3–15 psi signals are used on input and output, find the ratio of pivot distances that provides a 23% PB.

10.15 If the setpoint in Problem 10.14 is 7 psi and the zero error output is 9.2 psi, find the inputs yielding outputs of 3 psi and 15 psi.

Section 10.5

10.16 Explain how the setpoint in Example 10.10 can be changed. Implement such a change to provide a setpoint of 90 lux. Show all new component values for a proportional band of 48%.

10.17 Show how the circuit of Problem 10.10 can be modified to provide switched reset times of 0.02, 0.04, 0.06, and 0.08 minutes.

10.18 Design a proportional controller for a 4–20 mA, ground-based input, a 0- to 9-volt output, zero error output adjustable from 0 to 100%, and a K_P adjustable from 1 to 10. Design so the setpoint can be selected in the range 4 to 20 mA.

10.19 A sensor measures pressure as 22 mV/psi. Develop signal conditioning, error detection, and a PI controller for the following specifications: 0–300 psi measurement range; setpoint adjustable from 100 to 200 psi; K_P adjustable from 1.5 to 5.0; K_I switchable between 0.8, 1.6, 2.4, and 3.2 $\min^{-1}$; output range of 0 to 10 volts.

CHAPTER 11

DIGITAL CONTROLLERS

INSTRUCTIONAL OBJECTIVES

After a comprehensive study of this chapter, including the end-of-chapter problems, you should be able to

1. Give examples of how single- and multiple-variable alarms are implemented in process control.
2. Draw a diagram of a typical data logging system for use in process control.
3. Explain how computer supervisory process-control operations are used in an analog process-control loop.
4. Draw a diagram of a direct digital control system with identification of each element.
5. Explain the effect of ADC time and computer execution time on data sampling rate.
6. Contrast microcomputers and mainframe computers as applied to process control.
7. Define the effects of aliasing in data sampling systems.
8. Explain how controller modes are implemented in DDC.
9. Determine the computer flow diagram for typical DDC applications in process control.

11.1 INTRODUCTION

Digital electronics received its initial impetus to produce smaller, faster, and cheaper digital integrated circuits (ICs) from the computer industry. As ICs were developed, applications extended to many other areas, such as digital watches, TV tuners, and electronic calculators. In process control, there are numerous areas where digital circuits are used *directly*, such as alarms and multivariable control. A few examples of these applications will be presented in this chapter.

The evolution of digital computers, which have higher speed, higher reliability, smaller size, and reduced cost, brought about increased use of digital computers in process control. One of the first computer applications was *data logging*, where the computer is used to collect and store the vast amount of measurement data produced in a complex process and to display the data for review (by process engineers) to determine the condition of the process. Gradually, the computer performed certain kinds of reduction of this data using control equations and even indicated the type of action, if any, that should be taken to tune a process for maximum operating efficiency. All of the loops in the process were still analog and for the most part independent except for manual adjustment of setpoint under the guidance of a process engineer.

A natural extension of this concept led to the development of a technique where the computer itself performs adjustments of loop setpoints and provides a record of process parameters. The loops are still analog, but the setpoints that determine the overall process performance are set by a computer on the basis of equations solved by the computer using measured values of process parameters as inputs. Such a system is called *supervisory computer control.*

The programmable controller (PC) presented in Chapter 8 is one modern application of computers to control systems. A computer is used to provide control of systems that can be described in terms of discrete states. Control action consists of driving a process through a pattern of such states in time, where the progression of states is dependent on measurements of the present state of the system. In such an application, everything associated with the system under control is expressed digitally; that is, everything is an inherently two-state variable.

The ultimate result of computer applications in process control has been to use the computer to perform *continuous controller* functions. In such a computer-based system, often called *direct digital control (DDC),* the only analog elements left in the process-control loop are the measurement function and the final control element. This chapter gives an overall view of digital electronic applications and computer-based process control.

11.2 DIGITAL ELECTRONICS METHODS

There are some instances in process control where digital logic circuits provide the desired control. In general, the controller-mode equations studied in Chapter 9 are too complicated to be implemented by logic circuits, but simple on/off or two-position control are exceptions. In this section, we will see how logic circuits provide this type of control.

Of course, for very complex systems it is more practical to use a programmable controller to provide the control functions. There remain many cases, however, when simple digital electronic circuits offer an easy and more economical solution to required control.

11.2.1 Simple Alarms

One of the simplest digital applications to process control is the implementation of simple alarm circuits. These are very elementary binary processes because we are only concerned about whether a variable is *above* or *below* an alarm *level*. In industrial manufacturing operations, there are many variables, over and above the process-controlled variable, to be monitored. A system may operate without control of pressure, but if the pressure exceeds some preset limit, then an alarm is generated and some corrective action is taken. In this sense, the alarm is similar to two-position controller operation. In digital circuitry, a simple alarm is constructed from a *comparator*, where a voltage level indicates the alarm condition.

Example 11.1

Design an alarm that provides a logic high of 5 volts when a liquid level exceeds 4.2 m. The level has been linearly converted to a 0–10 volt signal for a 0–5.0 m level.

Solution The information given shows that level L and voltage V are related by 10 V per 5 m or 2 V/m, that is

$$V = 2L$$

so that the critical level of 4.2 m is $V_L = \dfrac{2 \text{ V}}{\text{m}}(4.2 \text{ m}) = 8.4$ volts.

A simple comparator with the level signal compared to 8.4 volts provides the desired alarm function. Above 8.4 volts, an alarm signal is generated.

Alarms generally are indicators of trouble and mean that other than normal control procedures are required.

11.2.2 Two-Position Control

It is possible to develop a two-position controller using digital electronics methods. Such a controller differs from the simple alarm comparator just presented by the deadband that exists between ON and OFF state transitions.

The hysteresis comparator of Chapter 3 (Figure 3.7) can be used for two-position control. The only problem is that the equations given for high and low trip points are only good if the deadband is small, which requires $R_f \gg R$. Otherwise, although the hysteresis effect is still present, the trip points and deadband depend on the resistances in a more complicated manner. This type of circuit for two-position control only should be used when the deadband is small compared to the trip voltages. It is really intended for noise suppression.

In a more general way, it is possible to develop two-position control using a combination of comparators and digital logic circuits. Figure 11.1 shows one such circuit. Two comparators and a D-flip/flop are used. One comparator determines the upper trip voltage and the other the lower.

Operation of the circuit can be described by noting the output changes as V_{in} ranges from a low to a high value.

1. $V_{in} < V_L$; B will be low and, thus, the output Q will be low because the F/F is in the clear state.
2. $V_{in} > V_L$ but $<V_H$; B will go high but the F/F output remains low because, although not in the clear state, the F/F has not been clocked to pass the D input through to the Q output.
3. $V_{in} > V_H$; The A comparator goes high. This clocks the F/F, passing the high D input through to the Q output. Thus, the output goes high.
4. On return, $V_{in} < V_H$ but $>V_L$; A will go LOW but this has no effect on the F/F output, which stays in the HIGH output state.
5. $V_{in} < V_L$; B goes low, which places the F/F in the clear state so that the output Q goes LOW.

Steps 1 through 5 describe a typical two-state controller. The deadband is determined by

$$\Delta V = V_H - V_L \qquad (11.1)$$

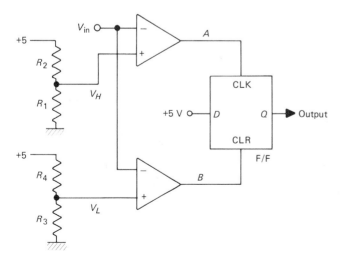

Figure 11.1 Comparators and a flip/flop (F/F) can be used to make a digital two-state controller.

where the high and low trip voltages can be obtained from dividers as shown.

$$V_H = R_1 V_0/(R_1 + R_2) \tag{11.2}$$

$$V_L = R_3 V_0/(R_3 + R_4) \tag{11.3}$$

Example 11.2

Temperature is measured with a response of 15 mV/°C. Devise a two-position controller that turns a 115-vac fan ON if the temperature reaches 70°C and OFF when it falls to 40°C.

Solution Let us use the two-position controller of Figure 11.1. The trip voltages are

$$V_H = (0.015 \text{ V/°C})(70°C) = 1.05 \text{ V}$$

and

$$V_L = (0.015 \text{ V/°C})(40°C) = 0.6 \text{ V}$$

Any combination of divider resistors will do. Practically, it is desirable to keep currents in the mA range, so let us make $R_2 = R_4 = 1$ kΩ. Then, assuming $V_0 = 5$ V, the other resistors are easily found to be $R_1 = 3.76$ kΩ and $R_3 = 7.33$ kΩ. A simple 5-volt relay is used to switch the fan ON and OFF. The circuit is shown in Figure 11.2 using a standard TTL 7474 F/F and a type LM319 dual comparator. The 500-Ω resistors on the comparator outputs are necessary because the LM319 has open-collector outputs.

11.2.3 Multivariable Alarms

Some alarms are not predicated on the state of one variable but on the relative states of *several* variables. If each of the variables is expressed through two states, that is, HIGH and LOW, then digital approaches are ideally suited for a multi-variable alarm. The general procedure is to express the variables as Boolean parameters and to find the Boolean equation between the variables that gives an alarm (HIGH output). This equation then is implemented using digital logic circuit elements.

Example 11.3

In Figure 11.3, a holding tank is shown for which liquid level, inflow A, and inflow B are monitored. These measurements are converted to voltages and then, with comparators, to digital signals that are high when some limit is exceeded. The flow variables FA and FB will be **0** for low flow and **1** for high flow. The level variables are such that L_2 is **1** if the level *exceeds* the lower limit and L_1 will be **1** if the level *exceeds* the upper limit. The alarm will be triggered if either of the following conditions occurs:

1. L_2 low and neither FA nor FB high
2. L_1 high and FA or FB or both high

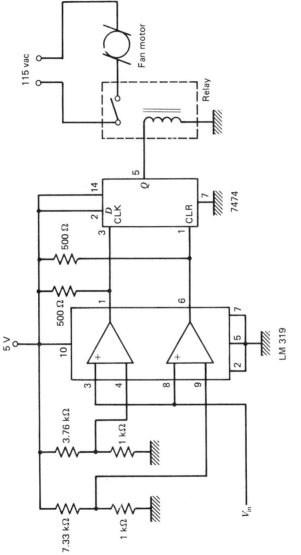

Figure 11.2 Solution for Example 11.2.

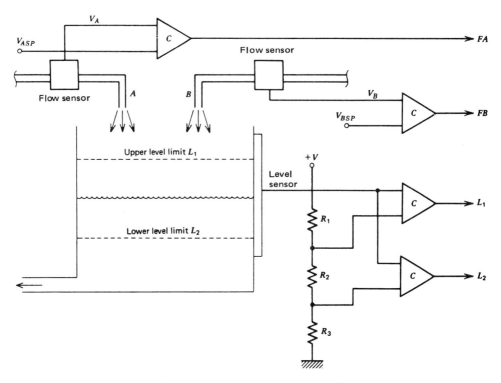

Figure 11.3 Holding tank level-control system for Example 11.3.

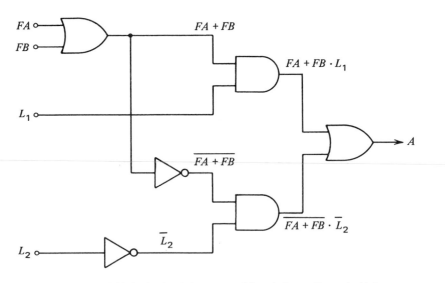

Figure 11.4 This logic circuit is one possible solution to Example 11.3.

Implement this problem with digital logic circuits.

Solution   The variables FA, FB, L_1, and L_2 are already Boolean in that they have values of logic **0** or **1**. We first write Boolean equations giving an alarm output $A =$ **1** for the given two conditions. This can be done directly as

1. $A = \overline{L_2} \cdot \overline{(FA + FB)}$
2. $A = L_1 \cdot (FA + FB)$

Now either of these conditions is provided by an OR operation

$$A = \overline{L_2} \cdot \overline{(FA + FB)} + L_1 \cdot (FA + FB)$$

Logic gates that can be used to directly implement this equation are shown in Figure 11.4.

11.3 COMPUTERS IN PROCESS CONTROL

There are a number of ways that computers are employed in the process industries. Of course, the continued reduction in size and price of computers coupled with the improvement in reliability and computing speed has caused increasing use of computers. In this section the nature of the common applications of computers in the process industries is summarized.

11.3.1 Programmable Controllers

You already learned in Chapter 8 about the extensive use of computers as programmable controllers (PCs). In this highly specialized application, the computer is used to control process operations that are two-state in nature. The computer used for such control is dedicated to this highly specialized application and is not recognized as a computer. This is true to the extent of using a peculiar "programming" language or technique, called the ladder diagram.

11.3.2 Data Logging

The efficient operation of a manufacturing process may involve the interplay of many factors, such as production rates, materials costs, and efficiencies of control. When the process requires implementation of many process-control loops, then the interaction of one stage of the system with another often can be analyzed in terms of the controlled variables of the loops. An example of this is the rate of production of one loop, expressed as a flow rate, which serves as a determining factor in the production rate of a following control system. Historically, an understanding of this type of interaction required analysis, after the fact, of strip-chart recordings taken from process parameters during a production run. Such analysis, carried out by trained personnel, may then dictate settings of operational limits of future production runs.

With the development of high-speed digital computers with mass digital storage, it became possible to record such data continuously and automatically, display the data on command, and perform calculations on the data to reduce it to a form suitable for evaluation by appropriate technical individuals.

Fixed loggers

The general features of a computer data logging system are shown in Figure 11.5. Let us assume the process is under the control of many analog process-control loops and there is provision for analog process variable measurements to be available as a commonly scaled voltage. Thus, some signal conditioning converts all measurements into a given range, often a specified voltage range as required by a data acquisition system. A brief accounting of the elements of the system is given later.

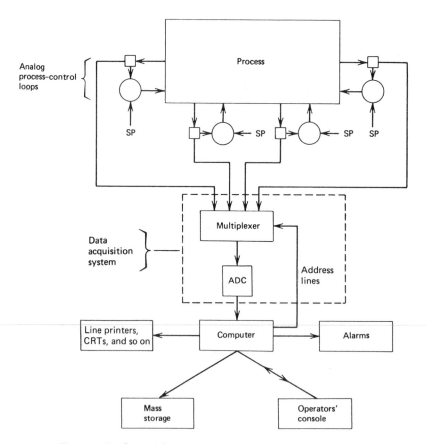

Figure 11.5   General features of a data logging system using a computer.

Data acquisition system (DAS)

The data acquisition system, discussed in detail in Chapter 3, is the switchyard by which the computer inputs samples of process variable values. The concept of "samples" of these values is an important topic that will be discussed in more detail later in this chapter. The reason for concern over this point is that there are situations where the sample rate can be such that erroneous information about the variable variations results. The rate at which samples of a process variable can be taken depends on how long it takes for the DAS to acquire a value, how long it takes the computer to process the value, and how many other variables are to be sampled. This is illustrated in Example 11.4.

Alarms

An important part of any data logging circuit is an alarm system that monitors inputs from excursions beyond some specific limits. With scan rates of the data as high as 5000 per second, it is possible for a computer to maintain a very tight vigilance over variable values. Every time the computer inputs a particular variable, the value is compared to its preset limits which, if exceeded, trigger an alarm.

Computer

The computer, of course, is the central element in the system. Through programming, the computer accepts inputs and performs prescribed reductions of the data through mathematical operations. The results are evaluated by further programmed tests to oversee the operation of the entire process from which the inputs are taken. Projections of future yields, evaluations of efficiency, deviation trends, and many other operations can be performed and made available to process personnel.

Peripheral units

The peripheral units are the *support* equipment to communicate computer operations to the outside world. These units include the *operator console* where the programs are entered and through which commands can be given to initiate specific actions, such as calculations and data outputs by the computer. The console usually has a CRT/keyboard and a typewriter unit for input and outputs. A mass storage system, such as magnetic tape, is used to store data, such as periodically sampled inputs from the process, that can be used in later, more detailed analysis of process performance.

Example 11.4

A data logging system such as that shown in Figure 11.5 must monitor 12 analog loops. A small computer requires 4 μs per instruction and 100 instructions to address a multiplexer line and to read in and process the data in that line. The ADC performs the conversion in 30 μs. The multiplexer requires 20 μs to select and capture the value of an input line. Calculate the maximum sampling rate of a particular line.

Solution   The 100 instructions require a time of (4 μs)(100) = 400 μs, and this must be done for 12 loops. Thus, the total instruction time is (12)(400 μs) = 4800 μs. The ADC converts in 30 μs so that for 12 conversions we have (12)(30 μs) = 360 μs, and the total time spent in multiplexer switching is 240 μs. Adding 4800 + 360 + 240 = **5400 μs** as the minimum time before a particular line can be readdressed. The maximum sampling rate is the reciprocal, or **185** samples per second.

Portable data loggers

There are many cases when data needs to be logged for a period of time from a loop and no fixed logger is provided. A portable data logger can be temporarily connected to the measurement output of the loop for this purpose.

A portable data logger has many of the same features as a fixed installation, such as a DAS, computer, or an operator's console. Of course, some portable data loggers do not use a computer; they are merely strip-chart recorders or magnetic tape recorders.

In general, the computer-based portable data loggers have some mechanism for saving the logged data for later analysis. Possible recording mediums are

1. Printed output
2. Digitized strip-chart recording
3. Magnetic tape
4. Magnetic floppy disks
5. Networked data communication

In network data communications, the data logger may be connected to a local area network (LAN). The data then can be transmitted over the network to another, fixed computer installation. The data can be stored on mass-storage facilities that are part of the network.

11.3.3 Supervisory Control

A natural extension of a computer data logging system involves computer feedback on the process through automatic adjustment of loop setpoints. As various loads in a process change, it is often advantageous to alter setpoints in certain loops to increase efficiency or to maintain the operation within certain precalculated limits. In general, the choice of setpoint is a function of many other parameters in the process. In fact, a decision to alter one setpoint may necessitate the alteration of many other loop setpoints as interactive effects are taken into account. Given the number of loops, interactions, and calculations required in such decisions, it is more natural and expedient to let a computer perform these operations under program control.

Such a system is represented in Figure 11.6, where the effect is shown by the addition of a data output system (DOS). Such a system assumes the controllers of analog loops have been designed to accept setpoint values as some properly scaled voltage. By proper switch addressing, the computer then outputs a signal through the multiplexer and DACs, representing a new setpoint to a controller connected to that output line.

It might be helpful in understanding use of this type of control if a hypothetical example is given. Study for a moment the reaction process shown in Figure 11.7. The process specifications are

1. Reactants A and B combine such that one part A to two parts B produce one part C.

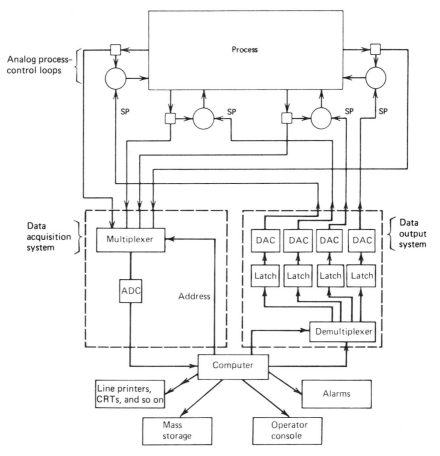

Figure 11.6 In computer supervisory control, the computer sets the setpoints of analog process-control loops.

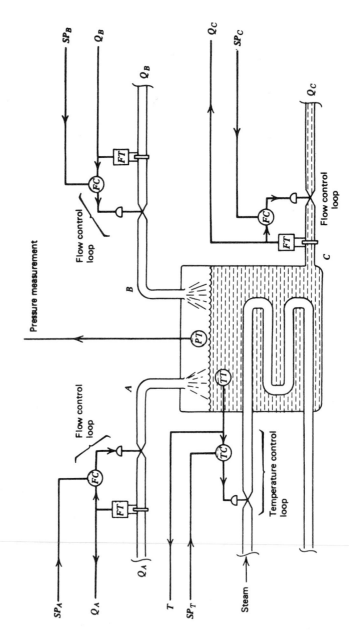

Figure 11.7 Computer supervisory control is ideally suited to strongly interacting, nonlinear control problems.

2. Volume production of C varies as the square root of the A and B flow rate product.

3. The operating temperature must be linearly decreased with C volume production rate.

4. For stability, the reaction must occur with the pressure maintained below a critical value.

A decision is made to increase production in this operation. The first step to accomplish this is to increase the flow rate of A by a change in setpoint. Let us see the consequences of this in the rest of the process.

1. The setpoint of B flow must be set to twice the A setpoint, keeping pressure below p_{max}.

2. The setpoint of C flow must be increased by the square root of the new A and B flow rates.

3. The temperature setpoint must be decreased by a proportion of the new C setpoint.

To accomplish this change in a purely analog system requires monitoring pressure constantly while the operations of the three steps are gradually performed and the new production rate is finally established. With each new setting of setpoint, we must wait until all parameters have adopted the new setpoints and wait for a safe pressure. To perform this manually requires constant human monitoring as the adjustments are made. In a supervisory control system, the computer performs these operations automatically while still performing other activities in the production.

Flowchart

To describe the steps a computer must go through to operate in some specified manner, we use an event flowchart. This is the same as the flow diagram used by computer programmers, and in fact some required programs can often be written (coded) directly from such a process-operation diagram. For purposes of illustration, we will use only three types of symbols to prepare a flowchart. These symbols are presented in Figure 11.8a. In using such a diagram, we do not have to get lost in the details of *how* the input, output, operations, and decisions are made, and we can thus better design the overall solution. The next step would be to consider these details.

The event flowchart by which a computer might accomplish monitoring is shown in Figure 11.8b. Remember, analog control loops are maintaining the control variables at the setpoint values. The boxes labeled INPUT refer to computer commands to address the input multiplexer to obtain the current values of these parameters. The OUTPUT boxes serve a similar function for the output multiplexer. The notes on the flow diagram indicate the function of each section. An

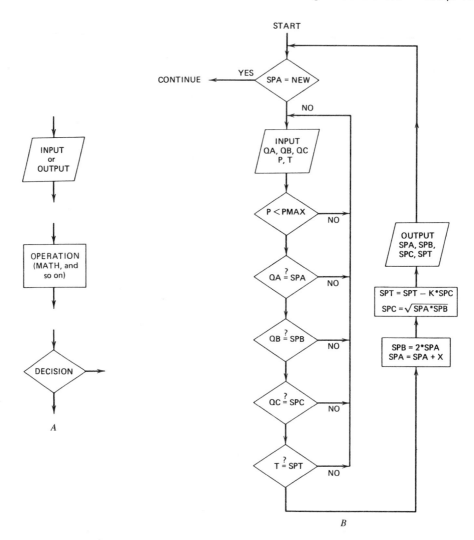

Figure 11.8 Flowchart symbols and an example for setpoint changes in supervisory process control.

important feature of a computer supervisory control system is that it produces the desired change in operation rate in the minimum possible time. The completion of one run through the instructions in Figure 11.8b might typically require less than 100 μs for an average computer. Most of the adjustment time is spent waiting for the loops to stabilize. The instant such stabilization occurs, the next increment of setpoints is made by the computer via the controller.

Example 11.5

Develop the supervisory control flowchart of a system to increase the temperature setpoint of a pressure reaction vessel to a new value, TSPNU. The temperature setpoint (TSP) is to be increased in steps of 0.2% with a 5-second delay between increases. If the pressure (P) rises above a critical value (PCR), the TSP should be decreased by 0.1% until P falls below PCR. Then setpoint increases can begin again.

Solution   Basically, we just build up a flowchart that satisfies each specification of the description. Setpoint increases are done by the operation of $1.002 * TSP \rightarrow TSP$ and decreases by $0.999 * TSP \rightarrow TSP$. It is assumed that either a hardware or software timer is available with its status available to the software. The result is shown in Figure 11.9.

11.3.4 Computer-Based Controller

The ultimate application of digital methods in process control is the use of a digital computer to perform the controller's functions. In this case the analog loop is gone and the computer becomes a fundamental part of the feedback system. The use of a computer for process control is not in and of itself new because an analog controller is actually an *analog computer*. What is new is the use of a *digital computer* for the controller, that is, as direct digital control (DDC).

Hardware

The hardware associated with computer-based control is essentially the same as that for supervisory control. In fact, the principal difference is that the analog controller is eliminated. Figure 11.10 shows a typical configuration for a DDC system. The measurement system, which consists of sensor and signal conditioning, is still present as well as the final control element.

Note that in Figure 11.10 a single computer is used to control four loops. It is quite common in computer-based control to let a computer control more than one loop. The speed with which the computer can compute control output usually exceeds typical process rates by many orders of magnitude. Thus, it is practical to let the computer control several loops.

In addition to the usual data acquisition and output, operator's console, and mass memory hardware, Figure 11.10 shows a *network interface*. In computer-based control applications it is often advantageous, particularly in large-scale installations, to interface the controller computer with other computers and facilities in the plant, that is, a *local area network* (LAN). With this interface it may be possible for a process engineer to monitor the values of dynamic variables of some part of the process from his or her office.

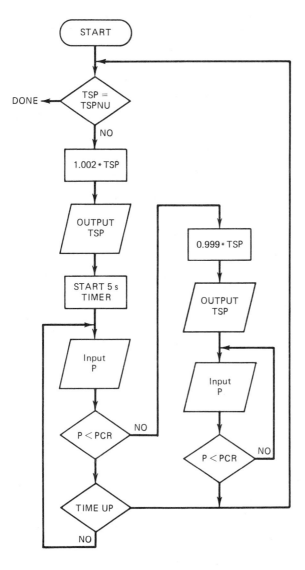

Figure 11.9 Flowchart solution of Example 11.5.

Software

For the computer to perform controller functions, the error must be determined and the controller-mode equations solved to determine the necessary output to the final control element. These operations are performed by software. The computer must have programs to solve the equations of the proportional, integral, and derivative modes introduced in Chapter 9. Later in this chapter, we will study the essential features of such software.

There are many types and sizes of computers available for industrial appli-

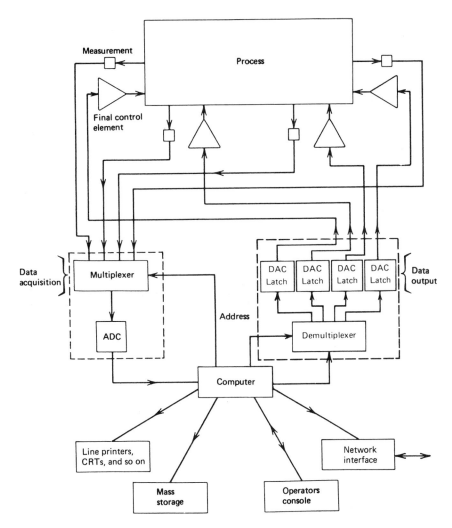

Figure 11.10 In computer-based control, the analog process-control loop is replaced by digital computer operations.

cation. The following paragraphs describe several types of computers employed as process controllers and in the process industries.

Single-board computers (SBCs)

A modern computer-based controller is often designed and built on a single printed circuit board, which is referred to as a single-board computer (SBC). When packaged for industrial use, such a controller often looks much like an analog controller such as that illustrated in Figure 10.1. Error, setpoint, and output readouts are usually digital, but this is also true of many modern analog controllers.

Figure 11.11 shows a block diagram of the elements of an SBC controller. The principal element is the microprocessor, which performs all actual data computations and manipulations required by the software. The controller program is contained in permanent memory, usually solid-state read-only memory (ROM). Temporary storage during calculations and other functions use the volatile random access memory (RAM). RAM means that this memory can be written to as well as read from, and *volatile* means that the memory is erased when power is lost.

This type of controller often is used for control of a single loop such as the analog controller it replaces. In addition, it is often designed to directly input and output the same 4–20 mA transmission signals in common use in control systems. In fact, it often can be used as a direct plug-in replacement for an analog controller.

Programmable controller

In Chapter 8 you learned about the programmable controller (PC) as a computer-based system to provide control of discrete-state systems. All data input and control output was described as two-state, either ON or OFF. The inputs and outputs are often 115-vac signals.

Because the PC already has an internal computer, it is a logical extension to building the software and hardware required to provide continuous mode controller functions. Thus, many PCs also have the capability of providing controller actions in the proportional, integral, and derivative modes. For these applications, PCs are much like the SBC controller because they have the facility to input and output dc signals such as the common 4–20 mA current.

Setpoint and gains are set as part of the ladder diagram program. When a

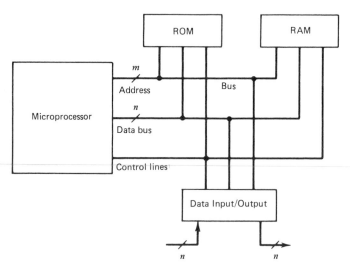

Figure 11.11 An SBC process controller. The programs are contained in the ROM.

rung is executed that contains the continuous mode controller function, the appropriate controller-mode equations are solved and a control signal to the final control element is output.

Personal computers

Many of the popular personal computers can be used as process controllers. We must be careful to distinguish between personal computers referred to as PCs and programmable controllers also referred to as PCs. We will restrict the use of PC to programmable controller in this text. In general, a personal computer will need special hardware in the form of I/O boards and software consisting of programs for data input/output and the controller modes. This approach to computer-based control is particularly suitable for small-scale or isolated control system needs.

Figure 11.12 shows the typical configuration for a personal computer application. Of course, the computer operation must be dedicated to the control system and thus cannot be used for other computing needs. It is possible to provide communication with other computers and remote terminals via the communication port. Such a computer also can be tied into a network.

Mainframes

A *mainframe* computer generally means a large-scale computing facility that requires special housing, serves many users, and has high computing speed. Such computers generally are not used for actual controller operations. They are more suitable as supervisors of an entire plant. As such, they receive measurement and status information from controller computers in the plant and issue updates of setpoints and other control commands.

The mainframe is the overseer of plant operation. It also is used for financial evaluation, engineering studies of plant operation, inventory control, personnel operations, and many other management activities. Figure 11.13 suggests how a fully integrated computer-based process-control installation might be configured.

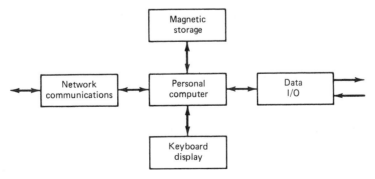

Figure 11.12   A personal computer process controller. Special software will also be required.

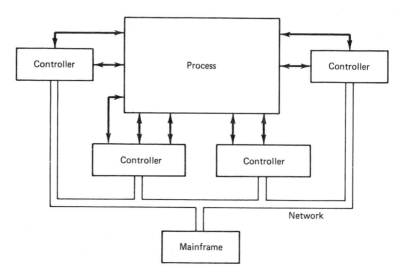

Figure 11.13 Mainframes are used with a network to manage SBC and personal computer process controllers.

A LAN links all computers so that information and commands can be exchanged. Each computer on the network operates according to its own program and performs its assigned duties. Network interface software is also resident that allows the computers to make and receive calls over the network.

Control output

The result of a solution to the mode equation for a particular variable is the updated setting of the final control element for that loop. This is provided by addressing the output multiplexer for that particular loop and outputting the calculated value. If the final control element is a digital device such as a stepping motor, then the output is direct. If it is an analog device, a DAC is used to convert the digital information into a properly scaled analog signal. If the final control element requires constant excitation to hold a particular state, as a control value, then a latch must be provided to hold the excitation when the computer is not outputting to that loop. In this instance, the output of the computer serves to update or refresh the state of the final control element.

Alarms and interrupts

An important feature of any computer controller system is its alarm and interrupt facilities. The *alarm* has the same meaning as in supervisory control where a signal is generated to notify process personnel that a parameter has exceeded some preset limits. Such an alarm is often both an audio and a visual signal to attract attention to the condition. Often, the computer will have a programmed set of operations to perform under an alarm condition, such as an orderly shutdown of some facet of the process. In any case, however, such an alarm in DDC is usually a computer's cry for help in a situation beyond its direct control.

The *interrupt* feature enables the operator or an external process condition to halt the normal computer operation and initiate some other procedure. Thus, boiler pressure may be constantly monitored by the computer on an interrupt line that goes HIGH when pressure exceeds a preset limit. The computer never directly addresses the line to see if the pressure is HIGH, but if it does happen, internal mechanisms cause the computer to immediately stop its present execution and perform some other operation, such as turning the heater off. In this case, it is similar to an alarm except the computer never addresses and examines the condition specifically. The interrupt also is used to terminate normal operations so that new data, setpoints, and so on can be input by an operator. Interrupts are often provided on a *priority* basis. The levels of priority may be such that high priority requires immediate cessation of present execution and low priority allows completion of a particular program instruction set before the interrupt condition is given attention.

11.4 CHARACTERISTICS OF DIGITAL DATA

There are many advantages to using computers for the controller function. For example, one computer can handle many loops, interaction between loops can be accounted for in the software, data is less susceptible to noise-induced errors, and linearization can be easily provided by software. There are other advantages that soon will be generally available, such as self-tuning, error correction, and automatic failure recovery.

It seems there should be a price for these advantages, and, indeed, there are some disadvantages to the use of computers in controller operations. A serious disadvantage is that conversion of analog data into digital data results in a *loss* of knowledge about the value of the variable. The nature and consequences of this will be considered.

11.4.1 Digitized Value

Chapter 3 discussed analog-to-digital conversion (ADC) of analog data into a digital format. The format of the ADC output was an n-bit binary representation of the data. With n-bits it is possible to represent 2^n values, including zero. There is a finite *resolution* of the physical data being represented of one part in 2^n, and that means we now are ignorant about the value of the variable after it has been converted into the binary representation.

In equation form we can write the relation between a physical variable and its n-bit digital representation as

$$N = \frac{(V - V_{min})}{(V_{max} - V_{min})} 2^n \qquad (11.4)$$

where

N = base 10 equivalent of binary representation
V = input value
V_{max} = maximum input value
V_{min} = minimum input value

Only the integer part of the right side of Equation (11.4) is used to determine N. Equation (11.4) assumes that the measurement system and ADC have been designed so that the binary output switches from the equivalent of $2^n - 1$ to 2^n just at V_{max}. The resolution of the measurement can be found by noting what change in V will produce a single integer (bit) change of N. This is easily seen to be

$$\Delta V = (V_{max} - V_{min})/2^n \qquad (11.5)$$

The following example illustrates some of the consequences of the digital conversion of data. A careful study of this example will help you to understand the limitations of digital representation.

Example 11.6

A temperature between 100°C and 300°C is converted into a 0- to 5.0-volt signal. This signal is fed to an 8-bit ADC with a 5.0-V reference. What is the actual measurement range of the system? (a) What is the resolution? (b) What hex output results from 169°C? (c) What temperature does a hex output of C5H represent?

Solution   The nature of the ADC is that the output will change to FFH at a voltage of $5.0 - 5/256 = 4.98$ V (299.22°C) and would change to 100H at exactly 5.0/V (300°C). Thus, FFH would seem to mean any temperature between 299.22°C and 300°C. Because there is no 100H or higher (only 8-bits), FFH actually means any temperature *greater* than 299.22°C. Similarly, the output will be 00H for any voltage less than $5/256 = 0.0195$ V, which is a temperature of 100.78°C so 00H is output for any temperature *less* than 100.78°C. Thus, the *actual* measurement range is from 100.78°C to 299.22°C.

(a) The temperature span of (300°C − 100°C) = 200°C is divided into $2^8 =$ 256 values. Therefore, the resolution is given by Equation (11.5).

$$\Delta T = 200°C/256 = 0.78°C/bit$$

Let us make sure we understand what this means. If the temperature is 100°C, the output will be 00H. It will *stay* at this value until the temperature reaches 100.78°C; then it will change to 01H. Thus, the resolution of 0.78°C means that we are ignorant about the value of the temperature by this amount with any reading.

(b) For every temperature within the range, there is one specific output value. Thus, for a temperature of 169°C, we can find that value by Equation (11.4).

We find the fraction of the measurement range 169°C represented by

$$N = \frac{(169 - 100)}{(300 - 100)} 2^8 = \frac{69}{200} 256 = (0.345)(256)$$

$$N = 88.32$$

But only the integer part is used, so $88_{10} \rightarrow$ 58H and, therefore, 169°C $\rightarrow$ 58H. Another way to get this result is to divide the quantity (169 − 100) by the resolution to find the fraction of the binary number, $69/0.78 = 88.46 \rightarrow$ 58H (round-off error accounts for the difference).

(c) Now our ignorance of value really shows up. What temperature does C5H represent? The procedure is quite straightforward. We simply solve Equation (11.4) for T knowing that C5H $\rightarrow 197_{10}$.

$$197 = \frac{(T - 100)}{(300 - 100)} 256$$

$$T = (197)(200)/256 + 100$$

$$T = 253.9°C$$

But wait! The fact is that the hex value will stay C5H until the temperature increases by the resolution, 0.78°C. So the actual answer only can be correctly stated as, "The temperature is *between* 253.9°C and 254.68°C."

One of the consequences of the digitizing resolution is that we cannot be expected to control a value any closer than this resolution. If we were supposed to control temperature to within ±0.2°C using the measurement system of Example 11.6, it would be impossible because we do not know its value within that tolerance.

Problems of resolution are reduced by using more bits in the digital word. With 16-bits, for example, the resolution is one part in 65536. With a 16-bit ADC using a 5.0-V reference, the least significant bit is toggled for voltage changes of only 5.0 V/65536 = 76.3 μV! Thus, noise becomes a severe problem in a typical industrial environment.

11.4.2 Sampled Data Systems

The previous section dealt with the consequences of having only discrete knowledge of the value of the physical variable. Consider also that we have only discrete knowledge of the value *in time*. That is, the computer control system takes only periodic samples of the variable value. Thus, we are ignorant of the value or variation of the variable between samples. For the control system to function correctly, certain conditions must be assumed about variations between samples. That is what sampled data systems are all about. In the following section we will consider the nature and consequences of having only periodic samples of the physical variable.

Sampling rate

The key issue with respect to sampling in a computer-based controller is the rate at which samples must be taken. The sample rate is expressed either through t_s, the time between samples, or $f_s = 1/t_s$, the sampling frequency.

There is a maximum sampling rate in any system, that is, the time required to take a sample (ADC conversion time) plus the time required to solve the controller equations to determine the appropriate output (program execution time).

There is a minimum sampling rate in any system that depends on the nature of the time variation of the sampled variable. Simply put, samples must be taken at a high enough rate so that the signal can be reconstructed from the samples. There are serious consequences to sampling at too small a rate. For example, the control system will not be able to correct variations of the controlled variable that are missed because too few samples were taken.

Figure 11.14 illustrates the consequences of sampling rate on knowledge of signal variation. The actual signal is shown in Figure 11.14a. Figures 11.14b, 11.14c, and 11.14d illustrate knowledge about signal variation deduced from various sample rates. Reconstructions of the original signal are indicated by the dashed lines between samples.

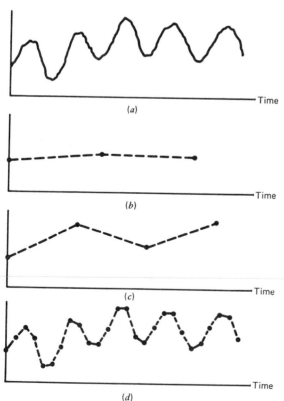

Figure 11.14 Samples must be taken fast enough to allow reconstruction of the data. The samples in (d) are just enough to allow crude representation of the real data in (a).

The sampling rate of Figure 11.14b is much too slow because little information about the actual signal variation is contained in the reconstruction from the samples.

For Figure 11.14c, the signal seems to possess a frequency of variation that is not in fact present in the actual signal. This is called *aliasing*, and it is one consequence of too small a sampling rate.

The sampling rate of Figure 11.14d shows that the essential features of the signal can be reconstructed from the samples. A general rule for the minimum sampling rate can be deduced from the maximum frequency of the signal. The rule is that for adequate reconstruction of the signal from samples, the samples must be taken at a frequency that is about 10 times the maximum frequency of the signal.

$$f_s = 10f_{max} \qquad\qquad (11.6)$$

where

f_s = sampling frequency
f_{max} = maximum signal frequency

This is, of course, equivalent to taking 10 samples within the shortest period of the signal.

To determine the minimum sampling rate, an estimate must be made of the highest possible frequency (shortest possible period) of the signal. The sampling frequency will be 10 times that value. For Figure 11.14d, six samples are taken instead of 10 in one basic period of the data. The reconstruction from samples is somewhat crude, but the basic structure is present.

Example 11.7

The plot of Figure 11.15 shows typical data taken from pressure variations in a reaction vessel. Determine the maximum time between samples for a computer control system to be used with this system.

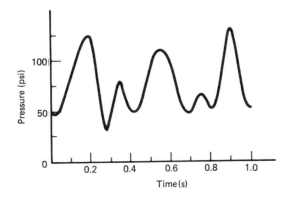

Figure 11.15 Pressure data for Example 11.7.

Solution An examination of the signal of Figure 11.15 shows that the shortest time between any two peaks is 0.15 seconds. This gives a maximum signal frequency of $f_{max} = 1/0.15 = 6.7$ Hz. From Equation (11.6), the minimum sampling frequency is given by $f_s = 10f_{max} = (10)(6.7\,\text{Hz}) = 67$ Hz. The maximum time between samples is $t_s = 1/f_s = 15$ ms.

11.5 CONTROLLER SOFTWARE

When the controller function is taken over by a computer, error determination and controller mode equation evaluation are done using software. In this section the nature of the software required to perform these functions is considered.

11.5.1 Software Format

The particular format of the software used for the computer controller depends on the computer and its programming language. Let us consider some formats in which the software may be described.

Flowchart

The most universal way of presenting a description of the required software is to use a flowchart. The basic flowchart concept was presented in Section 11.3.3, on supervisory control. The most common symbols were presented in Figure 11.8.

The flowchart defines the basic operations and decisions required to input a controlled variable value, process it, and output the required final control element drive signal. The flowchart will be used to define the operations necessary to support the three controller modes: proportional, integral, and derivative.

A flowchart is independent of the computer and programming language being used to describe a solution to some problem. Furthermore, once a flowchart solution to a problem is developed, it is quite easy to construct the required computer and language-dependent specific program.

Machine language

The internal operations of a computer are directed by a sequence of *instructions*. These instructions perform operations such as adding two numbers, fetching a number from memory, storing a number in memory, and so on. Most computers have from 50 to several hundred such instructions. Within the computer each of these instructions appears as a unique binary number. Thus, a binary number like 10010010_2 might be an instruction to add two numbers. *All* programs eventually end up as a combination of such binary numbers. This is referred to as a *machine-language* program and to build programs directly by such instructions is machine-language programming. This is very difficult to do.

A program is simply a sequence of such binary instructions in memory. The computer fetches each instruction from memory and executes it. A machine-language program to obtain two numbers from memory, add them, and place the result back into memory is shown in Figure 11.16. The two numbers are obtained from memory locations, A500H and A501H, and the sum is stored back into A502H.

Assembly language

The assembly programming language assigns an alphabetic code or mnemonic to each machine-language instruction of a computer. The code acts as a reminder of the function of the instruction. Thus, the addition of two numbers might be indicated by ADD B and storage of a number in memory might take STA A500H, where A500H is a memory address in hex. It is much easier for programmers to construct programs using these mnemonics.

A special program called an *assembler* then translates the assembly language program into machine language so that the program can be executed by the computer. This type of programming is still quite difficult and requires rather extensive knowledge of the architecture of the computer and all the instructions of the computer.

SBC controllers often are programmed initially in assembly language. The resulting debugged and error-free machine-language program is then stored in ROM for execution by the computer during controlling operations.

For comparison, Figure 11.17 shows the assembly language program for the same operations defined by the machine language program of Figure 11.16. Even

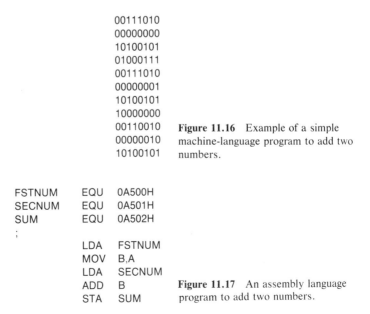

```
00111010
00000000
10100101
01000111
00111010
00000001
10100101
10000000
00110010
00000010
10100101
```

Figure 11.16   Example of a simple machine-language program to add two numbers.

```
FSTNUM    EQU    0A500H
SECNUM    EQU    0A501H
SUM       EQU    0A502H
;
          LDA    FSTNUM
          MOV    B,A
          LDA    SECNUM
          ADD    B
          STA    SUM
```

Figure 11.17   An assembly language program to add two numbers.

memory addresses can be referenced by mnemonics. In this case we call the memory locations of the numbers FSTNUM, SECNUM, and the result is stored in SUM. The EQU statements show the actual numeric values of the hex addresses.

Higher-level languages (HLL)

To facilitate the practical use of computers, special programming languages have been developed that do not require the programmer to have detailed knowledge of either the computer architecture or the instruction set of the computer. These languages use human language statements and common math symbols to describe the required operations.

Programs written in these languages are translated into machine-language programs by a variety of other programs called, for example, compilers and interpreters. In a higher-level language, the operations defined by Figures 11.16 and 11.17 might be written simply

$$SUM = FSTNUM + SECNUM$$

Higher-level language programming is supported by personal computers and mainframes. If it was necessary to write your own controller programs, it could be done most easily using one of these languages. Typical examples of higher-level languages include FORTRAN, BASIC, PASCAL, ADA, and C.

11.5.2 Input Data Operations

The first phase of software use involved in computer-based controllers is to input the data and put it into a form suitable for controller-mode evaluation.

Data input

An ADC has a certain conversion time required to develop the appropriate digital output for some analog input. Certain communications must pass between the computer and the ADC for a data sample to be taken.

In general, this involves the operations shown in the flowchart of Figure 11.18.

1. Program issues a start convert (SC) signal to ADC.
2. Program starts a timer.
3. Program enters a loop.
   a. If an end-of-convert (EOC) is issued by the ADC, control passes to 4.
   b. If no EOC is issued before the timer times out, control passes to an error routine.
4. Program inputs a data sample.

A certain length of time is allotted for the converter to indicate that the conversion is complete. If such a response is not issued within this time, an error

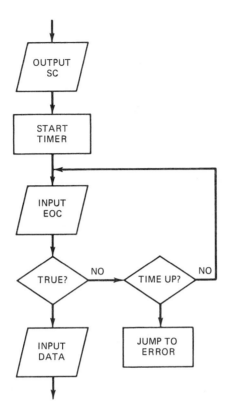

Figure 11.18   This flowchart shows the typical operations required to take a data sample from an ADC.

routine is initiated. Such a routine is necessary in case the ADC has failed and is unable to respond with a data sample. Without the error escape, the computer and control system would be "locked up" in a wait loop waiting for the EOC response from the ADC.

When a higher-level language is used, it is necessary that special software be provided showing the user how to construct the data input process. This is often done with user-defined functions (UDFs). In these cases the program sequence of Figure 11.18 would be written in assembly language and called up from the higher-level language with some command like $Y = UDF(1)$.

Linearization

In many cases the input binary number and the controlled variable are not linearly related. In such cases it is necessary to execute a program that will linearize the binary number so that it is proportional to the controlled variable value. There are two common approaches: equation inversion and table look-up.

Linearization by equation

When an equation is known that relates the value of the controlled variable and the binary number in the computer, an equation can be developed to determine the linearized value of the variable. For example, suppose that a transducer outputs a voltage related to pressure by

$$V = K[p]^{1/2} \tag{11.7}$$

This voltage is converted to a binary number DV by an ADC. Then it is also true that the binary number and pressure are still related by the square root

$$DV \, \alpha[p]^{1/2} \tag{11.8}$$

What we want is a binary number that is linearly related to pressure. The way to get this is to square DV

$$DP = DV * DV \, \alpha \, p \tag{11.9}$$

Thus, the program would input a sample DV and multiply it by itself. The resulting number would be linearly related to the pressure. Of course, there may have to be scale shifts and offsets before we have a number equal to the pressure, as the following example shows.

Example 11.8

Pressure from 50 to 400 psi is converted to voltage by the relation

$$V = 0.385[p]^{1/2} - 2.722$$

This is input to an ADC with a 5.0-V reference, which provides 00H to FFH over the pressure range. An HLL uses an instruction $DV = UDF(1)$ to input the data from the ADC as a base 10 number DV that varies from 0 to 255 over the pressure range. Develop a linearization equation to give a quantity p in the program that is equal to the actual pressure.

Solution We have enough information to work backwards through the ADC, signal conditioning, and measurement. Thus, we know the voltage is related to DV by the ADC transformation

$$DV = (V/V_{\text{ref}})256 = (V/5)256$$

or

$$V = (5/256)DV$$

Then, using the known relation for V in terms of p

$$0.385[p]^{1/2} - 2.722 = (5/256)DV$$

and solving for p

$$p = (0.0507DV + 7.071)^2$$

This number p in the program is equal to the actual pressure value, 50 to 400 psi.

Linearization by table look-up

There are many measurement processes where it is impossible to find a simple equation such as Equation (11.7) to relate the controlled variable and the binary number. Also, when the program must be written in assembly language, it may be difficult to evaluate even quite simple equations such as that of Example 11.8. In these cases it becomes much easier to use the look-up table approach.

This is really just what we humans do when we use thermocouple tables, for example. We measure the voltage, go to a table, and look up the temperature, and sometimes interpolate. It is the exact same thing with the software approach. The table of input values and corresponding physical variable values are stored in a table in memory. Following a measurement, the input value is looked up in the table and the correct measured value found.

Figure 11.19 shows a software approach for table look-up in flowchart format. It is assumed that the input values are stored in ascending value in N memory locations and the corresponding physical variable values are stored in the following N locations. Thus, if the input is found at the Ith memory location from the start of the table, then the actual variable value is found at the $N + I$ location.

Of course, there are many other methods of table construction and search. In many cases it is necessary to write interpolation routines using programming equations such as Equation (4.14) to refine the values between table values.

11.5.3 Controller modes

The operation of controller action is entirely taken over by software within the computer, that is, by programs. The process-controlled variable has been mea-

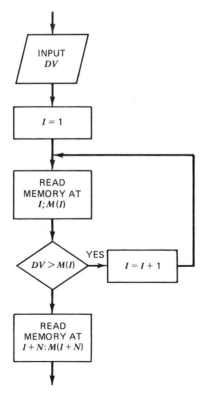

Figure 11.19 Linearization by table look-up can be accomplished by the operations of this flowchart.

sured in samples and provided to the computer, the value of the setpoint has been input by the computer operator via keyboard or other input device, and a program calculates the error and the required controller output to the final control element. It is interesting that the control algorithm, that is, the calculations required as controller operation, are the same three *modes* used in analog controllers: *proportional, integral (reset), and derivative (rate)*. In this section we discuss briefly what kinds of algorithms are used to solve the controller-mode equations by computer.

Error

In the computer, error can be represented by the same equation used to define error as percentage of span in Chapter 9.

$$e_p = \frac{r - b}{b_{\text{max}} - b_{\text{min}}} \cdot 100 \qquad (9.3)$$

It is important, even in computer-based controllers, to express both error and output as a percentage or fraction of a range. All the gains are determined on the basis that error is a percentage of measurement range and that this error determines some change in output, also expressed as the percent of full output. In the computer we often leave the error as a fraction of range. This is just Equation (9.3) without the 100.

In the computer software the error is expressed by an equation such as

$$DE = \frac{DSP - DV}{DMAX - DMIN} \qquad (11.10)$$

where

$$DV = \text{measured input}$$
$$DSP = \text{setpoint}$$
$$DMAX = \text{maximum of range}$$
$$DMIN = \text{minimum of range}$$

If the relation between the controller variable and the input is nonlinear, linearization will be necessary before the error can be computed. In these cases, when using an HLL, it is often convenient to express the measurement in terms of the actual variable being measured.

Suppose voltage input to an 8-bit ADC with a 10-volt reference is related to the physical variable, pressure, by the relation

$$V = 3.1[p + 10]^{1/2} - 9.8$$

The pressure p is in psi and the setpoint is 15 psi. Thus, the voltage and input digital word are nonlinearly related to the pressure. Assuming that an HLL is being used for the programming, we can easily construct equations to compute the actual pressure from the input digital word.

Suppose we call the input digital word *DIN*. This varies from 0 to 255_{10} as the voltage varies from 0 to 9.961 V. Thus, the voltage is found from

$$DV = DIN * 10/256$$

The pressure is now found by solving this equation for pressure in terms of voltage

$$DP = [(DV + 9.8)/3.1]^2 - 10$$

The error as a fraction of range can be found as

$$DE = (15 - DP)/30.6$$

The range of pressure is known to be 30.6 psi by working backward from the voltage range of 0 to 9.961 V.

Proportional mode

In Chapter 9 we saw that the proportional mode controller action is defined by a term that is directly proportional to the error. The equation is

$$p = K_P e_p + p_0 \qquad (9.14)$$

where

K_P = proportional gain
e_p = error
p_0 = controller output with no error

The gain is expressed as percent controller output per percent error. The concept of proportional band (*PB*) is defined as $1/K_P$ and represents the percentage error that will cause a 100% change in controller output. This mode is easily implemented by the computer in the form of an algorithm that simply calculates Equation (9.14) directly.

The proportional mode is provided through the software by an equation that is entirely like the analog equation. Because we are expressing the error as a fraction of range, what is calculated is the fraction of the maximum output.

$$P = P0 + KP * DE \qquad (11.11)$$

$$POUT = P * ROUT \qquad (11.12)$$

where

$P0$ = fraction of output with no error
KP = proportional gain (%/%)
P = fraction of output with error
$ROUT$ = maximum output
$POUT$ = output

Example 11.9

A proportional mode has $K_P = 2.4$, input range of 255, and a setpoint of 130. The output maximum is 180 and the output fraction with no error is 0.45. **(a)** Develop the control equations (what is the output for no error?) and **(b)** find the output for an input of 124.

Solution

(a) The equations are found simply from Equations (11.10), (11.11), and (11.12).

$$DE = (130 - DV)/255$$

$$P = 0.45 + 2.4 * DE$$

$$POUT = P * 180$$

When there is no error ($DE = 0$), $POUT = 0.45 * 180 = 81$.

(b) For an input of 124, we get an error fraction of

$$DE = (130 - 124)/255 = 0.024$$

$$P = 0.45 + 2.4 * (+0.024) = 0.45 + 0.0576$$

$$= 0.5076$$

$$POUT = 0.5076 * 180$$

$$= 91.4$$

When these control equations and those that follow are implemented in assembly language, it is usually necessary to express everything in hex. For the purposes of this text we will assume that an HLL is used so that a base 10 representation can be used.

Figure 11.20 shows a general flowchart for the proportional mode from which software can be developed. Of course, there may be added complications such as the need for linearization and specification of the input and output software.

Integral Mode

The integral or reset mode calculates a controller output that depends on the history of the controlled variable error. In a mathematical sense, history is measured by an integral of the error

$$p = K_I \int_0^t e_p dt + p(0) \tag{9.17}$$

where

K_I = integral gain in percent controller output per percent-second error (or, more commonly, per minute)

To use this mode in computer control, we need a way of evaluating the integral

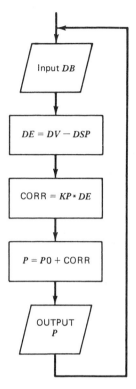

Figure 11.20 Proportional mode flowchart.

of error. There are many algorithms that have been developed to do this, all of which are only approximate, as only samples of the error in time are available. The simplest is called *rectangular* and is often accurate enough for use in process control. To see how this works, you should note that the integral in Equation (9.17) is merely the net area of the e_p curve from 0 to t. This is shown in Figure 11.21.

$$\int_0^t e_p dt = \text{net area} = (\text{area of } e_p > 0) - (\text{area of } e_p < 0)$$

In rectangular integration, we simply use the periodic samples of e_p to construct a series of rectangles of height equal to the sample error and width equal to the time between samples. The integral (or area) is then approximately equal to the sum of the rectangle areas. This is shown in Figure 11.21b. In an equation, rectangular integration specifies that

$$\int_0^t e_p dt \approx [S + e_{pi}] \Delta t$$

where

Δt = time between samples

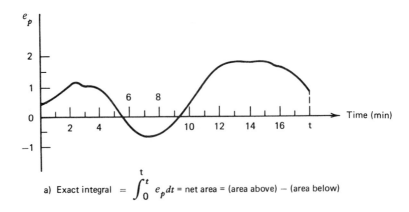

a) Exact integral $= \int_0^t e_p dt =$ net area = (area above) − (area below)

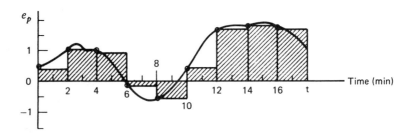

b) Approximate integral = sum of rectangle areas

Figure 11.21   Rectangular integration algorithm construction.

$S = e_{p1} + e_{p2} + \cdots$ (sum of errors calculated from previous variable samples)

$e_{pi} =$ last sample taken at time t specified in the integral

It should be clear that the smaller the time between samples, the more closely the approximate answer will approach the actual integral.

The issue of sampling time becomes important for the integral mode. If the time between samples is too large, the area will be in serious error and control will be compromised. If the criteria developed earlier of $f_s = 10 f_{max}$ is satisfied, then the integral term also will be of sufficient accuracy using the rectangular algorithm. The following example illustrates the effect of sampling time variation.

___

Example 11.10

Find the approximate integral of e_p in Figure 10.21a from 0 to 14 minutes. Do this for the sample time shown in the figure (2 minutes) and again for a sample time of

1 min. What percentage change in the value of the integral results from the difference in sample time?

Solution   For the sample time of 2 minutes, we find the integral from the rectangular integral procedure as

$$Integral_{2\ min} = 2(0.4 + 1.1 + 1 - 0.2 - 0.6 - 0.4 + 1.6)$$

$$Integral_{2\ min} = 7.4\% - min$$

For a sample time of 1 minute, we find the integral as

$$Integral_{1\ min} = 1(0.4 + 0.6 + 1.1 + 1.1 + 1 + 0.4 - 0.2 - 0.6$$

$$- 0.6 - 0.4 + 0.4 + 1.1 + 1.6 + 1.7)$$

$$Integral_{1\ min} = 7.6\% - min$$

This gives a percentage change in integral value between the two approaches as

$$Change = \frac{7.6 - 7.4}{7.6} \times 100 = +3.7\%$$

Implementation of this mode in software involves the following basic equations

$$SUM = SUM + DE \qquad (11.13)$$

$$PI = KI * DT * SUM \qquad (11.14)$$

$$POUT = PI * ROUT$$

where

SUM = a running sum of errors
KI = the integral gain
DT = time between samples
PI = fraction of maximum output

The flowchart of Figure 11.22 illustrates how such a mode can be programmed. The time delay routine must be built in to provide the required time between samples, because time appears as part of the mode equation [Equation (11.14)] and must therefore be known. Another important point is that the units of KI and DT must be the same.

An alternate way of writing the integral mode equations is to use the present error sample to modify the previous output. To see this, let us substitute Equation (11.13) into Equation (11.14).

$$PI = KI * DT * (SUM + DE)$$

$$PI = KI * DT * SUM + KI * DT * DE$$

The first term in this equation is simply the PI from the previous sample. There-

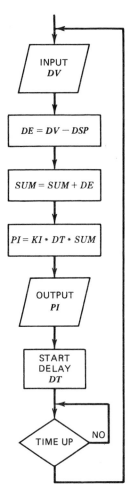

Figure 11.22 Flowchart of the integral mode by rectangular integration.

fore, we can write

$$PI = PI0 + KI * DT * DE$$

$$POUT = PI * ROUT \tag{11.15}$$

$$PI0 = PI$$

where

$PI0 = PI$ from previous sample

Derivative mode

The derivative controller mode, also called rate, derives a controller output that depends on the instantaneous rate of change of the error

$$p = K_D \frac{de_p}{dt} \tag{9.18}$$

where

K_D = derivative gain (% per %/s error)

$\frac{de_p}{dt}$ = rate of error change in percent per second (or minute)

The gain expresses the percent controller output for each percent/second change in error. This mode is implemented in computer control by calculating an approximate derivative of the error from the data samples. A derivative is defined as the rate at which a quantity is changing at an instant in time. We can calculate only the rate at which it is changing over the sample period Δt, which is therefore only an approximation. In terms of an equation, this is

$$\frac{de_{pi}}{dt} \approx \frac{e_{pi} - e_{pi-1}}{\Delta t}$$

where

e_{pi} = present error sample

e_{pi-1} = previous error sample

Δt = time between samples

Figure 11.23 shows that this process results in a derivative that is not the actual derivative. Notice that as the time between samples is made smaller, the error will become less.

Example 11.11

Determine an approximate value of the derivative of e_p at a time of 12 minutes from Figure 11.21a, using samples every 2 minutes and every 1 minute. Compare the results.

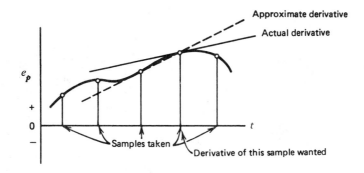

Figure 11.23 Approximate calculation of the derivative from sampled data.

Solution   For 2-minute samples, we get the derivative by using the sample at 10 minutes and at 12 minutes.

$$\text{Derivative}_{2 \text{ min}} = \frac{1.6 - 0.4}{2} = 0.6\%/\text{min}$$

For samples every minute, we use a sample at 11 minutes and at 12 minutes.

$$\text{Derivative}_{1 \text{ min}} = \frac{1.6 - 1.1}{1} = \textbf{0.5\%/min}$$

This means there is a difference of about 17%.

The set of equations for the derivative output can be developed directly from the definitions. We find

$$DDE = DE - DEO \tag{11.16}$$

$$DEO = DE \tag{11.17}$$

$$PD = KD * DDE/DT \tag{11.18}$$

The flowchart for this mode is presented in Figure 11.24. As in the case of the integral mode, it is very important that the units of KD and DT agree. In the following example, the PID control mode section illustrates this point.

PID control mode

The optimum control mode is a composite of the three previous modes: proportional, integral, and derivative. With computer-based control, a composite mode is developed by simply combining the three mode equations into the computation of the fractional output. According to the principles of PID control, the proportional gain should multiply all three terms. The control equations can be written

$$DDE = DE - DEO \tag{11.16}$$

$$DEO = DE \tag{11.17}$$

$$SUM = SUM + DE \tag{11.13}$$

$$PI = KP * KI * DT * SUM \tag{11.19}$$

$$PD = KP * KD * DDE/DT \tag{11.20}$$

$$P = KP * DE + PI + PD \tag{11.21}$$

$$POUT = P * ROUT$$

where all the terms have been previously defined. These equations then are programmed into the control software for determination of the required output.

An alternative expression for the PID output can be constructed by using errors to provide corrections to the current output. To develop this, let us adopt a convention that a subscript will denote a particular sample. Thus, DE_i is the

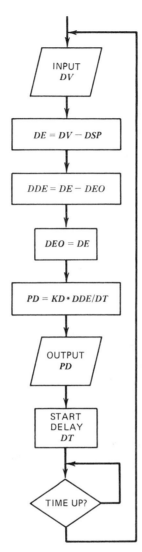

Figure 11.24 Flowchart of the derivative mode.

ith sample and P_i is the fractional output for that sample. The output for the P_{i-1} sample, according to Equation (11.21), can be written in the form

$$P_{i-1} = KP * DE_{i-1} + KP * KI * DT[SUM + DE_{i-1}]$$
$$+ KP * KD * [DE_{i-1} - DE_{i-2}]/DT$$

The result for P_i will be

$$P_i = KP * DE_i + KP * KI * DT * [SUM + DE_{i-1} + DE_i]$$
$$+ KP * KD * [DE_i - DE_{i-1}]/DT$$

Let us take the difference between these two expressions. This will give the correction to the previous output because of the present sample error. The result will be

$$P_i - P_{i-1} = KP[DE_i - DE_{i-1}] + KP * KI * DT * DE_i$$

$$+ KP * KD * [DE_i - 2DE_{i-1} + DE_{i-2}]/DT$$

This equation can be simplified to give

$$P_i = P_{i-1} + A * DE_i - B * DE_{i-1} + C * DE_{i-2} \qquad (11.22)$$

where

$$A = KP + KP * KI * DT + KP * KD/DT$$
$$B = KP + 2 * KP * KD/DT$$
$$C = KP * KD/DT$$

Then the result of Equation (11.22) is used to determine the output from

$$POUT = P_i * ROUT \qquad (11.23)$$

The following example illustrates how the controller equations are developed for a typical problem.

Example 11.12

A digital controller is to be developed with the following specifications: KP = 5%/%, KI = 0.5%/(% − min), KD = 0.08%/(%/min), time between samples = 5 seconds, input range 0 to 255, setpoint = 130. The output range is 0 to 255. Set up the control equations for PID control using both approaches given in the text.

Solution   By the first approach it is merely necessary to express the equations given in the mode description. The first set of equations is given easily by

$$DE = (130 - DV)/(255 - 0)$$

$$DDE = DE - DEO$$

$$DEO = DE$$

$$SUM = SUM + DE$$

To compute the gains we need to set the units the same.

$$KP * KI * DT = (5)(0.5 \text{ min}^{-1})(5 \text{ s})(1 \text{ min}/60 \text{ s})$$

$$= 0.208$$

$$KP * KD/DT = (5)(0.08 \text{ min})/(5 \text{ s})(1 \text{ min}/60 \text{ s})$$

$$= 4.8$$

Thus, the remaining equations become

$$PI = 0.208 * SUM$$

$$PD = 4.8 * DDE$$

$$P = 5 * DE + PI + PD$$

$$POUT = P * 255$$

For the second approach we need to evaluate the three constants of Equation (11.22).

$$A = 5 + 0.208 + 4.8 = 10.008$$

$$B = 5 + 2 * 4.8 = 14.6$$

$$C = 4.8$$

Then Equation (11.22) can be written

$$P = P1 + 10.008 * DE - 14.6 * DE1 + 4.8 * DE2$$

$$DE2 = DE1$$

$$DE1 = DE$$

The last two equations are necessary because the next sample $DE1$ and $DE2$ contains the previous two samples. Then the previous output must be updated.

$$P1 = P$$

11.6 COMPUTER CONTROLLER EXAMPLES

The following examples illustrate some of the difficulties that arise from seemingly simple control problems implemented digitally. It is not the purpose of this chapter or text to give you all the expertise to fully design a DDC system, but merely to give you an understanding of the steps of such a design and how some of the features of the design are actually built into the system.

Example 11.13

A proportional control system with a 50% *PB* controls conveyor speed by using weight measurement. Specifications are

1. Weight range: 40 to 90 lb
2. Weight setpoint: 65 lb
3. Signal conditioning already available converts 0 to 100 lb to a voltage scaled at 0.05 volts/lb, that is, 0 to 5 V
4. An 8-bit ADC converts an input of 0 to 5 volts to 00H to FFH, where 5 volts input just produces FFH
5. Conveyor speed is regulated by a motor control circuit that operates directly from the output of an 8-bit DAC. There are 256 speeds selected according to the following: 00H is off, FFH is full speed, and 80H is the speed corresponding to the setpoint of 75 lb, that is, zero error.

Find the digital controller equations. Express all numbers in their hex form.

Solution   The proportional mode equations are given by Equations (11.10), (11.11), and (11.12). We must define the constants that appear in these equations.

W_{min} = 40 lb so

$$DMIN = (40/100)256 = 102.4 \rightarrow 66H$$

W_{max} = 90 lb

$$DMAX = (90/100)256 = 230.4 \rightarrow E6H$$

W_{sp} = 65 lb

$$DSP = (65/100)256 = 166.4 \rightarrow A6H$$

The error equation becomes

$$DE = (DW - A6H)/(E6H - 66H)$$

$$DE = (DW - A6H)/80H$$

Because the zero error output is given to be 80H, $P0$ = 80H. A 50% PB means K_P = 2%/%. Both the error and output are given as fractions of range so that this value of K_P can be used directly. Thus

$$P = 80H + 2 * DE$$

This is the fraction of full scale output. The final output would be

$$POUT = P * FFH$$

Actually, multiplying an 8-bit number by FFH is effectively equivalent to a shift of 8 bits, so that in terms of numbers

$$POUT = P$$

You can see from this example that the development of actual coding from the original specification involves consideration of many factors including the encoding of numerical data. For more complicated operations, such as finding averages, linearization, integration, and finding derivatives, the math operations in binary become even more complicated. In general, routines are written in a top-down fashion, starting with a general statement of the problems and then working into the details as shown in this example. The following example illustrates a more complicated problem. The problem takes the control algorithm only to the stage of a rather general flow diagram, from which the detailed flow diagram would be developed.

Example 11.14

A DDC system, shown in Figure 11.25, will use PI control with a 40% PB, 0.5%/ min integration time, and a 0.75-minute sample time. The input is to be the average of five temperatures, and the output will be a control signal to a valve. Specifications of the system are as follows:

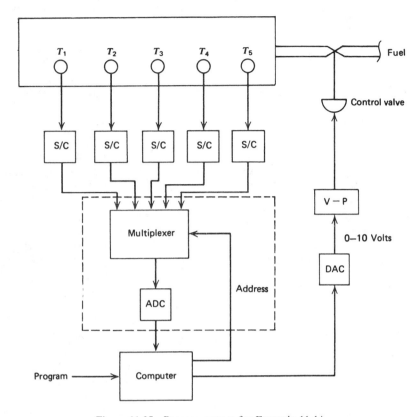

Figure 11.25   Process system for Example 11.14.

1. A temperature measurement system is available that provides a voltage from the temperature transducer with the following relationship

$$V = 0.075T + 0.0004T^2 \qquad (11.24)$$

2. Temperature range is 0°C to 90°C with a 57°C setpoint.

3. For initial startup, the output signal to the valve should be 5 volts. The valve is driven directly from a DAC, adjusted so that 0H is 0 volts and FFH is 10 volts.

4. A data acquisition system is available with five channels and an ADC that converts 0 volts to 0H and 10 volts to FFH.

5. The computer has floating-point hardware so that inputs are converted automatically to floating-point numbers for use in the programs. Thus, 0H input produces 0.0, and FFH produces 225.0_{10}.

Construct the general flow diagram for the control algorithm using rectangular integration.

Solution   We find the solution by working through the system from input to output in steps. For convenience, we will express the temperature error in terms of percent.

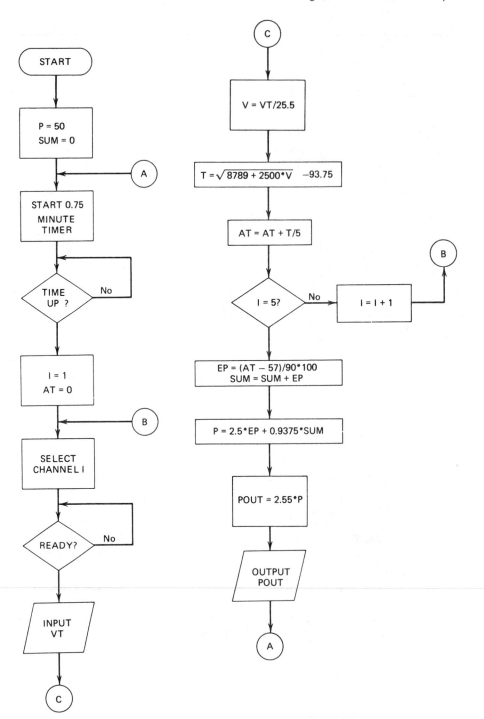

Figure 11.26 Flowchart solution of Example 11.14.

1. *Find the temperature*   The input section will provide a value from 0 to 255 that is the voltage representing the temperature. We can find the voltage by dividing by 25.5. Knowing the voltage, we find the temperature by inversion of the relationship between voltage and temperature, Equation (10.9). First we write it in a quadratic form

$$0.0004T^2 + 0.075T - V = 0$$

From the quadratic root equation we find

$$T = \frac{-0.075 \pm \sqrt{(0.075)^2 - 4(-V)(0.0004)}}{2(0.0004)}$$

This relation reduces to

$$T = -93.75 + \sqrt{8789 + 2500V} \tag{11.25}$$

where we have used the positive sign because we know T must be 0 when the voltage is 0.

2. *Find the average temperature*   We can find the average by simply inputting all five temperatures, by addressing the proper input channels, and then summing and dividing by 5.

3. *Find the error*   The error will be found in percent, as usual, using Equation (9.3)

$$e_p = \left(\frac{AT - 57}{90 - 0}\right) \times 100 \tag{11.26}$$

where AT = temperature average.

4. *Proportional gain*   A 40% PB means that when the error changes by 40% the output changes by 100%. Because e_p is expressed in percent, we have $K_P = 100/40 = 2.5\%/\%$.

5. *Integral term*   The integral gain is the 0.5%/min given in the specification of the problem. For a 1% error the term contributes 0.5% change in proportional-mode controller output every minute. Because the samples from which the integral is constructed are taken every 0.75 minutes, the contribution actually will be (0.5%/min)(0.75 min/sample) or 0.375%/sample.

6. *Output*   The output signal can be constructed from Equation (9.19) along with Equations (11.13) and (11.14), adjusted so that 100% corresponds to 255 so that an output of FFH is sent to the DAC when $P = 100\%$.

The final flow diagram for the controller is given in Figure 11.26. It has been assumed here that the input channels are identified by a single integer, 1 through 5.

SUMMARY

This chapter presented a discussion of the important features of digital and computer applications in process control. General features are as follows:

1. Application of digital methods started with the simple application of digital logic methods to simple control problems and the use of computer installations to provide data logging services in process control.

2. In data logging, advantages of the computer to oversee the operation of a process through high-speed sampling of loop data proved the computer's value in process control. ADC and multiplexer systems allow rates of 5000 samples per second to be taken.

3. Supervisory control lets the computer adjust process-loop setpoints for optimum process performance even though standard analog control loops are still used for control.

4. Computer-based control eliminates the analog controller and replaces the mode implementation by programs within the computer.

5. Computer control can implement the proportional, integral, and derivative modes using approximation algorithms. Sampling time effects must be considered.

PROBLEMS

Section 11.2

11.1 A force transducer has a transfer function of 2.2 mV/N. Design an alarm using a comparator that triggers at 1050 N.

11.2 A type J TC with a 0°C reference monitors a process temperature. Design a two-alarm system turning on one LED at 100°C and another at 150°C.

11.3 Design a two-position digital controller using the force transducer of Problem 11.1. The output should go high at 500 N, and the deadband should be 75 N.

11.4 Assume that a conveyor system speed S, load L, and loading rate R have been converted to binary signals by comparators. What Boolean equation provides an alarm A when either the loading and loading rate are high when the speed is low, or when the speed is high when the load is low?

11.5 Show how digital circuit elements can implement the alarm of Problem 11.4.

11.6 A multiple-variable digital control uses a reaction vessel for which binary inputs of level, temperature, and pressure drive binary outputs to the input valve and the output valve. A digital HIGH opens a valve. The following conditions must be satisfied:

Pressure	Temperature	Level	Input Valve	Output Valve
H	H	H	L	H
L	H	H	L	H
L	H	L	L	L
L	L	L	H	L
H	L	L	H	H
H	L	H	L	L

Devise a Boolean equation for this controller and the digital logic circuit.

Section 11.3

11.7 A data logging system must take samples of 40 variables at 10 samples per second each. What is the maximum signal acquisition and processing time in microseconds?

11.8 A computer must sequentially sample 100 process parameters. It requires 14 instructions at 5.3 μs/instruction for the computer to address and process one line of data. The multiplexer switching time is 2.3 μs, and the ADC conversion time is 34 μs. Find the maximum sampling rate for a line.

11.9 Develop a block diagram similar to Figure 11.6 of a supervisory computer control system for the process illustrated in Figure 11.27. This is a firing operation for ceramic articles. The conveyor motor speed setpoint is determined by the feed rate of articles.

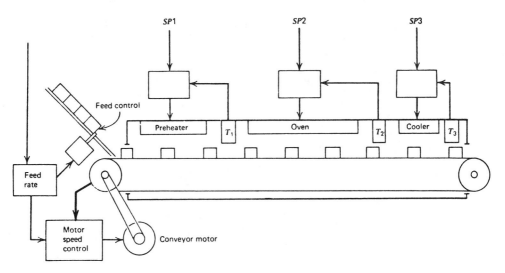

Figure 11.27 Process used for problems.

11.10 For the process of Figure 11.27 under supervisory control, to obtain an increase in conveyor speed by 5% requires that the preheat setpoint be increased by 3.9%, the oven setpoint by 7.2%, and the cooler by 4.4%. The speed increase must be performed in 1% steps with reacquisition of each setpoint before continuing. Prepare a flowchart showing how computer control can provide this increase.

Section 11.4

11.11 A digital control system is to provide regulation of pressure within 1.2 kPa in the range 30 to 780 kPa. How many bits must be used for the data acquisition?

11.12 A computer will be used to control flow through 10 pumping stations. The pumps exhibit a surging effect with a period of 2.2 seconds. What is the minimum sampling rate to assure quality data? How much time can be spent processing each station's data? Data acquisition hardware and software takes 200 μs for a channel.

Section 11.5

11.13 The output voltage of a silicon photovoltaic cell is given by $V = 0.11 \log_e(I_L)$, where I_L is the light intensity in W/m^2. Intensity is to be measured from 100 to 400 W/m^2, and input to a computer is via an 8-bit ADC with a 5.00-volt reference.
  (a) Develop signal conditioning to interface the cell to the ADC so that 00H to 01H occurs at 100 W/m^2 and FEH to FFH occurs at 400 W/m^2.
  (b) Develop software in the language of your choice to linearize the input data so that a program variable is the intensity in W/m^2.

11.14 A Type K TC with a 20°C reference will be used to measure temperature from 200°C to 280°C in a computer controller application. The data acquisition hardware and software makes the actual measured voltage, in mV, available in software as a variable VTC. Devise a flowchart and/or program in the language of your choice which makes the reference correction and performs linearization by table look-up and interpolation. Use table values in 5°C increments for a Type K TC with a 0°C reference.

11.15 A strain gauge is used in a bridge circuit as in Figure 5.14a with $R_1 = R_2 = 120 \, \Omega$, $V = 10.0$ volts, and specifications $R = 120 \, \Omega$, GF = 2.05. Assume the bridge offset voltage is provided as input to a computer and appears in a variable DELTAV. Prepare a flowchart and/or program which will obtain the strain in μs from this voltage.

11.16 The setpoint for the measurement of Figure 11.15 is 75 psi. Find the approximate integral term response at 0.8 seconds assuming a 0.05-s sample time and an integral gain of 2.45 min^{-1}. The control range is 25 to 100 psi.

11.17 A process is to operate under PID with a 60% PB, 1.2-min integration time, and 0.05-min derivative time. If the error is available as percent of span, develop the control equations and show a flowchart of computer controller action with all constants evaluated. The sample time is 0.8 minutes.

Section 11.6

11.18 A PI controller is required for the following control problem:
  (a) Temperature is to be controlled from 50 to 150°C with a setpoint of 100°C.
  (b) A sensor provides temperature-dependent voltage as

$$V = (T - 100)/10 + 5$$

  (c) An 8-bit ADC is used for which 00H is 0 volts and FEH to FFH occurs at 10.0 volts.
  (d) The PB is 70% and the integral gain is 1.5 min^{-1}.
  (e) Sample time is 0.5 minutes.
  (f) Output is through an 8-bit DAC with a 10.0-volt reference.

Develop a flowchart for the control process, control equations, and a program in the language of your choice.

11.19 Measurement of position L in mm is provided in terms of voltage by the relation

$$V = (e^{0.02L} - 1)/2$$

This if fed directly into an 8-bit ADC with a 5.00-volt reference. Develop the control equations for the following controller specifications: $K_P = 3.7$, $K_I = 0.9$ min^{-1}, 2-second sample time, setpoint of 50 mm, 8-bit output.

11.20 Add derivative action to Problem 11.19 with $K_D = 0.1$ s.

CHAPTER 12

CONTROL LOOP CHARACTERISTICS

INSTRUCTIONAL OBJECTIVES

This chapter provides process-control technologists with a general background in the practical considerations of process-control loop implementations. After reading this chapter, you should be able to

1. Explain the characteristics of single-variable, compound, cascade, and multivariable control.
2. Define three standard measures of quality in a control system.
3. Describe the control loop stability criteria with respect to a Bode plot.
4. Describe the open loop transient disturbance method of loop tuning.
5. Describe the Ziegler–Nichols method process-control loop tuning.
6. Define phase and gain margin.
7. Explain how the frequency response method can be used to tune a process-control loop.

12.1 INTRODUCTION

In the previous chapters of this text, the detailed elements of process control have been described. The instrumentation used to perform the process-control function has been presented in some detail. At this point, the reader is similar to an individual who has acquired detailed knowledge of all the elements of an airplane.

Such a person, however knowledgeable about the airplane parts, is certainly *not* considered competent to *pilot* the aircraft. In process control, there also is a remarkable difference between knowing the elements of a process-control loop and being able to tune and operate a process-control loop installation.

In this chapter, consideration is given to the various ways of setting up a process-control loop system and tuning this system for optimum performance. A word of caution: A complete understanding of such *optimization* requires detailed mathematical study of stability theory, a substantial depth of knowledge of the process, and many years of experience. The general intent of this chapter is merely to make the reader familiar with the general concepts.

The objectives of this chapter have been carefully selected to provide the reader with as much background as can be expected without a more extensive mathematical background. The emphasis is not on the theoretical development of process-control loop characteristics but on the interpretation and employment of the results of such theory. Many volumes are devoted to studies of loop stability, optimization techniques, multiloop DDC operations theory, and a host of other topics that could never be included in this brief chapter. In short, this chapter provides the process-control technologist with a general background in the practical considerations of process-control loop implementations.

12.2 CONTROL SYSTEM CONFIGURATIONS

We consider first the types of control loop configurations that are encountered in typical industrial applications. The control loop concept as outlined in the previous chapters is a correct one but necessarily oversimplifies many of the actual configurations used in industry. The decision of the arrangement to select is made by the process designers based on the goal of the process, relative to production requirements, and the physical characteristics of operations under control.

12.2.1 Single Variable

The elementary process-control loop (which has formed the basis for most of our discussion in previous chapters) is a *single-variable* loop. The loop is designed to maintain control of a given process variable by manipulation of a controlling variable, regardless of the other process parameters.

Independent single variable

In many process-control applications, certain regulations are required regardless of other parameters in the process. In these cases a setpoint is established, controller action is started, and the system is left alone. Thus, in Figure 12.1, a flow-control system is used to regulate flow into a tank at a fixed rate determined by the setpoint. This system then makes adjustments in valve positions as necessary following a load change to maintain *flow rate* at the setpoint value.

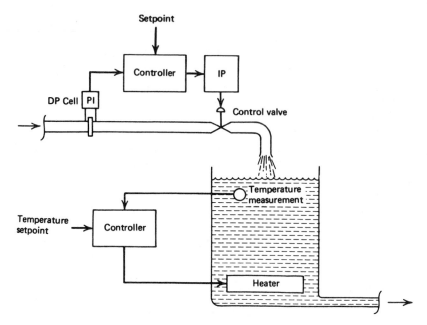

Figure 12.1   Two variable process-control loops with interaction.

Interactive single variable

A second single-variable control loop, shown in Figure 12.1, regulates the *temperature* of liquid in the tank by adjustment of heat input. This also is a single-variable loop that maintains the liquid temperature at the setpoint value. Under nominal conditions, the flow into the tank is held constant and the temperature is also held constant, both at their respective setpoint values. Note, however, that a change in the setpoint of the flow-control system appears as a *load change* to the temperature-control system because the fluid level in the tank or rate of passage through the tank must change. The temperature system now responds by resetting the heat flux to accommodate the new load and bring the temperature back to the setpoint.

We say then that these two loops *interact*. Almost *any* process where several variables are under control shows such *interactive behavior*. Any cycling or other instability of the flow control loop *causes* cycling in the temperature system because of this interaction.

Compound variable

In some cases, a *single* process-control loop is used to provide control of the relationship between *two or more* variables. This can be accomplished by using measurements from, say, two sensors as input to the process controller. A signal conditioning system must *scale* the two measurements and add them prior to input to the controller for evaluation and action. The analysis of such systems can become quite complicated.

A common example is when the *ratio* of two reactants must be controlled. In this case, one of the flow rates is measured but allowed to float, that is, not regulated, and the other is both measured and adjusted to provide the specified constant ratio. An example of this system is shown in Figure 12.2. The flow rate of reactant *A* is measured and added, with appropriate scaling, to the measurement of flow rate *B*. The controller reacts to the resulting input signal by adjustment of the control valve in the reactant *B* input line.

Example 12.1

In a compound control system, the ratio between two variables is to be maintained at 3.5 to 1. If each has been converted to a 0–5-volt range signal, devise a signal conditioning system that will output a zero signal to the controller when the ratio is correct.

Solution   In this case, we use a summing amplifier (Section 2.5.2). Then the output is related to the input by

$$V_{out} = -\frac{R_f}{R_1} V_1 - \frac{R_f}{R_2} V_2$$

If we make $V_{out} = 0$, then the voltage ratio is

$$\frac{V_1}{V_2} = -\frac{R_1}{R_2}$$

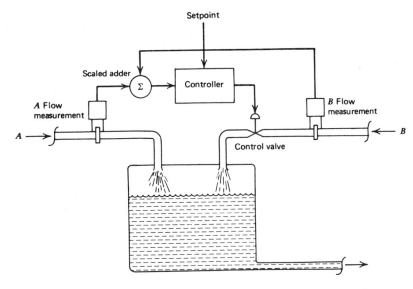

Figure 12.2   A compound system for which the ratio of two flow rates is controlled.

One input voltage must be negative (because we cannot use negative resistance!) and because the ratio of the resistance should also be 3.5 to 1

$$R_1 = 3.5R_2$$

Because the gain is unspecified, we can use $R_f = R_1$, for example. Then the circuit of Figure 12.3 accomplishes the desired function, where

$$V_{out} = -\left(\frac{3.5 \text{ k}\Omega}{3.5 \text{ k}\Omega}\right)(-V_1) - \left(\frac{3.5 \text{ k}\Omega}{1 \text{ k}\Omega}\right)V_2$$

or

$$V_{out} = V_1 - 3.5V_2$$

Thus, whenever $V_1 = 3.5V_2$, the output is zero.

In some cases, the compound control system is known as a cascade, although it differs from true cascade systems which will be described next.

12.2.2 Cascade Control

The inherent interaction that occurs between two control systems in many applications is sometimes used to provide better overall control. One method of accomplishing this is for the *setpoint* in one control loop to be determined by the *measurement* of a different variable for which the interaction exists. A block diagram of such a system is shown in Figure 12.4. Two measurements are taken from the system and each used in its own control loop. In the outer loop, however, the controller output is the *setpoint* of the inner loop. Thus, if the outer loop

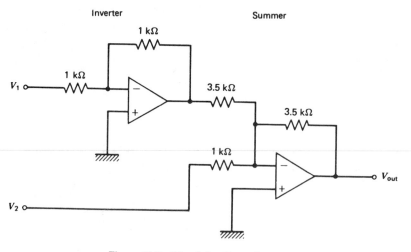

Figure 12.3 Circuit for Example 12.1.

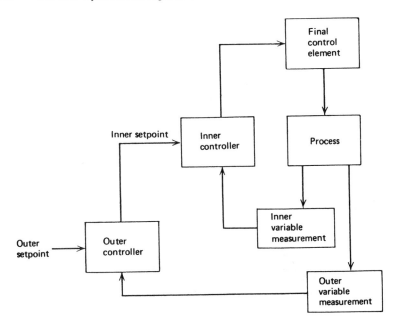

Figure 12.4   General features of a cascade process-control system.

controlled variable changes, the error signal that is input to the controller effects a change in setpoint of the inner loop. Even though the measured value of the inner loop has not changed, the inner loop experiences an error signal and thus new output by virtue of the setpoint change. Cascade control generally provides better control of the outer loop variable than is accomplished through a single-variable system.

An example of a cascade control system not only shows how it works, but suggests how control is improved. Consider the problem of controlling the level of liquid in a tank through regulation of the input flow rate. A single-variable system to accomplish this is shown in Figure 12.5a. A level measurement is used to adjust a flow-control valve as a final control element. The setpoint to the controller establishes the desired level. In this system, upstream load changes cause changes in flow rate that result in level changes. The level change is, however, a second stage effect here. Consequently, the system cannot respond until the level has actually *been changed* by the flow rate change.

Figure 12.5b shows the same control problem solved by a *cascade* system. The flow loop is a single-variable system as described earlier, but the setpoint is determined by a measurement of *level*. Upstream load changes are *never seen* in the level of liquid in the tank because the flow-control system regulates such changes before they appear as substantial changes in level.

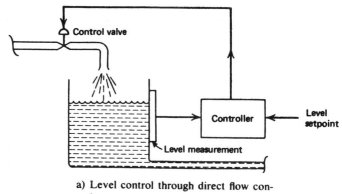

a) Level control through direct flow control

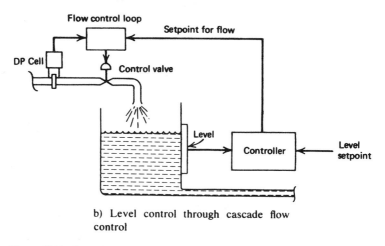

b) Level control through cascade flow control

Figure 12.5   Cascade control often provides better control than direct methods.

12.3 MULTIVARIABLE CONTROL SYSTEMS

One could correctly say that any reasonably complex industrial process is multivariable because many variables exist in the process and must be regulated. In general, however, many of these are either noninteracting or the interaction is not a serious problem in maintaining the desired control functions. In such cases, either single-variable or cascade loops suffice to effect satisfactory control of the overall process. The use of the word *multivariable* in this section refers to those processes wherein many, *strongly* interacting variables are involved. Such a multivariable system can have such a complex interaction pattern that the adjustment of a single setpoint causes a profound influence on many other control loops in the process. In some cases, instabilities, cycling, or even runaway result from the indiscriminate adjustment of a few setpoints.

12.3.1 Analog Control

When analog control loops are used in multivariable systems, a carefully prepared instructional set must be provided to the process personnel regarding the procedure for adjustment of setpoints. Generally, such adjustments are carried out in very small increments to avoid instabilities that may result from large changes. Let us try to give you an idea of the kind of situation under consideration. Suppose we have a reaction vessel in which two reactants are mixed, react, and the product is drawn from the bottom. We are now concerned with controlling the reaction rate. It is also important, however, to keep the reaction temperature and vessel pressure below certain limits and, finally, the level is to be controlled at some nominal value. If the reactions are exothermic, that is, self-sustaining and heat-producing, then the relation between all of these parameters can be critical. If the temperature is low, then an indiscriminant increase in steam-flow setpoint could cause an unstable runaway of the reaction. Perhaps, in this case, the level and reaction flow rates must be altered as the steam-flow rate is increased to maintain control. The necessary steps are often empirically determined or from numerical solutions of complicated control equations.

12.3.2 Supervisory and Direct Digital Control

The computer is ideally suited to the type of control problem presented by the multivariable control system. The computer can make any necessary adjustments of system operating points in an *incremental* fashion, according to a predetermined sequence, while monitoring process parameters for interactive effects. The problem in such a system is determining the *algorithm* that the computer must follow to provide the control function of the setpoint change sequence. In some cases, control equations are used. Usually, in complex interactions, these relations are not analytically known. In some cases *self-adapting* algorithms are used, causing the computer to sequence through a set of operations and letting the result of one operation determine the next operation. As an example, if the temperature is slightly raised and the pressure rises, then drop the temperature, and so on. The computer can sequence through a myriad of such microadjustments of setpoints looking for the optimum adjustment path.

Example 12.2

A process requires adjustment of setpoints to increase production. A particular sequence must be followed to provide the increase. SP1, SP2, and SP3 are the setpoints, P and PCR are the pressure and a critical pressure, respectively, and T and TCR are the temperature and critical temperature, respectively. Develop a flowchart that increases the setpoints as follows:

1. Increase SP1 by 1%.
2. Wait 10 s, test for pressure compared to critical.

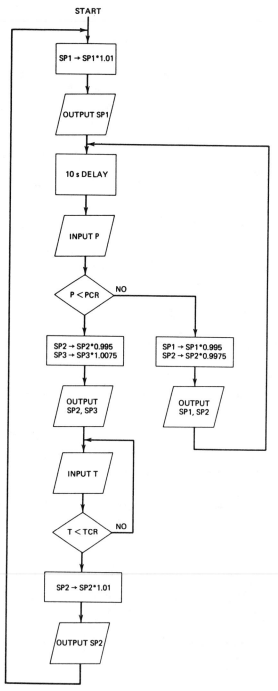

Figure 12.6 Flowchart for an interactive control problem.

3. If the pressure is less than critical, then
    a. decrease SP2 by $\frac{1}{2}\%$
    b. increase SP3 by $\frac{3}{4}\%$
    c. wait for T < TCR
    d. increase SP2 by 1%
    e. go to step 1
4. If the pressure is above critical,
    a. decrease SP1 by $\frac{1}{2}\%$
    b. decrease SP2 by $\frac{1}{4}\%$
    c. go to step 2

Solution   To implement this with either DDC or supervisory systems requires a flowchart from which the program instructions are developed. Such a flowchart is shown in Figure 12.6. The computer sequences through these steps in an optimum fashion according to the stipulations of the problem. If the sequence is performed manually, several operators would be required to monitor equipment and make necessary adjustments.

12.4 CONTROL SYSTEM QUALITY

When a manufacturing concept is to be implemented, the ultimate goal is to develop a product that satisfies present "design" criteria. If the product is crackers, they should have certain color, flavor, salinity, size, and so forth. If it is gasoline, it should possess certain octane, lead, viscosity, and so forth. The manufacturing process depends on the operation of a set of process-control loops to impart the desired characteristics to the product. The ultimate gauge of control system quality is, therefore, whether such control provides a product that is within specifications. The operation cannot even get started unless that *level* of quality is provided. We assume that in any practical situation it *is* satisfied. There are, however, other levels of quality that represent a deeper study of the process-control system, and these are the subject of the present discussion. In brief, given that a control system *can* provide a product that meets specifications, we ask how well does it perform this job, what variation in parameters exists, what percentage of rejected product occurs, and so on. To answer these questions, we must first describe measures of quality in a control system and then analyze how the loop characteristics affect these measures. It is most important to note that no *absolute* answers exist. What is considered good control in one manufacturing process may be unsatisfactory in another. Some of the general criteria that are applied are discussed later.

12.4.1 Definition of Quality

Let us consider for a moment one control loop in a manufacturing process. We need a control loop because the variable under control is dynamic, changes from many influences, and therefore needs regulatory operations. It is impossible for a control loop to regulate this variable to *exactly* the setpoint. Let's face it, the variable must *change* before the loop can generate a corrective action to oppose the change. Then, in considering quality, we must accept from the outset that *perfect* control is impossible and that some inevitable deviations of variables from the optimum values will occur. In fact, then, a definition of quality is concerned with these deviations and their interpretation in terms of the ultimate product. A process-control technologist is not necessarily in a position to evaluate how deviations of a loop variable from the setpoint actually (1) affect the specific properties of the final product, (2) cause it to be rejected, or (3) may affect the cost efficiency of the manufacturing process. To provide a link of communication between the process experts and the product experts, a set of measures or criteria have been devised. These criteria provide a *common language* so that a product can be evaluated in terms of the dynamic characteristics of a specific loop and serve, therefore, as our measures of quality. To understand the measures, we must first define quality in terms of the process-control loop.

Loop disturbance

The process-control system is supposed to provide regulation so that disturbances in the system will cause minimum deviation of the controlled variable from the setpoint value. The quality of the control system is defined by the degree to which the deviations that result from the disturbances are minimized.

There are three basic types of disturbances that can occur in a process-control system.

1. Transient
2. Setpoint changes
3. Load changes

A transient disturbance results from a temporary change of some parameter in the system that affects the controlled variable. It is impractical to use transient disturbances to define control quality because the nature of a transient cannot be well defined. That is, the transient can vary in duration, peak amplitude, and shape. For definition of quality, we need a more regular type of disturbance. To be specific, the two other disturbance conditions are used, both of which introduce a step-function change into the loop. A step-function change in *setpoint*, as shown in Figure 12.7a, is an *instantaneous* change of the loop setpoint from an old value to a new value. The second possible disturbance is a step-function change in *process load*, as suggested in Figure 12.7b, that also occurs instantaneously in time. The load change can come from the sudden permanent change of any of the process parameters that constitute the process load.

To provide measures of quality, we evaluate how the system responds to either of these sudden changes.

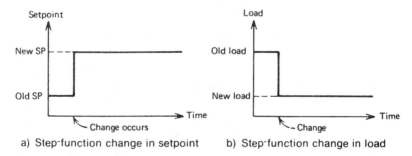

Figure 12.7   Loop disturbances can occur from intentional setpoint changes or changes in process load.

Optimum control

The most universal definition of quality in a control system is that the system provides *optimum control*, that is, the best control possible. If anything is modified in the system, then the deviation of controlled variable from a load or setpoint change is always worse. So the overall settings of the system are at an optimum. This does not mean that the control is "perfect" or even very good; it simply means that it is the best it can be.

Optimum control, and therefore control quality, can be defined in terms of the three effects resulting from a load or setpoint change.

1. Stability
2. Minimum deviation
3. Minimum duration

Stability

The most basic characteristic in defining process-loop quality is that it provides stable regulation of the dynamic variable. Stable regulation means that the dynamic variable does not grow without limit. In Figure 12.8, two types of unstable responses are shown. In one case, a disturbance causes the dynamic variable to simply increase without limit. In the other, the variable begins to execute growing oscillations, where the amplitude is increasing without limit. In both cases, some nonlinear breakdown (such as an explosion or other malfunction) eventually terminates the increase. We will consider stability in the next section, but for a measure of quality we assume stable operation has been achieved. A controlled variable in some process may be stable and still cyclic. This is the case, for example, in two-position control where the controlled variable oscillates between two limits under nominal load conditions. A change in load may change the period of oscillations, but the amplitude swing remains essentially the same; hence, the variable is under stable control.

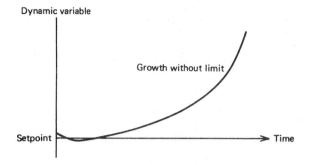

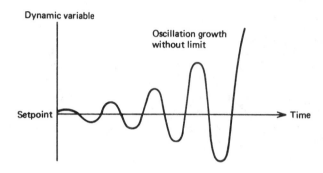

Figure 12.8 Instability in a process-control loop refers to the uncontrolled growth of the controlled variable.

Minimum deviation

If a process-control loop has been adjusted to regulate a variable at some setpoint value, then an obvious definition of quality is the extent to which a disturbance causes a deviation from that setpoint. Where a disturbance is a change in setpoint, this can be considered as any *overshoot* or *undershoot* of the variable in achieving the new setpoint. In general, we want to *minimize* any deviation of the dynamic variable from the setpoint value.

Minimum duration

If a disturbance occurs, we can conclude that some deviation will occur. Another definition of quality is the *length of time* before the controlled variable regains or adopts the setpoint value or at least falls within the acceptable limits of that value.

Thus, the quality of a process-control loop is defined through an evaluation of *stability*, *minimum deviation*, and *minimum duration* following a disturbance of the dynamic variable.

12.4.2 Measure of Quality

In general, it is not enough to simply state that we will design or adjust the process-control loop to provide for stable, minimum deviation, minimum duration operation. For example, the achievement of minimum deviation may result in a less than minimum duration (it usually occurs). The final product also may favor a less than absolute minimum adjustment to provide for a faster production rate at an acceptable degradation in product specification. To accommodate such circumstances, we distinguish several *measures of quality* by which we can convey the degree to which we have approached the ideals.

Assume stable operation has been achieved. There are three possible responses to a disturbance that a dynamic variable in a process-control loop can execute. The specific response depends on the controller gains and lags in the process. Referring to Figure 12.9a for a load change and Figure 12.9b for a setpoint change, we have the following definitions.

Overdamped

The loop is *overdamped* in case *A* of Figure 12.9; the deviation approaches the setpoint value smoothly (following a disturbance) with no oscillations. The duration is *not* a minimum in such a case. For that matter, the deviation itself usually is not a minimum either. Such a response is safe, however, in assuring that no instabilities occur and that certain maximum deviations never occur.

Critically damped

Careful adjustment of the process-control loop brings about curve *B* of Figure 12.9. In this case, the duration is a minimum for a noncycling response. This is

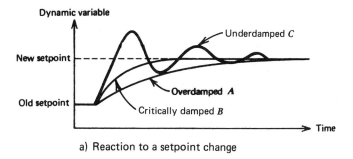

a) Reaction to a setpoint change

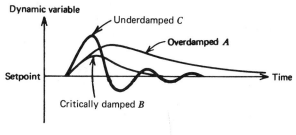

b) Reaction to a load change

Figure 12.9   Tuning determines the reaction of a variable to changes.

the optimum response for a condition where no overshoot or undershoot is desired in a setpoint change or no cycling, in general, is desired.

Underdamped

The natural result of further adjustments of the process-control loops is cyclic response, where the deviation executes a number of oscillations about the setpoint. This is shown in Curve C of Figure 12.9. It is possible that this response gives minimum deviation and minimum duration in some cases. If the cycling can be tolerated, then such a response is preferred.

Two specialized measures of control are used when *none* of these conditions serves to define the measure of control desired in a process.

Quarter amplitude

When a process-control loop has a damped cyclic response to a disturbance, a criteria is sometimes used that is *neither* minimum deviation nor minimum duration. This measure of quality is found by adjusting the loop until the deviation from a disturbance is such that each deviation peak is down by one quarter from the preceding peak, as shown in Figure 12.10. In this case, the *actual* magnitude of the deviation is *not* included in the measure, nor is the time between each peak. In this sense, neither duration nor magnitude of the deviation is directly involved in a quarter-amplitude criterion.

Example 12.3

Suppose the deviation following a step-function disturbance is a 4.7% error in the controlled variable. If a quarter-amplitude criterion is used to evaluate the response, find the error of the second and third peak.

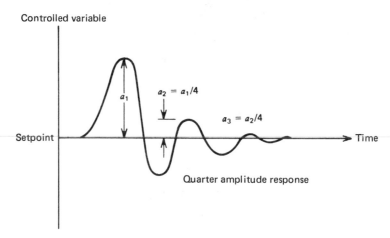

Figure 12.10 In one type of cyclic response, the system is adjusted to make each peak down by one quarter of the previous peak.

Solution   The amplitude of each peak must be one quarter of the previous peak. In this case the second peak is 4.7%/4 or 1.18%. The third peak is an error of 1.18%/4 or 0.30%.

Minimum area

In cases of *cyclic* or underdamped response, the most critical element is sometimes a combination of *duration* and *deviation*, which must be minimized. Thus, if minimum deviation occurs at one loop setting and minimum duration at another, then *neither* is optimum. One type of optimum measure of quality in these cases is to minimize the net area of the deviation as a function of time. In Figure 12.11 this is shown as the sum of the shaded areas. Analytically, this can be expressed as

$$A = \int | r - b | \, dt \qquad (12.1)$$

where

A = area of deviation
b = measured value
r = setpoint value

Another way of representing this is to use the full-scale percent error (Section 9.3.1) to write

$$A_p = \int | e_p | \, dt \qquad (12.2)$$

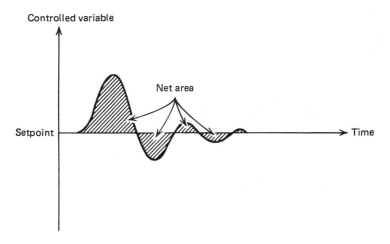

Figure 12.11   The minimum area criterion for cyclic response adjusts the process-control loop until the net area is a minimum.

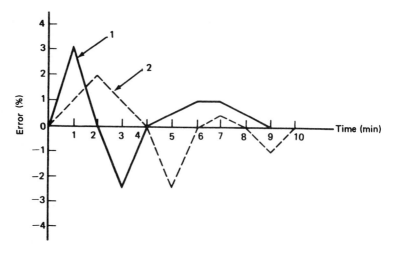

Figure 12.12 Error versus time for Example 12.4.

where

e_p = full-scale error in %
A_p = area as % − s

By adopting this criteria of loop response, we are keeping the extent and duration of deviation, and thus product rejections, to a minimum.

Example 12.4

Given the two response curves of Figure 12.12 for deviation versus time, following an initial disturbance, find the response preferred using the minimum area criterion.

Solution   We find the area by application of

$$A_p = \int |e_p| \, dt \tag{12.2}$$

In these cases we find the areas geometrically by finding the net area of each curve. For curve 1 we have

$$A_1 = 13\% - \text{minute}$$

and for curve 2 we get

$$A_2 = 10.5\% - \text{minute}$$

Thus, it is clear that curve 2 is preferred under the minimum area test.

12.5 STABILITY

Earlier in this chapter, we assumed that the loop response is stable as a prerequisite for use in a practical control application. In fact, a great deal of effort is expended in the design and development of process-control loops to achieve this stability. Although a detailed treatment of stability theory in process control is beyond the intent of this book, it is of value to discuss some of the important considerations of such studies.

12.5.1 Transfer Function Frequency Dependence

The static transfer function of an element in a process-control loop tells how the output is determined from the input when the input is constant in time. The dynamic transfer function of an element tells how the output is determined from the input when the input varies in time. For the study of stability, we are interested in the particular time variation which is sinusoidal (i.e., the dynamic transfer function when the input is oscillating at some frequency f).

Consider some element block as shown in Figure 12.13 with a transfer function $T(\omega)$ and where the input is a sinusoidal given by

$$r = a \sin(\omega t)$$

The frequency has been expressed in terms of the angular frequency, $\omega = 2\pi f$, measured in radians/second. There are only two things which can happen, in a linear study at least: The amplitude can change, and there can be a phase shift. Thus the output can be described as

$$c = b \sin(\omega t + \phi)$$

The ratio of the amplitudes is called the gain, gain $= b/a$, and the phase shift is called the phase lag, ϕ. In general, both the gain and amount of phase lag of an element vary with frequency. The gain decreases and the phase lag becomes larger.

The whole issue of stability is tied up with the frequency variation of gain and phase of all elements in a control loop.

Source of instability

To see how a process-control loop can cause instability, consider the open loop block diagram of Figure 12.14, for which an oscillating transient disturbance has been imposed at r. Notice that the feedback line has been broken at the error

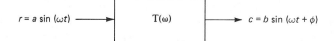

$r = a \sin (\omega t) \longrightarrow \boxed{T(\omega)} \longrightarrow c = b \sin (\omega t + \phi)$

Figure 12.13 A transfer function changes the amplitude and phase of a sinusoidal input.

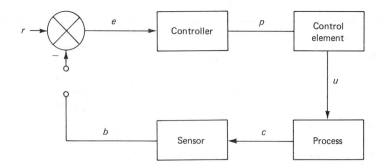

Figure 12.14   This control loop has been "opened" so the effect of the loop on a disturbance can be traced.

detector so no actual feedback occurs. Each element of the loop has a gain and phase lag, including the process itself. The net gain is the product of all gains and the net phase lag is the sum of all phase lags.

In a perfect world, the feedback *b* would be an exact replica of the disturbance at every frequency. This would mean a loop gain of unity and no phase shift. Then the − 180° phase shift (lag) of the error detector would subtract the feedback from the disturbance, and it would be cancelled. In reality, there are gain variations and extra phase shifts, and both vary with the disturbance frequency.

If the gain for the disturbance frequency is greater than one and the system phase lag small, the disturbance is cancelled within a few cycles of oscillation. However, as the phase lag becomes greater with increasing frequency, the effectiveness of the feedback is reduced and the oscillation will persist longer. This is the cyclic response to a transient discussed elsewhere in this book.

Similarly, as the gain becomes smaller with frequency, the effectiveness of the feedback to cancel error is reduced. But the control system is still working and stable.

Consider, however, the particular case of a frequency where the phase lag of the system reaches − 180° while the gain remains unity or greater. When combined with the error detector phase shift, the net shift around the loop will be 360° and so the feedback will be summed instead of subtracted. If the gain at that frequency is just unity, then the disturbance will persist, forever, with constant amplitude. If the gain is greater than unity, the disturbance will grow in amplitude. This is the instability caused by the control system.

Instability illustration

Instability is caused by a condition where for some frequency the transfer function is such that feedback to the error summer actually *increases* the error because of the gain and phase shift. Now, if there is *any* frequency for which this condition exists, then oscillations will always start and grow at that frequency. To illustrate this, we have assumed that a small oscillation is introduced from external sources as a disturbance in *r*. Assume that at this frequency the gain is greater than one, say two, and the phase shift is $-180°$, that is, a lag of $180°$. Let us study the result, frozen instant by instant. In Figure 12.15a, the original input and the feedback *b* is shown. Note that we are using the model of Figure 12.14 for the process-control loop but with a closed loop. The summer subtracts the feedback (*b*) from the input, giving the signal *e* of the next instant in Figure 12.15b, and the output with the gain of two and phase shifted by $-180°$. Again, we subtract output from

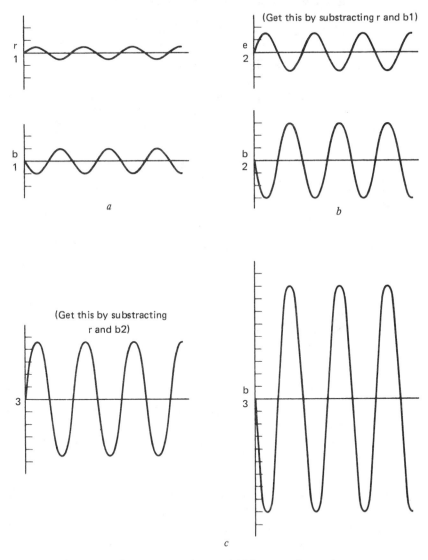

Figure 12.15 This figure suggests how an initial, one-cycle transient can grow under the appropriate feedback conditions.

input, getting the input of the next instant, as in Figure 12.15c, and the corresponding output. Thus, you can see that the error is actually growing! The control system is forcing the oscillating error to increase, instant by instant. Of course, in actuality, it happens smoothly and the output would look something like Figure 12.8, for oscillating growth. If there is *any* frequency where such conditions for growth exist, the system is unstable and something like random noise will eventually set the system into growing oscillation. When a process-control installation is designed, one has the objective of regulating the controlled variable *without* instability in the loop. Stability can be assured by designing the controller gains so that oscillation growth is never favorable, according to certain criteria.

12.5.2 Stability Criteria

We can derive stability criteria by determining just what conditions of system gain and phase lag can lead to an enhancement of error. We are led to two conclusions:

1. The gain must be greater than one.
2. The phase shift must be $-180°$ (lag).

Thus, if there is any frequency for which the gain is greater than one and the phase is $-180°$, then the system is unstable. From this argument, we develop two ways of specifying when a system is *stable*. These rules are as follows:

- **Rule 1**   A system is stable if the phase lag is less than 180° at the frequency for which the gain is unity (one).
- **Rule 2**   A system is stable if the gain is less than one (unity) at the frequency for which the phase lag is 180°.

The application of these rules to an actual process requires evaluation of the gain and phase shift of the system for all frequencies to see if Rules 1 and 2 are satisfied. This is easier to do if a plot of gain and phase versus frequency is used.

Bode plot

A particular type of graph, called a Bode plot or diagram, normally is used to plot the gain and phase of the control system versus frequency. In this case, the frequency is plotted along the abscissa (horizontal) on a log scale and expressed commonly as rad/s. Remember, if you want to know the actual frequency in Hz, you just divide the rad/s by 2π. Use of a log scale permits a very large range of frequency to be displayed. The ordinate actually consists of two parts: Gain is plotted on a log scale in one part, and phase is plotted linearly as degrees in another part. Figure 12.16 shows the Bode plot with a gain/phase plotted for an example. We have labeled the gain as *open loop gain*, since we do not include the effect of actually feeding that signal back into the system.

Example 12.5

Determine if the system of Figure 12.16 is stable according to the rules given earlier.

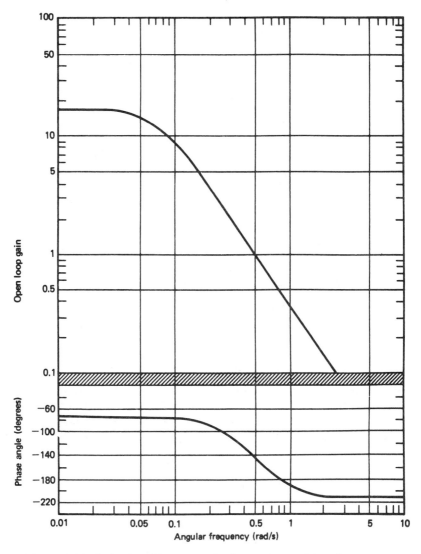

Figure 12.16   Bode plot with an example of process-control loop frequency response.

Solution   By *Rule 1*, we find the frequency for which the gain is unity (one). The system is stable if the phase lag is less than 180°. Unity gain occurs at 0.5 rad/s; this is $0.5/2\pi$ = 0.08 Hz, or about five cycles per minute. Such frequencies are not uncommon in many industrial processes. For this frequency, the phase is about −140° so the lag is less than 180°, and the system is **stable** by *Rule 1*. For *Rule 2*, we find the frequency for which the phase lag is 180° and check the gain, which should be less than one for stability. The phase lag is 180° at 0.8 rad/s, for which the gain is 0.5. Thus, the system is **stable** by *Rule 2*.

In some cases, a modified form of the Bode plot given in Figure 12.16 is used, in which the gain is plotted in dB. This gain is then plotted on a linear scale where

$$\text{Gain (dB)} = 20 \log_{10}(|\ T(\omega)\ |) \tag{12.3}$$

Thus, a gain of 100 gives 40 dB, unity gain is 0 dB, and a gain less than one gives a negative dB, as 0.1 gives −20 dB. The stability rules would be revised to read 0 dB instead of unit gain.

If the transfer functions of several elements are known, the transfer functions of a series of such cascaded elements is found by a *product* of the *magnitudes* and a *sum* of the *phases* or

$$|\ T(\omega)| = |\ T_1(\omega)\ |\ \cdot\ |\ T_2(\omega)\ |\ \cdot\ |\ T_3(\omega)| \tag{12.4}$$

$$\angle T(\omega) = \angle T_1(\omega) + \angle T_2(\omega) + \angle T_3(\omega) \tag{12.5}$$

where

$$T(\omega) = \text{system transfer function}$$
$$T_1, T_2, T_3 = \text{element transfer function}$$

On log-log scales the product of Equation (12.4) becomes addition. The transfer function for an entire process-control loop therefore involves summation of the transfer functions of all the elements of the loop, *including the process itself*.

$$\log|\ T(\omega)| = \log|\ T_1(\omega)| + \log|\ T_2(\omega)| + \log|\ T_3(\omega)| \tag{12.6}$$

In most cases, an exact solution cannot be found using Bode plots because the *process transfer function* is too complicated or unknown. Stability can still be assured through adjustment of the proportional gain, derivative time, and integral time. The final adjustment of these quantities *tunes* the system for stable operation.

12.6 PROCESS LOOP TUNING

The last aspect of process-control technology we consider refers to the actual startup and adjustment of a process-control loop. We have seen how the various settings of the controller can have a profound effect on loop performance. Now, the most natural question is how to select these settings. There are in fact *many* methods for determination of the optimum mode gains, depending on the nature and complexity of the process. We consider here three common tuning methods to give a basic idea of how optimum adjustments are found. Two of the methods given are semiempirical in that they depend on measurements made on the system to determine factors used in the adjustment formulas. The last is more analytical in that it is based on both a known transfer function of the process and loop.

12.6.1 Open Loop Transient Response Method

This method of finding controller settings was developed by Ziegler and Nichols and is sometimes referred to as a *process reaction* method. The basic approach is to open the process-control loop so that no control action (feedback) occurs. This usually is done by disconnecting the controller output from the final control element. All of the process parameters are held at their nominal values. This method can only be used for systems with self-regulation.

At some time, a transient disturbance is introduced by a small, manual change of the controlling variable using the final control elements. This change should be as small as practical for making necessary measurements. The controlled variable is measured (recorded) versus time at the instant of and following the disturbance.

A typical open loop controller response is shown in Figure 12.17, where the disturbance is applied at t_1. We have expressed the deviation as a percent of range as usual, and we assume the final control element disturbance also is expressed as a percentage change. A tangent line, shown as a dashed line, is drawn at the *inflection point* of the curve. The inflection point is defined as that point on the curve where the slope stops increasing and begins to decrease. Where the tangent

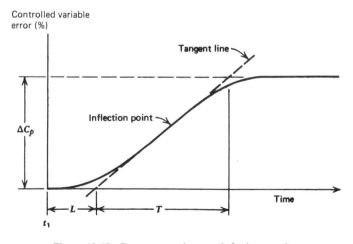

Figure 12.17 Process reaction graph for loop tuning.

line crosses the origin, we get

$$L = \text{lag time in minutes} \qquad (12.7)$$

as the time from disturbance application to the tangent line intersection as shown in Figure 12.17. Also from the graph we get T, the process reaction time and

$$N = \frac{\Delta C_p}{T} \qquad (12.8)$$

where

$$N = \text{reaction rate in \%/min}$$
$$\Delta C_p = \text{variable change in \%}$$
$$T = \text{process reaction time in minutes}$$

The quantities defined by Equations (12.7) and (12.8) are used with the controlling variable change ΔP to find the controller settings. The following paragraphs give the stable control definitions for the various modes as developed by Ziegler and Nichols and corrections developed by Cohen and Coon (when the quarter-amplitude response criterion is indicated). In the latter case, a log ratio is used, defined by

$$R = \frac{NL}{\Delta C_p} \qquad (12.9)$$

where

$$R = \text{log ratio (unitless)}$$

Proportional mode

For the proportional mode, the proportional gain setting K_P is found from

$$K_P = \frac{\Delta P}{NL} \qquad (12.10)$$

Corrections to the value of K_P are sometimes used to obtain the quarter-amplitude criterion of response. One given by Cohen and Coon is shown bracketed as

$$K_P = \frac{\Delta P}{NL}\left[1 + \frac{1}{3}\frac{NL}{\Delta C_p}\right] \qquad (12.11)$$

Proportional-integral mode

When the controller mode is proportional-integral, the appropriate settings for proportional gain and integration time are

$$\left.\begin{array}{l} K_P = 0.9\,\dfrac{\Delta P}{NL} \\[2mm] 1/K_I = T_I = 3.33L \end{array}\right\} \qquad (12.12)$$

If the quarter-amplitude criterion is used, the gain is

$$K_P = \frac{\Delta P}{NL}\left[0.9 + \frac{1}{12}R\right] \left.\rule{0pt}{24pt}\right\}$$
$$T_I = \left[\frac{30 + 3R}{9 + 20R}\right]L \qquad\qquad (12.13)$$

Three mode

For the three-mode controller, we find the appropriate proportional gain, integration time, and derivative time from

$$K_P = 1.2\frac{\Delta P}{NL} \left.\rule{0pt}{50pt}\right\}$$
$$T_I = 2L \qquad\qquad (12.14)$$
$$T_D = 0.5L$$

If the quarter-amplitude criterion is used, these equations are corrected by

$$K_P = \frac{P}{NL}[1.33 + R/4] \left.\rule{0pt}{60pt}\right\}$$
$$T_I = 1\left[\frac{32 + 6R}{13 + 8R}\right]L \qquad\qquad (12.15)$$
$$T_D = L\left(\frac{4}{11 + 2R}\right)$$

Example 12.6

A transient disturbance test is run on a process loop. The results of a 9% controlling variable change give a process reaction graph as shown in Figure 12.18. Find settings for three-mode action.

Solution   By drawing the inflection point tangent on the graph, we find a lag $L = 2.4$ minutes, and a process reaction time of 4.8 minutes. The reaction rate is

$$N = \frac{\Delta C_p}{T} = \frac{3.9\%}{4.8 \text{ min}} = 0.8125 \text{ \%/min} \qquad\qquad (12.8)$$

The controller settings are found from Equations (12.14)

$$K_P = 1.2\frac{\Delta P}{NL} = 1.2\frac{9\%}{(0.8125)(2.4)}, \qquad \text{yielding } K_P = 5.54$$

or a proportional band of

$$\frac{100}{K_P} = \frac{100}{5.54} = 18\%$$

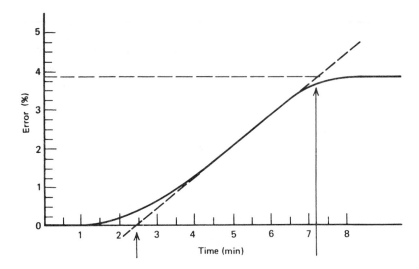

Figure 12.18 Process reaction graph for Example 12.7.

$$T_I = 2L = (2)(2.4 \text{ min}) = \textbf{4.8 min}$$

$$T_D = 0.5L = (0.5)(2.4 \text{ min}) = \textbf{1.2 min}$$

Example 12.7

For example 12.6, find the three-mode settings for a quarter-amplitude response.

Solution   We use Equations (12.15) together with the lag ratio

$$R = \frac{NL}{\Delta C_p} = \frac{(0.8125)(2.4)}{3.9}$$

(12.9)

$$R = \textbf{0.50}$$

Then we get

$$K_P = \frac{\Delta P}{NL} (1.33 + R/4)$$

$$K_P = \frac{9}{(0.8125)(2.4)} (1.33 + 0.50/4)$$

(12.15)

$$K_P = \textbf{6.72}$$

or a **14.9%** proportional band.

$$T_I = L\left(\frac{32 + 6R}{13 + 8R}\right)$$

$$T_I = 2.4\left(\frac{32 + 6(0.5)}{13 + 8(0.5)}\right)$$

(12.15)

$$T_I = \textbf{4.94 min}$$

$$T_D = L\left(\frac{4}{11 + 2R}\right)$$

$$T_D = (2.4)\frac{4}{11 + 2(0.5)} \tag{12.15}$$

$$T_D = \textbf{0.80 min}$$

12.6.2 Ziegler–Nichols Method

Ziegler and Nichols also developed another method of controller setting assignment that has come to be associated with their name. This technique, also called the *ultimate cycle method*, is based on adjusting a closed loop until steady oscillations occur. Controller settings are then based on the conditions that generate the cycling.

The particular method is accomplished through the following steps:

1. Reduce any integral and derivative actions to their minimum effect.
2. Gradually begin to increase the proportional gain while providing periodic small disturbances to the process. (These are to "jar" the system into oscillations.)
3. Note the critical gain K_c at which the dynamic variable just begins to exhibit steady cycling, that is, oscillations about the setpoint.
4. Note the critical period T_c of these oscillations measured in minutes.

This method can be used for systems without self-regulation. Now, from the critical gain and period, the settings of the controller are assigned as follows:

Proportional

For the proportional mode alone, the proportional gain is

$$K_P = 0.5K_c \tag{12.16}$$

A modification of this relation is often used when the quarter-amplitude criterion is applied. In this case, the gain is simply adjusted until the dynamic response pattern to a step change in setpoint obeys the quarter-amplitude criterion. This also results in some gain less than K_c.

Proportional-integral

If proportional-integral action is used in the process-control loop, then the settings are determined from

$$\left.\begin{aligned} K_P &= 0.45K_c \\ T_I &= T_c/1.2 \end{aligned}\right\} \tag{12.17}$$

In case the quarter-amplitude criterion is desired, we make

$$T_I = T_c \qquad (12.18)$$

and adjust the gain for that necessary to obtain the quarter-amplitude response.

Three mode

The three-mode controller requires proportional gain, integral time, and derivative time. These are determined for nominal response as

$$\left. \begin{array}{c} K_P = 0.6K_c \\[6pt] T_I = T_c/2.0 \\[6pt] T_D = T_c/8 \end{array} \right\} \qquad (12.19)$$

For adjustment to give quarter-amplitude response, we set

$$\left. \begin{array}{c} T_I = T_c/1.5 \\[6pt] T_D = T_c/6 \end{array} \right\} \qquad (12.20)$$

and adjust the proportional gain for satisfaction of the quarter-amplitude response.

Example 12.8

In an application of the Ziegler–Nichols method, a process begins oscillation with a 30% proportional band in an 11.5-min period. Find the nominal three-mode controller settings.

Solution   First, a 30% proportional band means the gain is (Section 9.5.1)

$$K_c = \frac{100}{PB} = \frac{100}{30} = 3.33$$

Then, from Equation (12.19) we find settings of

$$K_P = 0.6K_c = (0.6)(3.33) \qquad (12.19)$$

$$K_P = 2$$

$$T_I = T_c/2 = 11.5/2 \qquad (12.19)$$

$$T_I = 5.75 \text{ min}$$

$$T_D = T_c/8 = 11.5/8 \qquad (12.19)$$

$$T_D = 1.44 \text{ min}$$

12.6.3 Frequency Response Methods

The frequency response method of process controller tuning involves use of Bode plots for the process and control loops. The method is based on an application of the Bode plot stability criteria given in Section 12.5.2 and the effects that proportional gain, integral time, and derivative time have on the Bode plot.

Gain and phase margin

The stability criteria given in Section 12.5.2 represent *limits* of stability. For example, if the gain is slightly less than one when the phase lag is 180°, the system is *stable*. But if the gain is slightly greater than one at 180°, the system is *unstable*. It would be well to design a system with a margin of safety from such limits to allow for variation in components and other unknown factors. This consideration leads to the *revised stability criteria*, or more properly, *a margin of safety* provided to each condition. The exact terminology is in terms of a *gain margin* and *phase margin* from the limiting values quoted. Although no standards exist, a common condition is

---

1. If the phase lag is less than 140° at the unity gain frequency, the system is stable. This, then, is a 40° *phase margin* from the limiting value of 180°.
2. If the gain is 5 dB below unity (or a gain of about 0.56) when the phase lag is 180°, the system is stable. This is a 5 dB *gain margin*.

Example 12.9

Determine if the system with a Bode plot of Figure 12.19 satisfies both the gain and phase margin conditions.

Solution   We first examine the phase at unity gain. Unity gain occurs at an angular frequency of 0.15 rad/s, and the phase is −120°. Thus, the first stability condition is satisfied with a 60° phase margin.

Now a 180° phase occurs at 0.3 rad/s angular frequency where the gain is 0.7, which is too high. Thus, the 0.56 or less gain margin is *not* satisfied, and the controller gain will have to be reduced slightly.

Tuning

The operations of tuning using the frequency response method involve adjustments of the controller parameters until the stability is proved by the appropriate phase and gain margins. If the process and control elements' transfer functions are known, the correct settings can be determined analytically. If not, the Bode plot can be determined experimentally by opening the loop and providing a variable frequency disturbance of the controlling variable. If measurements of phase and gain are made, then the Bode plot can be constructed. From this the proper settings can be determined. The significance of the unity gain crossover in frequency is that the system can correct any disturbances of frequency less than that of the unity gain frequency. Any disturbance of *higher* frequency has no effect on the control system.

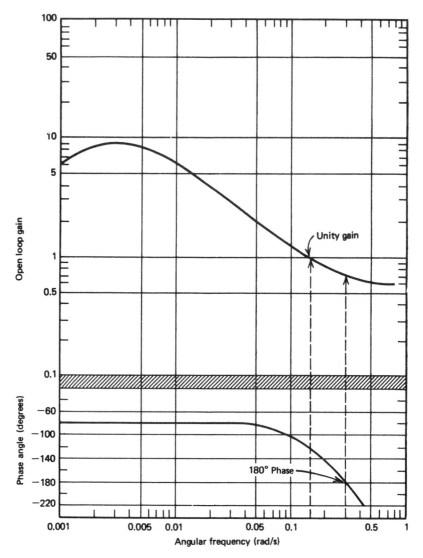

Figure 12.19 Bode plot for Example 12.10.

The tuning operation is based on the fact that the gains of each mode have a particular effect on the system Bode plot. By adjusting these gains, we can alter the Bode plot until it satisfies the gain and phase margins of safety for a stable system. Remember that on the Bode plot, the gains appear as products and phases of the modes simply and algebraically.

Proportional action

This mode of the controller simply multiplies the gain curve by a constant independent of frequency and has no phase effect at all. Thus, if a system gain curve is found for a proportional gain of 2 and we increase this to 4, then the entire gain curve is multiplied by 2. This term can be used to move the intact gain curve up (increased gain) or down (decreased gain). Generally, moving the curve up extends the range of frequency that the system can control, provided the stability margins are maintained.

Integral action

In its pure form, the integral mode contributes

$$\text{Integral Gain} = \frac{K_I}{\omega}$$

$$\text{Integral Phase} = -90° \text{ (lag)}$$

Thus, we see that the integral mode contributes more gain at lower frequency and less gain at higher frequency because of the inverse dependence on the frequency ω. The phase shift is made more lagging by 90°. The integral *mode* gain K_I can increase the gain at lower frequencies. In practical form, the integral mode has some lower frequency (called the *breakpoint*) at which its effects cease, and the gain curve becomes flat while the phase lag goes to zero. This is shown in Figure 12.20, where $K_I = 1$ has been assumed.

Derivative action

Because of problems of instability at higher frequencies, the derivative mode is almost never used in its pure form. If it were, the gain and phase would be

$$\text{Derivative Gain} = K_D\omega$$

$$\text{Derivative Phase} = +90° \text{ lead}$$

Because the gain increases with frequency, it must be limited at some upper frequency. In Figure 12.20, this mode action is shown (for a mode gain $K_D = 10$) with the characteristic upper-limit frequency at which the gain becomes constant, and the phase shift is zero. In general then, this mode can be used to increase the gain at higher frequency, but more importantly it can be used to drive the phase away from the $-180°$ (lag) shift where instability can begin.

Example 12.10

Find the fractional decrease in proportional gain that would make the system of Example 12.10 stable.

Solution   The system failed to satisfy the 5 dB gain margin at $-180°$, because the gain was 0.7 instead of 0.56 or less. Thus, if the proportional gain is reduced by a factor of 0.7/0.56 or about 0.8, then the gain margin will be satisfied. The phase margin will not be affected.

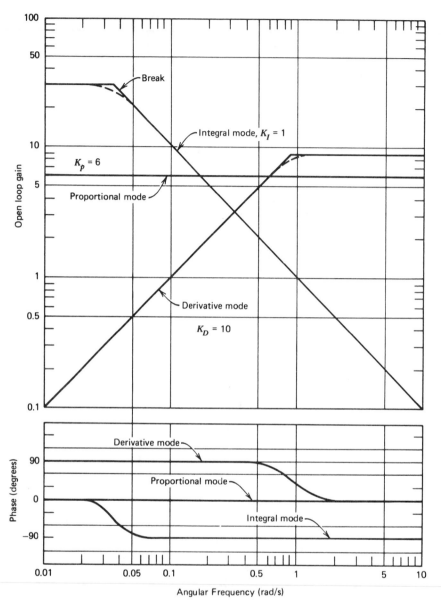

Figure 12.20 Bode plots of proportional, integral, and derivative controller modes.

Example 12.11

Consider the Bode plot of Figure 12.21 for some hypothetical process and controller with a proportional gain of 10. Show that the system is unstable, and find a proportional gain for satisfying gain and phase margins.

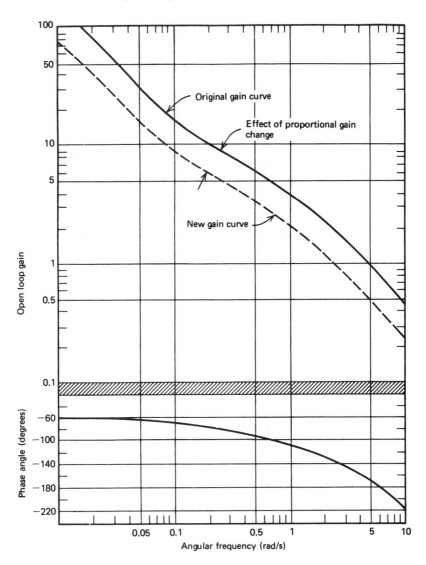

Figure 12.21 Bode plots for Example 12.11. Note that changes in proportional gain do not change the phase.

Solution Evaluation of the curve shows that the phase lag at unity gain is 170°, which, although stable, does not satisfy the required 40° phase margin. The gain at 180° phase lag is approximately 0.8, which, although less than one, does not satisfy the stipulated gain margin. Notice that if the entire gain curve is reduced or moved down by a constant amount, as indicated, with no effect on the phase lag, then the stability criteria can be satisfied. This is done by changing the *proportional gain* because this affects gain only. In particular, the gain must be derated by a factor of

½ (note difference at 5 rad/s) to achieve both margins. Thus, the new proportional gain should be 10/2 or 5.

In general, the derivative, integral, and proportional gains are adjusted until the system satisfies the stability criteria and the specified unity gain frequency.

SUMMARY

This chapter presented an overview of operational characteristics of the entire process-control loop. The following items were specifically considered:

1. Although the single-variable control loop is widely used in process control, we often must account for interactive effects between variables.

2. In many cases, a cascade control loop is used where the setpoint of one loop is determined by the controller output of another loop.

3. Multivariable control systems are used where strong interactions between variables exist. In many ways, supervisory or DDC techniques are best suited for these control problems.

4. The quality of control system response is defined by characterizing the response of the system to a disturbance. The stability, deviation magnitude, and duration are important to this quality.

5. The *minimum area* and *quarter-amplitude* criteria are often used to measure quality together with damping characteristics.

6. The stability of a process-control loop can be studied with a Bode plot, consisting of gain and phase lag versus frequency.

7. The tuning of a process-control loop consists of finding the optimum settings of controller gains for good control.

8. The open loop transient method of tuning relies on a semiempirical technique using the reaction of the open loop to a transient disturbance.

9. The Ziegler–Nichols method finds the conditions that generate steady oscillation and derates the settings for optimization.

10. The frequency-response method relies on the stability conditions determined from a Bode plot. Adjustments of gains provide for the phase and gain margin and frequency band width.

PROBLEMS

Section 12.2

12.1 A compound control system specifies that the ratio of pressure in Pa to temperature in K be held to 0.39.
(a) Diagram a control loop to provide this control.

(**b**) Identify the function of each element in the loop necessary to accomplish the control.

12.2 Suppose that for Problem 12.1 the temperature is available at 2.75 mV/K and the pressure at 11.5 mV/Pa. Design signal conditioning that outputs zero when the ratio is correct. *Hint:* First convert each to scales of 1 mV/K and 1 mV/Pa.

12.3 Draw a diagram of a cascade control system that has an inner loop of temperature input to a flow system and an outer loop of viscosity to determine the flow controller setpoint.

Section 12.3

12.4 Draw a block diagram of an operational flow diagram to show how Problem 12.1 is implemented by DDC.

12.5 Draw a block diagram and flowchart to show how the cascade system of Problem 12.3 can be implemented by DDC.

Section 12.4

12.6 Assume we want an electronic means of finding the area of the error-time graph following a disturbance.
  (**a**) Show that the circuit of Figure 12.22 does this.
  (**b**) Find the values of R_3 and R_4 so that the output is the area of Equation 12.2 in V-s.

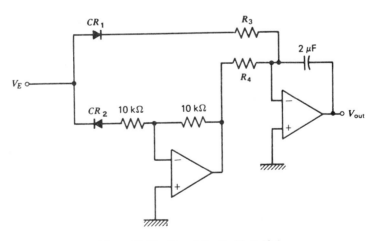

Figure 12.22 Circuit for Problem 12.6.

Section 12.5

12.7 A process and control system has components with transfer functions at 10 rad/s
of 20 $\angle 0°$, 1.4 $\angle -90°$, 0.05 $\angle 90°$, and 3.2 $\angle 85°$. Find the total system transfer function.

Section 12.6

12.8 An open loop transient test provides the process reaction graph given in Figure
12.23 for a 7.5% disturbance.
  (a) Find the standard proportional-integral gain settings.
  (b) Find the three-mode quarter-amplitude settings.

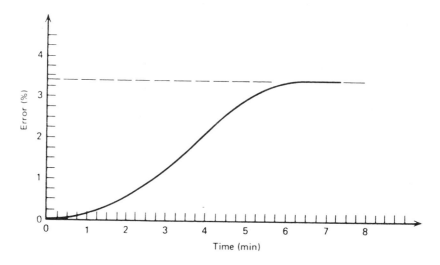

Figure 12.23 Process reaction graph for Problem 12.8.

12.9 In the Ziegler–Nichols method, the critical gain was found to be 4.2 and the critical
period was 2.21 minutes. Find the standard settings for (a) proportional mode control, (b) PI control, and (c) PID control.

12.10 Specify the gain and phase margins for Figure 12.24. Is the system stable?

12.11 If the nominal proportional gain for the process in Figure 12.24 was 11.5, determine
the gain that will just satisfy a gain margin of 0.56 and phase margin of 40°.

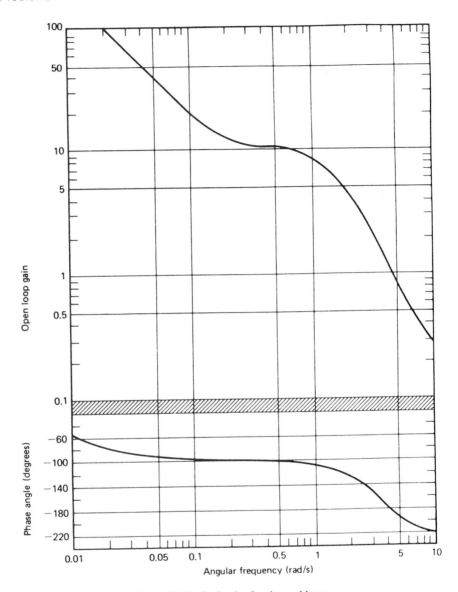

Figure 12.24 Bode plot for the problems.

12.12 Draw the resultant Bode plot of the combination of the proportional, integral, and derivative modes shown in Figure 12.20.

12.13 Figure 12.25 shows the Bode plots of a process, proportional mode, and integral mode of an associated controller. Find the composite transfer function Bode plot and evaluate the stability.

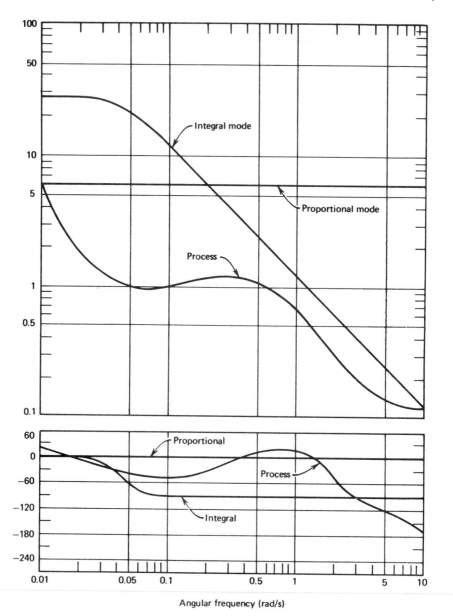

Figure 12.25   Bode plot for Problem 12.13.

12.14 Show how adding derivative action to the system of Problem 12.13 would change the composite Bode plot. Discuss the effect on stability of adding this action.

12.15 The composite Bode plot of an open loop system is shown in Figure 12.26. This is for proportional action only with $K_P = 1$. How much proportional gain will drive the system into instability? The critical gain of the ultimate cycle method is that

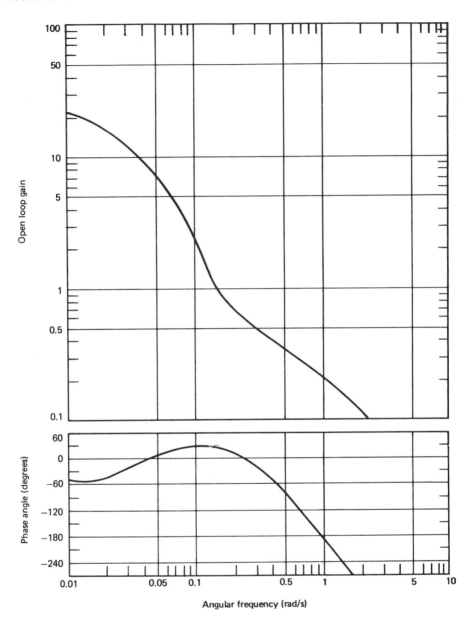

Figure 12.26 Bode plot for Problem 12.15.

which will drive the system to the margin of instability. What would be the proper gain setting for ultimate cycle tuning? What phase and gain margin does this provide?

12.16 A process is to be tuned by transient response to quarter amplitude using a PID controller. The system control temperature varies from 140 to 300°C with a 220°C setpoint. The output is a heater control voltage from 0 to 24 V. The test is started with the system having a 14-V output. The output is increased suddenly to 16.5 V. The resulting temperature graph is shown in Figure 12.27. Find the proper PID gain.

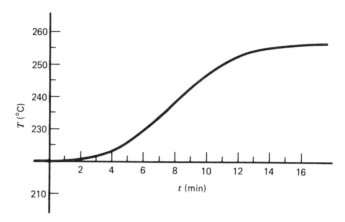

Figure 12.27   Process reaction graph for Problem 12.16.

12.17 A process Bode plot is shown in Figure 12.28. Prove the system is unstable to the 5-dB gain margin and 48° phase margin. The plot represents a proportional gain of 5.0. What gain will make the system stable?

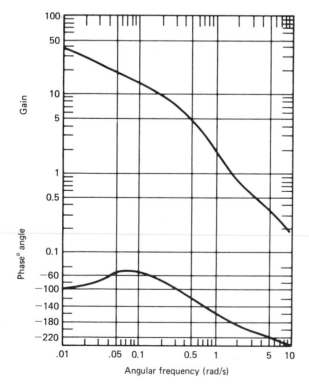

Figure 12.28   Bode plot for Problem 12.17.

APPENDIX 1

UNITS

A.1.1 SI UNITS

The system of units that is gaining acceptance throughout the world is called the Systéme International d'Unités (SI). This system is maintained by the Conférence Genérale des Poids et Measures. The basic set of these units is given in Section 1.6.1. The description of other physical quantities can be provided in terms of *derived* units that are expressible in terms of the basic set. The following list shows the more common derived units and their descriptors.

Quantity Symbol	Quantity Definition	Unit Name	Unit Symbol	Unit Definition
f	Frequency	Hertz	Hz	s^{-1}
W	Energy	Joule	J	$kg \cdot m^2/s^2$
F	Force	Newton	N	$kg \cdot m/s^2$
R	Resistance	Ohm	Ω	$kg \cdot m^2/(s^3 \cdot A^2)$
V	Voltage	Volt	V	$A \cdot \Omega$
p	Pressure	Pascal	Pa	N/m^2
ω	Angular frequency	Radians per second	rad/s	rad/s
E	Illuminance	Lux	lx	lm/m^2
Q	Charge	Coulomb	C	$A \cdot s$
L	Inductance	Henry	H	$kg \cdot m^2/(s^2 \cdot A^2)$
C	Capacity	Farad	F	$s^4 \cdot A^2/kg\text{-}m^2$
G	Conductance	Siemen	S	Ω^{-1}
Φ	Luminous flux	Lumen	lm	cd/sr
	Luminous efficacy	Lumen per watt	lm/w	lm/W
P	Power	Watt	W	J/S

A.1.2 OTHER UNITS

Many systems of units are employed in the world, although a strong effort is being made to adopt the SI units universally. Two widely employed unit systems are the English and centimeter-gram-seconds.

English System

The system of units that has been historically employed in the United Kingdom and United States is the English system. There are actually several types of English systems, but the most common one for technological practices is the engineering system. It is inevitable that some quantity in a design specification that has been employed in several countries for many years may be expressed in English units. Until this awkward system is eliminated, it will be necessary to perform conversion between the SI and English systems of units. The following table of conversion factors is provided to show the equivalent number of English units for particular SI units.

The measure of force and mass in the English system of units is somewhat confusing because of differences in definitions among users of the system. We will define and use what are considered the engineering definitions, although it is well to note that these are not the legal definitions.

The English unit of *force* is the *pound* (lb_f). The pound is best defined by its relation to the SI unit of force through the conversion 0.2248 lb/N. Thus, a force of 16 lb becomes

$$(16 \text{ lb}) \left(\frac{N}{0.2248 \text{ lb}} \right) = 71.2 \text{ N}$$

Measurement of *mass* in the English system is in the form of a unit called the *slug*. A slug is defined to be that mass that accelerates at 1 ft/s² when acted on by a force of 1 lb. The conversion between the slug and SI unit of mass is 14.59 kg/slug.

To complete the picture of English units, we note several other definitions that are sometimes employed that differ from those already presented.

(1) Pound mass

The pound was originally designed as a mass measurement. When used in this sense, it is best defined in relation to the SI kg from the conversion 1 lb (mass) = 0.454 kg.

(2) Poundal (pdl)

The poundal is a unit of force defined as that force that gives the pound mass, defined previously, an acceleration of 1 ft/s². In terms of the SI force unit, the poundal conversion is 1 pdl = 0.1383 N.

Centimeter-Gram-Second (CGS) System

The CGS system of units is often employed in scientific work and appears in journal articles and on some scientific measurement equipment. This system is based on the centimeter (cm) of length, the gram (g) of mass, and second (s) of time. Transformations may be made using the scale factors listed in the following table.

Quantity	SI	CGS	English
Length	1 meter	100 centimeters	3.28 feet
Mass	1 kilogram	10^3 grams	0.0685 slugs
Time	1 second	1 second	1 second
Force	1 newton	10^5 dynes	0.2248 pounds
Energy	1 joule	10^7 ergs	0.7376 ft-lb
Pressure	1 pascal	10 dyne/cm^2	1.45×10^{-4} lb/in^2

A.1.3 STANDARD PREFIXES

The SI decimal multiple and submultiple designations are shown in the following table:

Multiple	SI Prefix	Symbol
10^{12}	tera	T
10^9	giga	G
10^6	mega	M
10^3	kilo	k
10^2	hecto	h
10	deka	da
10^{-1}	deci	d
10^{-2}	centi	c
10^{-3}	milli	m
10^{-6}	micro	μ
10^{-9}	nano	n
10^{-12}	pico	p
10^{-15}	femto	f
10^{-18}	atto	a

APPENDIX 2

DIGITAL REVIEW

A.2.1 NUMBER SYSTEMS

The base 10 or decimal number system is normally employed in the analog description of information. In digital applications, a base 2 or binary counting system is often used with the octal or base 8 system for writing numbers. The hexadecimal system or base 16 also has become popular. It is important that an individual involved in digital applications be prepared to translate numbers between these systems. This can be accomplished by the following operations and equations.

Binary

A mathematical relationship can be constructed between a base 10 number and its binary equivalent using the equation

$$N_{10} = a_n 2^n + a_{n-1} 2^{n-1} + \cdots a_3 2^3 + a_2 2^2 + a_1 2^1 + a_0 2^0 \quad \text{(A.2.1)}$$

where

$$N_{10} = \text{base 10 number}$$
$$a_n a_{n-1} \cdots a_3 a_2 a_1 a_0 = \text{base 2 number}$$
$$n + 1 = \text{number of digits in the binary number}$$

The reverse procedure for finding the binary equivalent of a decimal number can be represented in several ways. The easiest way is to use a set of conversion tables. Lacking that, a process of successive division by 2 is employed. In any division, the remainder will always be zero or one half, and this determines the value of that binary digit. The first division yields the value of the least significant bit (LSB), which is the a_0 term, and the last remainder gives the most significant bit (MSB), which is a_n.

Example A.2.1

Find the binary equivalent of 259_{10}.

Solution   We use successive division by 2 as follows:

$$259/2 = 129 + 1/2 \qquad a_0 = \mathbf{1}$$

$$129/2 = 64 + 1/2 \qquad a_1 = \mathbf{1}$$

$$64/2 = 32 + 0/2 \qquad a_2 = \mathbf{0}$$

$$32/2 = 16 + 0/2 \qquad a_3 = \mathbf{0}$$

$$16/2 = 8 + 0/2 \qquad a_4 = \mathbf{0}$$

$$8/2 = 4 + 0/2 \qquad a_5 = \mathbf{0}$$

$$4/2 = 2 + 0/2 \qquad a_6 = \mathbf{0}$$

$$2/2 = 1 + 0/2 \qquad a_7 = \mathbf{0}$$

$$1/2 = 0 + 1/2 \qquad a_8 = \mathbf{1}$$

Thus, the answer is that 259_{10} is $\mathbf{100000011_2}$.

Octal

Because it is very frequently employed in digital work, we need to consider a number base formed of eight counting states. A principal reason for using this numbering system is that it becomes both easy and convenient to represent binary numbers in the octal system. This can be done by grouping the binary numbers in groups of three digits from the right and noting that these three digits represent the full range of octal numbers. Thus, $\mathbf{000_2}$ corresponds to 0_8, and $\mathbf{111_2}$ corresponds to 7_8. The small subscript 2 and 8 denote base 2 and base 8, respectively. Thus, 101111_2 can be grouped as $\mathbf{101_2}$ followed by $\mathbf{111_2}$ and written 57_8 in the octal system. The correspondence between octal and decimal could be accomplished through conversion first to binary. The direct method, however, is to use the equation

$$N_{10} = d_n 8^n + d_{n-1} 8^{n-1} + \cdots + d_2 8^2 + d_1 8^1 + d_0 8^0 \qquad \text{(A.2.2)}$$

where

$$N_{10} = \text{decimal number}$$
$$d_n d_{n-1} \cdots d_1 d_0 = \text{octal number digits}$$
$$n = \text{one less than the number of digits in the octal number}$$

Example A.2.2

Find the binary and decimal equivalent of the octal number 33_8.

Solution   We can find the binary most easily by

$$3_8 = \mathbf{011_2} \text{ and thus}$$

$$33_8 = 011011_2 = \mathbf{11011_2}$$

The decimal number could be found by either of Equations (A.2.1) or (A.2.2). Using

$$N_{10} = a_n 8^n + a_{n-1} 8^{n-1} + \cdots + a_1 8^1 + a_0 8^0 \qquad \text{(A.2.2)}$$

we have $n = 1$, so

$$N_{10} = (3)8^1 + (3)(8^0)$$

$$N_{10} = \mathbf{27_{10}}$$

To convert from decimal to octal, we can either convert first to binary and then to octal, use tables, or use the procedure of successive division where the division will be eight instead of two. As before, the least significant value is found by the first division.

Example A.2.3

Find the decimal number 356_{10} as both octal and binary numbers.

Solution   Let us use the successive division procedure to find the octal number first and then find the binary from that.

$$\frac{356}{8} = 44 + \frac{4}{8} \qquad \text{so } d_0 = 4$$

$$\frac{44}{8} = 5 + \frac{4}{8} \qquad \text{so } d_1 = 4$$

and

$$d_2 = 5$$

We find then that 356_{10} is $\mathbf{544_8}$ and from the octal to decimal conversion of 5_8, 4_8, and 4_8 we get a binary number of $\mathbf{101100100_2}$.

Hexadecimal

The rapid development of 4-bit, 8-bit, and 16-bit microcomputers has brought about the common use of base 16 number representations. This is called the *hexadecimal* or *hex system*. Any binary system with 4 or a multiple of 4 bits to the word can be conveniently expressed in hex because a hex number is formed from the counting states of 4 binary numbers. Because we must go beyond 9 before the hex set is complete, it is necessary to have new symbols to complete the set. The accepted standard is to use the letters A through F for the remaining counts. Instead of a subscript 16 to denote hex, it also has become common practice to use a cap letter H to indicate that a number is hex. Thus, from 0_{10} through 9_{10} we have the hex numbers 0H through 9H, and from 10_{10} through 15_{10} we have the hex equivalents of AH, BH, CH, DH, EH, FH. Conversion of a hex number into the decimal equivalent can be made through the equation

$$N_{10} = c_n 16^n + c_{n-1} 16^{n-1} + \cdots + c_2 16^2 + c_1 16^1 + c_0 16^0 \quad (A.2.3)$$

where

$$N_{10} = \text{decimal number (base ten)}$$
$$c_n c_{n-1} \cdots c_1 c_0 = \text{hex number}$$
$$n = \text{one less than the number of hex digits}$$

Conversion of a decimal number to hex can be made by successive division by 16, as was done for base 2 and base 8 systems. The following examples illustrate these conversions.

Example A.2.4

Find the decimal equivalents of 47H, 30DH, and A2FH.

Solution   We use Equation (A.2.3) directly, where for the actual evaluation the individual hex digits are expressed in their respective base 10 equivalents. For 47H, we have $c_0 = 7$ and $c_1 = 4$ so that

$$N_{10} = 4 \times 16^1 + 7 \times 16^0 = 64 + 7$$

$$N_{10} = 71_{10}$$

For 30DH, we have $c_0 = \text{DH} = 13_{10}$, $c_1 = 0$, and $c_2 = 3$

$$N_{10} = 3 \times 16^2 + 0 \times 16^1 + 13 \times 16^0 = 3 \times 256 + 13$$

$$N_{10} = \mathbf{781_{10}}$$

For A2FH, we have $c_0 = \text{FH} = 15$, $c_1 = 2$, and $c_2 = \text{AH} = 10$

$$N_{10} = 10 \times 16^2 + 2 \times 16^1 + 15 \times 16^0$$

$$N_{10} = \mathbf{2607_{10}}$$

Example A.2.5

Find the hex equivalent of binary $\mathbf{10110101_2}$ and octal 422_8.

Solution   The simplest way to convert binary to hex is to group the binary number into 4-bit sets and find the hex of each set. Thus, $\mathbf{10110101_2}$ is $\mathbf{1011_2}$ and $\mathbf{0101_2}$. But $\mathbf{1011_2}$ is 11_{10}, which is BH, and $\mathbf{0101_2}$ is 5_{10}, which is 5H. Thus, the hex is B5H. The octal number can be converted easily to binary and then to hex. Thus, 422_8 is $\mathbf{100010010_2}$, which can be grouped into three groups of 4 bits (using leading zeros) as $\mathbf{0001_2}$, $\mathbf{0001_2}$, and $\mathbf{0010_2}$. The hex is then 112H.

Example A.2.6

Find the hex equivalents of 29_{10}, 175_{10}, and 3412_{10}.

Solution   Conversion of decimal to hex is accomplished by successive division by 16. This finds the least significant digit first.

$$29/16 = 1 + 13/16 \qquad c_0 = 13 = \text{DH}$$

$$1/16 = 0 + 1/16 \qquad c_1 = 1 = \text{1H}$$

Thus, the hex is 1DH.
For the next decimal number, we have

$$175/16 = 10 + 15/16 \qquad c_0 = 15 = \text{FH}$$

$$10/16 = 0 + 10/16 \qquad c_1 = 10 = \text{AH}$$

Thus, the answer is AFH.
The last decimal number is converted as follows:

$$3412/16 = 213 + 4/16 \qquad c_0 = 4$$

$$213/16 = 13 + 5/16 \qquad c_1 = 5$$

$$13/16 = 0 + 13/16 \qquad c_2 = 13 = \text{DH}$$

Thus, the hex number is D54H.

Negative Numbers

The representation of negative numbers can be accomplished in both binary and octal by preceding the number with a negative sign as in decimal. In many cases, however, it is desirable to let a binary number itself carry information about the sign. This is accomplished in binary by one of several artificial techniques such as the following:

Sign and magnitude

In this method, the binary number has another term added to it. There is no convention here, and in some cases, a **1** represents a negative and a **0** represents a negative. Thus, we would write the binary number $-\mathbf{1011}$ as **11011** if our convention choice was to make the highest bit **1** for a negative number.

2s complement

In this method, an extra digit is (again) provided that is always **1** for a negative number, but the code of the number is altered. This method is very popular because of the ease with which math operations can be performed. In this method, a negative number is formed by the relation

$$(-N_2) = (2^{n+1})_2 - N_2 \tag{3.3}$$

where

$(-N_2)$ = the 2s complement binary number
n = highest digit in the number
$(2^{n+1})_2$ = a binary number 2 digits larger than N_2

Thus, the number -1011_2 becomes in 2s complement with $n = 4$.

$$100000_2 - 1011_2 = 10101_2 \text{ (2s complement)}$$

A simple rule for finding the 2s complement of a binary number can be stated as follows:

To find the 2s complement of a binary number, copy all zeros from LSB up to and including the first **1** and then replace all remaining **0**s by **1**s and **1**s by **0**s. Finally, place a **1** before the MSB.

Example A.2.7

Find the 2s complement of 1101010_2.

Solution   Using the aforementioned rule, we leave the **0** and **1** on the right alone, change all the rest, and include a final **1**.

$$10010110 \text{ (2s complement)}$$

A.2.2 BOOLEAN ALGEBRA

The application of digital techniques to the solution of control problems often involves the use of complex logical operations. Boolean algebra is a mathematical method that is very useful for setting up and solving logical equations.

The (only) values of variables in the logical math of Boolean algebra are those of true or false. Thus, the variables are binary in that they can take on only two values that are assigned as **1** for a true state and **0** for a false state. As an example, if the level of liquid in a tank is the variable, we let the variable A denote this level. Then, if the level is *higher* than that desired, we set $A = 1$ (a logical true) and if *lower*, $A = 0$ (a logical false).

There are four math operations in Boolean algebra that relate how combinations of variables may operate on one another and be related to one another.

Equality "="   **Equality between two variables is defined by the same symbol employed in traditional algebra and has the same meaning. If we write $A = B$ and know $A = 1$, then $B = 1$ also.**

Complement "$\overline{A}$"

The complement operation is denoted by a bar over a variable and indicates a variable of the opposite value as the original. If we know $B = 1$, then $\overline{B} = 0$. The complement operation is sometimes referred to as a *NOT*.

AND "·"

The AND operation is denoted by a dot "·" between two variables. If $C = A \cdot B$, then C will be **1** if and only if $A = 1$ and $B = 1$; otherwise $C = 0$. In some publications the dot will be omitted such that $A \cdot B = AB$.

OR "+"

The plus symbol is employed to define the OR operation between two variables. Thus, we note that if $C = A + B$, the C will be **1** if A or B or both are **1** and will be **0** otherwise.

These math operations may be combined into many combinations for the description of some required logical operation. Thus, we may have $C = \overline{A \cdot B}$, which would mean C was equal to the complement of A AND B or NOT A AND B. In many cases, a particular logical combination can be represented by a different logical combination but giving the same result. Thus, we can show that the previous $C = \overline{A \cdot B}$ is exactly the same as the operation $\overline{A} + \overline{B}$, that is, NOT A OR NOT B. In both cases, C takes on the same logic state for the same logic states of A and B. There are several theorems that aid in changing the form of Boolean math operations without resorting to a truth table in each case.

The basic theorems of Boolean algebra are separated into single variable and multivariable categories. Table A.2.1 shows the single variable theorems and Table A.2.2 the multivariable theorems.

The theorems can be of great value in reducing a logical equation into another equivalent form that may not be intuitively obvious and may be much simpler.

Example A.2.8

Find a simpler logic expression for the equation

$$D = A \cdot B + C \cdot (A \cdot B + \overline{C})$$

Solution   Let us apply the theorems in steps to simplify this equation. First, by distribution, we can write

$$D = A \cdot B + C \cdot A \cdot B + C \cdot \overline{C}$$

TABLE A.2.1

$A \cdot 0 = 0$	$A + A = A$
$A + 0 = A$	$A \cdot A = A$
$A \cdot 1 = A$	$A + \overline{A} = 1$
$A + 1 = 1$	$A \cdot \overline{A} = 0$

TABLE A.2.2

$A + B = B + A$	Commutation
$A \cdot B = B \cdot A$	
$A + (B + C) = (A + B) + C$	Association
$A \cdot (B \cdot C) = (A \cdot B) \cdot C$	
$A \cdot (B + C) = A \cdot B + A \cdot C$	Distribution
$A + B \cdot C = (A + B) \cdot (A + C)$	
$A + A \cdot B = A$	Absorption
$A \cdot (A + B) = A$	
$\overline{A \cdot B} = \overline{A} + \overline{B}$	DeMorgan
$\overline{A + B} = \overline{A} \cdot \overline{B}$	

But $C \cdot \overline{C} = 0$ by Table A.2.1. By applying association, we can write

$$D = A \cdot B + A \cdot B \cdot C$$

Then, from distribution again

$$D = A \cdot B \cdot (1 + C)$$

But $1 + C = 1$, so that

$$D = A \cdot B + C \cdot (A + B + \overline{C}) = A \cdot B$$

So the rather complex expression becomes a simple AND between A and B and is independent of C.

A.2.3 DIGITAL ELECTRONIC BUILDING BLOCKS

The building blocks for digital electronic circuits are a set of black boxes that implement the basic Boolean math operations for electrical signals. These elements are referred to as *gates*, and each has a characteristic symbol that has been adopted for universal indication of the device.

AND Gate

The AND gate accepts as input two signals A and B and outputs a signal that is given by $C = A \cdot B$. The symbol for this element is given in Figure A.2.1.

OR Gate

The OR gate takes as input two signals A and B and outputs a signal that is given by $C = A + B$. The symbol for this device is given in Figure A.2.1.

Inverter

(NOT) the inverter, which is sometimes referred to as a NOT operation, accepts as an input a signal A and outputs an inverted version of this signal $\overline{A}$. The symbol

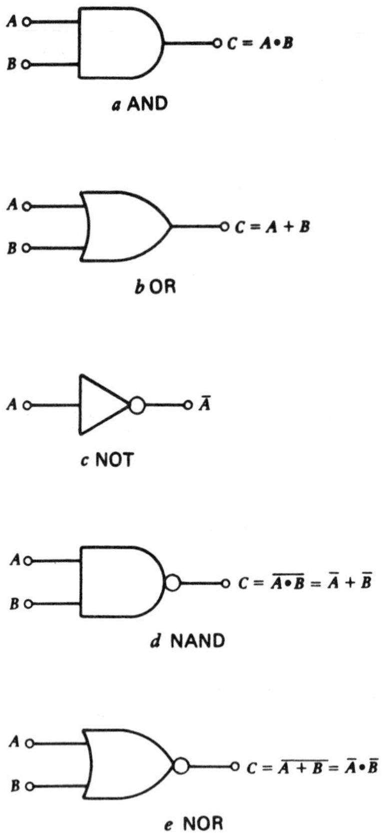

$C = A \cdot B$

a AND

$C = A + B$

b OR

$\bar{A}$

c NOT

$C = \overline{A \cdot B} = \bar{A} + \bar{B}$

d NAND

$C = \overline{A + B} = \bar{A} \cdot \bar{B}$

e NOR

Figure A.2.1   Digital gate symbols.

for this element is given in Figure A.2.1. This is equivalent to providing the complement of a variable.

There are two additional building blocks that are frequently employed in electronic digital testing. These elements are used because of the ease with which they may be implemented using the popular circuit families.

NAND Gate

The NAND gate is essentially a device that performs the AND of two inputs and then inverts the output of that operation. Thus, if the inputs are A and B, the output will be $C = \overline{A \cdot B}$. We note from De Morgan's theorem that this operation is identical to $\bar{A} + \bar{B}$. The symbol for this device is shown in Figure A.2.1.

NOR Gate

As in the previous case, the NOR gate performs an OR operation between two inputs and then inverts the results. Thus, if the inputs are A and B, the output will be $C = \overline{A + B}$, which we know again from De Morgan's theorem can be expressed as $C = \overline{A} \cdot \overline{B}$. The symbol for this device is shown in Figure A.2.1. Using the five logic gates described, most of the basic logic equations developed from problems and using Boolean algebra can be implemented.

APPENDIX 3

THERMOCOUPLE TABLES

The following tables give the output voltage of several thermocouple (TC) types over a range of temperature in 5°C increments. In each case, the TC reference temperature is 0°C. The first-named material will be the positive terminal, as *iron-constantan*; the iron will be the positive lead when the reference temperature is lower than the measurement. The temperature is in °C and the output is mV. Each column is in 5°C increments from the temperature of that row.

Type J: Iron-Constantan

	0	5	10	15	20	25	30	35	40	45
− 150	− 6.50	− 6.66	− 6.82	− 6.97	− 7.12	− 7.27	− 7.40	− 7.54	− 7.66	− 7.78
− 100	− 4.63	− 4.83	− 5.03	− 5.23	− 5.42	− 5.61	− 5.80	− 5.98	− 6.16	− 6.33
− 50	− 2.43	− 2.66	− 2.89	− 3.12	− 3.34	− 3.56	− 3.78	− 4.00	− 4.21	− 4.42
− 0	0.00	− 0.25	− 0.50	− 0.75	− 1.00	− 1.24	− 1.48	− 1.72	− 1.96	− 2.20
+ 0	0.00	0.25	0.50	0.76	1.02	1.28	1.54	1.80	2.06	2.32
50	2.58	2.85	3.11	3.38	3.65	3.92	4.19	4.46	4.73	5.00
100	5.27	5.54	5.81	6.08	6.36	6.63	6.90	7.18	7.45	7.73
150	8.00	8.28	8.56	8.84	9.11	9.39	9.67	9.95	10.22	10.50
200	10.78	11.06	11.34	11.62	11.89	12.17	12.45	12.73	13.01	13.28
250	13.56	13.84	14.12	14.39	14.67	14.94	15.22	15.50	15.77	16.05
300	16.33	16.60	16.88	17.15	17.43	17.71	17.98	18.26	18.54	18.81
350	19.09	19.37	19.64	19.92	20.20	20.47	20.75	21.02	21.30	21.57
400	21.85	22.13	22.40	22.68	22.95	23.23	23.50	23.78	24.06	24.33
450	24.61	24.88	25.16	25.44	25.72	25.99	26.27	26.55	26.83	27.11
500	27.39	27.67	27.95	28.23	28.52	28.80	29.08	29.37	29.65	29.94
550	30.22	30.51	30.80	31.08	31.37	31.66	31.95	32.24	32.53	32.82
600	33.11	33.41	33.70	33.99	34.29	34.58	34.88	35.18	35.48	35.78
650	36.08	36.38	36.69	36.99	37.30	37.60	37.91	38.22	38.53	38.84
700	39.15	39.47	39.78	40.10	40.41	40.73	41.05	41.36	41.68	42.00

Type K: Chromel-Alumel

	0	5	10	15	20	25	30	35	40	45
− 150	−4.81	−4.92	−5.03	−5.14	−5.24	−5.34	−5.43	−5.52	−5.60	−5.68
− 100	−3.49	−3.64	−3.78	−3.92	−4.06	−4.19	−4.32	−4.45	−4.58	−4.70
− 50	−1.86	−2.03	−2.20	−2.37	−2.54	−2.71	−2.87	−3.03	−3.19	−3.34
− 0	0.00	−0.19	−0.39	−0.58	−0.77	−0.95	−1.14	−1.32	−1.50	−1.68
+ 0	0.00	0.20	0.40	0.60	0.80	1.00	1.20	1.40	1.61	1.81
50	2.02	2.23	2.43	2.64	2.85	3.05	3.26	3.47	3.68	3.89
100	4.10	4.31	4.51	4.72	4.92	5.13	5.33	5.53	5.73	5.93
150	6.13	6.33	6.53	6.73	6.93	7.13	7.33	7.53	7.73	7.93
200	8.13	8.33	8.54	8.74	8.94	9.14	9.34	9.54	9.75	9.95
250	10.16	10.36	10.57	10.77	10.98	11.18	11.39	11.59	11.80	12.01
300	12.21	12.42	12.63	12.83	13.04	13.25	13.46	13.67	13.88	14.09
350	14.29	14.50	14.71	14.92	15.13	15.34	15.55	15.76	15.98	16.19
400	16.40	16.61	16.82	17.03	17.24	17.46	17.67	17.88	18.09	18.30
450	18.51	18.73	18.94	19.15	19.36	19.58	19.79	20.01	20.22	20.43
500	20.65	20.86	21.07	21.28	21.50	21.71	21.92	22.14	22.35	22.56
550	22.78	22.99	23.20	23.42	23.63	23.84	24.06	24.27	24.49	24.70
600	24.91	25.12	25.34	25.55	25.76	25.98	26.19	26.40	26.61	26.82
650	27.03	27.24	27.45	27.66	27.87	28.08	28.29	28.50	28.72	28.93
700	29.14	29.35	29.56	29.77	29.97	30.18	30.39	30.60	30.81	31.02
750	31.23	31.44	31.65	31.85	32.06	32.27	32.48	32.68	32.89	33.09
800	33.30	33.50	33.71	33.91	34.12	34.32	34.53	34.73	34.93	35.14
850	35.34	35.54	35.75	35.95	36.15	36.35	36.55	36.76	39.96	37.16
900	37.36	37.56	37.76	37.96	38.16	38.36	38.56	38.76	38.95	39.15
950	39.35	39.55	39.75	39.94	40.14	40.34	40.53	40.73	40.92	41.12
1000	41.31	41.51	41.70	41.90	42.09	42.29	42.48	42.67	42.87	43.06
1050	43.25	43.44	43.63	43.83	44.02	44.21	44.40	44.59	44.78	44.97
1100	45.16	45.35	45.54	45.73	45.92	46.11	46.29	46.48	46.67	46.85
1150	47.04	47.23	47.41	47.60	47.78	47.97	48.15	48.34	48.52	48.70
1200	48.89	49.07	49.25	49.43	49.62	49.80	49.98	50.16	50.34	50.52
1250	50.69	50.87	51.05	51.23	51.41	51.58	51.76	51.94	52.11	52.29
1300	52.46	52.64	52.81	52.99	53.16	53.34	53.51	53.68	53.85	54.03
1350	54.20	54.37	54.54	54.71	54.88					

Type T: Copper-Constantan

	0	5	10	15	20	25	30	35	40	45
− 150	− 4.603	− 4.712	− 4.817	− 4.919	− 5.018	− 5.113	− 5.205	− 5.294	− 5.379	
− 100	− 3.349	− 3.488	− 3.624	− 3.757	− 3.887	− 4.014	− 4.138	− 4.259	− 4.377	− 4.492
− 50	− 1.804	− 1.971	− 2.135	− 2.296	− 2.455	− 2.611	− 2.764	− 2.914	− 3.062	− 3.207
− 0	0.000	− 0.191	− 0.380	− 0.567	− 0.751	− 0.933	− 1.112	− 1.289	− 1.463	− 1.635
+ 0	0.000	0.193	0.389	0.587	0.787	0.990	1.194	1.401	1.610	1.821
50	2.035	2.250	2.467	2.687	2.908	3.132	3.357	3.584	3.813	4.044
100	4.277	4.512	4.749	4.987	5.227	5.469	5.712	5.957	6.204	6.453
150	6.703	6.954	7.208	7.462	7.719	7.987	8.236	8.497	8.759	9.023
200	9.288	9.555	9.823	10.093	10.363	10.635	10.909	11.183	11.459	11.735
250	12.015	12.294	12.575	12.857	13.140	13.425	13.710	13.997	14.285	14.573
300	14.864	15.155	15.447	15.740	16.035	16.330	16.626	16.924	17.222	17.521
350	17.821	18.123	18.425	18.727	19.032	19.337	19.642	19.949	20.257	20.565
400										
450										
500										
550										
600										
650										
700										

Type S: Platinum-Platinum/10% Rhodium

	0	5	10	15	20	25	30	35	40	45
− 150										
− 100										
− 50										
− 0										
+ 0	0.000	0.028	0.056	0.084	0.113	0.143	0.173	0.204	0.235	0.266
50	0.299	0.331	0.364	0.397	0.431	0.466	0.500	0.535	0.571	0.607
100	0.643	0.680	0.717	0.754	0.792	0.830	0.869	0.907	0.946	0.986
150	1.025	1.065	1.166	1.146	1.187	1.228	1.269	1.311	1.352	1.394
200	1.436	1.479	1.521	1.564	1.607	1.650	1.693	1.736	1.780	1.824
250	1.868	1.912	1.956	2.001	2.045	2.090	2.135	2.180	2.225	2.271
300	2.316	2.362	2.408	2.453	2.499	2.546	2.592	2.638	2.685	2.731
350	2.778	2.825	2.872	2.919	2.966	3.014	3.061	3.108	3.156	3.203
400	3.251	3.299	3.347	3.394	3.442	3.490	3.539	3.587	3.635	3.683
450	3.732	3.780	3.829	3.878	3.926	3.975	4.024	4.073	4.122	4.171
500	4.221	4.270	4.319	4.369	4.419	4.468	4.518	4.568	4.618	4.668
550	4.718	4.768	4.818	4.869	4.919	4.970	5.020	5.071	5.122	5.173
600	5.224	5.275	5.326	5.377	5.429	5.480	5.532	5.583	5.635	5.686
650	5.738	5.790	5.842	5.894	5.946	5.998	6.050	6.102	6.155	6.207
700	6.260	6.312	6.365	6.418	6.471	6.524	6.577	6.630	6.683	6.737
750	6.790	6.844	6.897	6.951	7.005	7.058	7.112	7.166	7.220	7.275
800	7.329	7.383	7.438	7.492	7.547	7.602	7.656	7.711	7.766	7.821
850	7.876	7.932	7.987	8.042	8.098	8.153	8.209	8.265	8.320	8.376
900	8.432	8.488	8.545	8.601	8.657	8.714	8.770	8.827	8.883	8.940
950	8.997	9.054	9.111	9.168	9.225	9.282	9.340	9.397	9.455	9.512
1000	9.570	9.628	9.686	9.744	9.802	9.860	9.918	9.976	10.035	10.093
1050	10.152	10.210	10.269	10.328	10.387	10.446	10.505	10.564	10.623	10.682
1100	10.741	10.801	10.860	10.919	10.979	11.038	11.098	11.157	11.217	11.277
1150	11.336	11.396	11.456	11.516	11.575	11.635	11.695	11.755	11.815	11.875
1200	11.935	11.995	12.055	12.115	12.175	12.236	12.296	12.356	12.416	12.476
1250	12.536	12.597	12.657	12.717	12.777	12.837	12.897	12.957	13.018	13.078
1300	13.138	13.198	13.258	13.318	13.378	13.438	13.498	13.558	13.618	13.678
1350	13.738	13.798	13.858	13.918	13.978	14.038	14.098	14.157	14.217	14.277
1400	14.337	14.397	14.457	14.516	14.576	14.636	14.696	14.755	14.815	14.875
1450	14.935	14.994	15.054	15.113	15.173	15.233	15.292	15.352	15.411	15.471
1500	15.530	15.590	15.649	15.709	15.768	15.827	15.887	15.946	16.006	16.065
1550	16.124	16.183	16.243	16.302	16.361	16.420	16.479	16.538	16.597	16.657
1600	16.716	16.775	16.834	16.893	16.952	17.010	17.069	17.128	17.187	17.246
1650	17.305	17.363	17.422	17.481	17.539	17.598	17.657	17.715	17.774	17.832
1700	17.891	17.949	18.008	18.066	18.124	18.183	18.241	18.299	18.358	18.416
1750	18.474	18.532	18.590	18.648						

A P P E N D I X   4

MECHANICAL REVIEW

A.4.1 MOTION

The mechanical description of motion is based on a specification of the position, velocity, and acceleration of an object relative to some reference section. If the object moves in an arbitrary manner with respect to the reference, we speak of linear or rectilinear motion. If the object is constrained to move *about* the reference, as in rotation, we speak of angular motion. The relationships between position, velocity, and acceleration are given in Table A.4.1. Linear position is measured in meters (m) with velocity in m/s and acceleration m/s^2. Angular motion is measured in radians (rad), angular velocity in rad/s, and angular acceleration in rad/s^2.

In some cases, a special unit of measure is used for linear acceleration. This unit is the **g** that is defined as the acceleration of a freely falling body at the earth's surface. The value of **g** is given by **g** $= 9.8$ m/s^2. In this context an acceleration of 50 **g** in any direction would be

$$a = (50 \text{ } \mathbf{g})(9.8 \text{ m/s}^2)$$

$$a = 490 \text{ m/s}^2$$

Special cases of motion frequently occur under constant speed or constant acceleration. In these cases, the motion can be described in terms of the speed or acceleration as shown in Table A.4.2.

TABLE A.4.1   Position, Velocity, and Acceleration

Motion	Position	Velocity	Acceleration
Linear	x	$v = \dfrac{dx}{dt}$	$a = \dfrac{d^2x}{dt^2}$
Angular	θ	$\omega = \dfrac{d\theta}{dt}$	$\alpha = \dfrac{d^2\theta}{dt^2}$

TABLE A.4.2   Special Motion Cases

Motion	Linear	Angular
Constant speed position	v $x = vt + x_0$	ω $\theta = \theta_0 + \omega t$
Constant acceleration speed	a $v = v_0 + at$	α $\omega = \omega_0 + \alpha t$
Position	$x = x_0 + v_0 t + \dfrac{at^2}{2}$	$\theta = \theta_0 + \omega_0 t + \dfrac{\alpha t^2}{2}$

x_0, θ_0, v_0, ω_0 are initial positions and speed

A.4.2 FORCE

If an object is in uniform motion, it is found that a change in this motion (acceleration) can be effected through the action of some external agent with appropriate exchange of energy. The action that is required is defined as a *force*, and the resultant energy exchange is defined as *work*. The unit of *force* is the newton (N), a derived quantity, equal to 1 kg-m/s². The measurement of force is a significant part of the process-control technology. The relationship of force to other mechanical variables can be formulated only after definition of a property of objects called *mass*.

Mass

Any object at rest or in uniform motion resists attempts to cause a change of its state. This resistance is indeed a reflection of the need for work to be done before a state change can occur. Even in a zero gravity condition (0 **g**) in deep space, we would find it much more difficult to move a locomotive than a ball bearing. This resistance or *inertia* is reflected in a characteristics of an object called *mass*. Mass is a pure property, independent of the location or state of motion (neglecting relativity) of the object. Because a locomotive has much more mass than a ball bearing, it is more difficult to move. Mass is measured in the SI unit of kilogram (kg).

Newton's Law

The relationship between the force applied to an object, its mass, and the resultant change in motion expressed in terms of acceleration is formulated as *Newton's law*.

$$F = ma \qquad\qquad (A.4.1)$$

where

F = net force in N
m = total mass in kg
a = acceleration in m/s²

Example A.4.1

An accelerometer indicates that a 2-kg object is accelerating at 2.2 m/s². Find the net force that acts on the object in newtons and pounds.

Solution   We can find this force by using

$$F = ma$$

$$F = (2 \text{ kg})(2.2 \text{ m/s}^2) \times \frac{1 \text{ N}}{\text{kg-m/s}^2} \qquad\qquad (A.4.1)$$

$$F = \textbf{4.4 N}$$

Note we have used the fact that 1 kg-m/s² = 1 N, and Appendix 1 shows that 1 lb = 4.448 N, so that

$$F = (4.4 \text{ N})\left(\frac{1 \text{ lb}}{4.448 \text{ N}}\right)$$

$$F = 0.989 \text{ lb}$$

Gravity and Weight

A fundamental law of nature states that all objects with finite mass exert a force of attraction on each other. In the case of the massive earth and objects close to the surface, this force is (approximately)

$$F_g = mg \qquad\qquad (A.4.2)$$

where

F_g = force of gravity in N
m = mass in kg
g = acceleration because of gravity defined earlier (9.8 m/s²)

Equation (A.4.2) expresses the *weight* an object possesses, measured as the *force* with which the earth attracts an object close to its surface. The proper measure of weight is also the newton because it is a force. Scales used for *weighing* read weight. These are, of course, calibrated to read the force of the earth on a mass and thus *do not* measure mass. A metric scale may indicate the weighed object mass in kg in terms of F_g/g. This is a true measure of mass because the scale has been calibrated for the earth's surface. An English system scale may indicate a weight in pounds, and this is actually, again, the force of the earth's attraction of the object.

Hooke's Law and Springs

A very important relationship exists between force and displacement for the special case of objects defined as *springs*. A traditional spring of coiled steel wire serves as a good descriptive example. Such a spring is shown in Figure A.4.1a in the equilibrium or relaxed state. Here no forces act on the spring, and its length is x_0 as shown. In Figure A.4.1b a force F is applied to the spring. Application of this force caused the spring to extend from x_0 to a new length x, where the system has again become stationary. The net force must be zero and hence the spring must be pulling back with an equal force F. *Hooke's law* states that the equilibrium force of a spring under compression or extension is given by

$$F = -k\Delta x \qquad\qquad (A.4.3)$$

where

F = force in N
k = spring constant in N/m

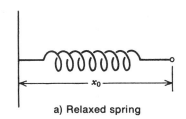

a) Relaxed spring

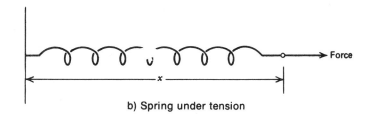

b) Spring under tension

Figure A.4.1   A force applied to a spring extends the length from a relaxed x_0 to a new x.

$$\Delta x = x - x_0 = \text{change in length}$$

The significance of the negative sign is to indicate that the spring force is *opposite* the applied force so that they balance. The spring constant k is a specification of the spring itself and is found from calibration or from a manufacturer specification sheet. The term *spring* is used loosely here. The devices described under Hooke's law could be a bellows, diaphragm, coil, spring, leaf spring, or indeed any number of configurations that relate displacement to force.

Example A.4.2

Find the extension that a spring with a spring constant of 5 N/m experiences when a force of 14 N is applied.

Solution   The equation for Hooke's law is

$$F = k\Delta x$$

$$\Delta x = \frac{14\ \text{N}}{5\ \text{N/m}} \tag{A.4.3}$$

$$\Delta x = \textbf{2.8 m}$$

A P P E N D I X   5

P&ID SYMBOLS

A.5.1 INTRODUCTION

Just as electronics has standard symbols to represent components in circuit schematics, process control has symbols to represent the elements of a process-control system. Instead of a schematic, we call the process-control diagram a piping and instrumentation drawing, or simply P&ID. The symbols used are standards that have been developed through the years and are accepted by the Instrument Society of America (ISA). In this appendix the principal symbols used in P&ID are presented.

A.5.2 INTERCONNECTIONS

Interconnections in a P&ID can involve many different types of signals and the flow of the process itself. For this reason, a simple drawn line, such as is used in an electronic schematic, is insufficient. Instead, we use the symbol of a line to denote the nature of the signal, as shown in Figure A.5.1. Further information may be necessary on the P&ID document itself to indicate the specific nature of the signal, such as a 4–20 mA signal as opposed to a 10–50 mA signal. Both would have the dashed line on the drawing.

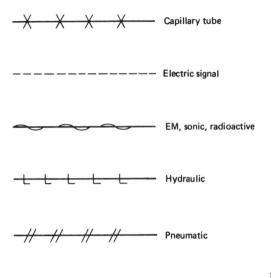

Figure A.5.1 P&ID signal and process lines.

A.5.3 BALLOON SYMBOLS

To denote the actual instruments used in process control, we use a circle called a balloon. Inside the balloon will be a code that tells the function and nature of the device represented. The code is given by letters, such as FC, and numbers, such as 117. The letters tell what the device is—for example, FC would mean flow controller and the number identifies the process-control loop that that element serves. The balloon also tells where the instrument is located, that is, local or control room. In Figure A.5.2, the balloon symbol is shown for the three cases

Figure A.5.2 Letter codes and balloon symbols.

	First Letter	Second Letter
A	Analysis	Alarm
B	Burner	
C	Conductivity	Control
D	Density	
E	Voltage	Primary element
F	Flow	
G	Gaging	Glass (sight tube)
H	Hand	
I	Current (electric)	Indicate
J	Power	
K	Time	Control station
L	Level	Light
M	Moisture	
O		Orifice
P	Pressure	Point
Q	Quantity	
R	Radioactivity	Record
S	Speed	Switch
T	Temperature	Transmit
U	Multivariable	Multifunction
V	Viscosity	Valve
W	Weight	Well
Y		Relay (transformation)
Z	Position	Drive

Figure A.5.3 P&ID symbols for sensors and elements.

of local (in the plant), control room board, and control room, but behind the board. Figure A.5.3 summarizes the meaning of the various letters that may appear in the balloon.

A.5.4 INSTRUMENT SYMBOLS

The transducer employed in the process-control loop is often broken down into a primary element (second letter E in balloon) and a transmitter (second letter T in balloon) to account for the actual transducer and the signal conditioning. The final control element may also be broken down into several elements, but the actual final control element itself will be in the process line. Figure A.5.4 shows some of the special symbols used for the transducers and final control elements.

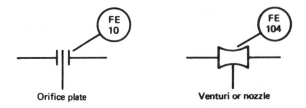

Orifice plate   Venturi or nozzle

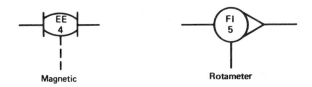

Magnetic   Rotameter

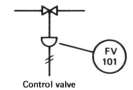

Control valve

Figure A.5.4 P&ID symbols for
transducers and elements.

REFERENCES

Andrew, *Applied Instrumentation in the Process Industries,* Gulf Publishing Co., 1974 (three volumes).

Bateson, *Introduction to Control System Technology,* Merrill, 1973.

Byran, *Control Systems for Technicians,* University of London Press, 1967.

Cerni and **Foster,** *Instrumentation for Engineering Measurement,* Wiley, 1962.

Considine, *Process Instrumentation and Control Handbook,* McGraw-Hill, 1957.

Considine and **Ross,** *Handbook of Applied Instrumentation,* McGraw-Hill, 1964.

Eckman, *Automatic Process Control,* Wiley, 1958.

Grabbe, Ramo, and **Wooldridge,** *Handbook of Automatic Computation and Control,* Wiley, 1958 (three volumes).

Lee, Adams, and **Gaines,** *Computer Process Control,* Wiley, 1968.

Liptak, *Instrument Engineers Handbook,* Volumes I and II, Chilton Book Company, 1970.

Norton, *Handbook of Transducers for Electronic Measuring Systems,* Prentice-Hall, 1969.

Oliver, *Practical Instrumentation Transducers,* Hayden Book Co., 1971.

GLOSSARY

Accelerometer. A transducer that measures the acceleration of the object to which it is attached.

Actuator. A part of the final control element that translates the control signal into action of the final control device in the process.

ADC. An analog-to-digital converter that converts an analog input of electric voltage or current into a proportional digital signal.

Alarm. In process control, an indicator that some process variable has exceeded preset limits.

Bellows. A pressure transducer that converts pressure into a nearly linear displacement.

Binary. A number representation system of base 2. This is the working system of digital computers.

Bode plot. A graph of transfer function versus frequency where the gain (often in decibels) and phase (in degrees) are plotted against the frequency on a log scale.

Bourdon tube. A pressure transducer that converts pressure to a displacement. The device is essentially a coiled, flattened tube that tends to straighten when pressure is applied.

Cascade control. A control system composed of two loops where the setpoint of one loop (the inner loop) is the output of the controller of the other loop (the outer loop).

Controlled variable. The process variable regulated by the process-control loop.

Controller. The element in a process-control loop that evaluates error of the controlled variable and initiates corrective action by a signal to the controlling variable.

Controlling variable. The process variable changed by the final control element under command of the controller to effect regulation of the controlled variable.

Cyclic. A condition of either steady-state or transient oscillation of a signal about the nominal value.

DAC. A digital-to-analog converter that converts a digital signal, often from a computer, into a proportional analog voltage or current.

DAS. A data acquisition system that interfaces many analog signals, called channels, to a computer. All switches, controls, and the ADC are included in the system.

DDC. Direct digital control, where a computer performs all the functions of error detection and controller action.

Derivative control mode. A controller mode in which controller output is directly proportional to the rate of change of controlled variable error.

DP cell. A pressure transducer that responds to the difference in pressure between two sources. Most often used to measure flow by the pressure difference across a restriction in the flow line.

Dynamic variable. Process variables that can change from moment to moment because of unspecified or unknown sources.

Error. The algebraic difference between the measured value of a variable and the ideal value. A positive error indicates that the measured value is greater than the ideal value.

Floating-control mode. A controller mode in which an error in the controlled variable causes the output of the controller to change at a constant rate. The error must exceed preset limits before controller change starts.

Foreground/background. A control system that uses two computers, one performing the control functions and the other used for data logging, offline evaluation of performance, financial operations, and so on. Either computer is able to perform the control functions.

Frequency response method. A method of tuning a process-control loop for optimum operation by proper selection of controller settings. This method is based on a study of the frequency response of the open process-control loop.

Gas thermometer. A temperature transducer that converts temperature to pressure of gas in a closed system. The relation between temperature and pressure is based on the gas laws at constant volume.

Hardware. When used in the context of computers hardware, refers to the physical equipment associated with the computer (e.g., ICs, printed circuit boards, cables, and so on).

Hex. A number of representation system of base 16. The hex number system is very useful in cases where computer words are composed of multiples of 4 bits (i.e., 4-bit words, 8-bit words, 16-bit words, and so on).

Hysteresis. The tendency of an instrument to give a different output for a given input, depending on whether the input results from an increase or decrease from the previous value.

Integral control mode. A controller mode in which the controller output increases at a rate proportional to the controlled variable error. Thus, the controller output is the integral of the error over time with a gain factor called the integral gain.

Ionization gauge. A pressure transducer based on conduction of electric current through ionized gas of the system whose pressure is to be measured. Useful only for very low pressures (e.g., below 10^{-3} atm).

I/P converter. A device that linearly converts electric current into gas pressure (e.g., 4–20 mA into 3–15 psi).

LASER. Stands for light amplification by stimulated emission of radiation. It is a source of EM radiation generally in the IR, visible, or UV bands and is characterized by small divergence, coherence, monochromaticity, and high colimation.

Linearity. The closeness to which the curve relating two variables approximates a straight line. It is usually expressed as the maximum deviation between the actual curve and the best-fit straight line.

Load. The process load is a term to denote the nominal values of all variables in a process that affect the controlled variable.

Load cell. A transducer for the measurement of force or weight. Action is based on strain gauges mounted within the cell on a force beam.

LVDT. A linear variable differential transformer that measures displacement by conversion to a linearly proportional voltage.

Microcomputer. A computer based on the use of a microprocessor integrated circuit. The entire computer often fits on a small printed circuit board and works with a data word of 4, 8, or 16 bits.

Microprocessor. A large-scale integrated circuit that has all the functions of a computer, except memory and input/output systems. The IC thus includes the instruction set, ALU, registers, and control functions.

Multiplexer. This device allows selection of one of many input channels of analog data under computer control. The device is often an integral part of a DAS.

Nozzle. A particular type of restriction used in flow systems to facilitate flow measurement by pressure drop across a restriction.

Nozzle/flapper. A fundamental part of pneumatic signal processing and pneumatic control operations. Basically, the device converts a displacement of the flapper to a pressure signal.

Octal. A number representation system of base 8. The octal system is most useful for computer systems with digital words of multiples of three bits (i.e., 3-bits, 6-bits, 9-bits, 12-bits, and so on).

Orifice plate. A type of restriction used in flow systems to facilitate measurement of flow by the pressure drop across a restriction.

PD. The designation of a controller operating in the proportional-derivative mode combination.

Photoconductive cell. A transducer that converts the intensity of EM radiation, usually in the IR or visible bands, into a change of cell resistance.

Photovoltaic cell. A transducer that converts the intensity of EM radiation, usually in the IR or visible bands, into a voltage.

PI. The designation of a controller operating in the proportional-integral mode combination.

P/I. A pressure-to-current converter linearly converts a signal pressure range into a signal current range (e.g., 3–15 psi into 4–20 mA).

PID. The designation of a controller operating in the proportional-integral-derivative mode combination. The PID is also called a three-mode controller.

Pirani gauge. A pressure transducer based on measurement of the resistance of a heated wire as a function of wire temperature that is a function of the gas pressure. Useful primarily for pressures less than one atmosphere.

P&ID. Stands for piping and instrumentation drawing, which is the primary schematic drawing used for laying out a process-control installation.

Pneumatic. Systems that employ gas, usually air, as the carrier of information and the medium to process and evaluate information.

Process. Any system comprised of dynamic variables, usually involved in manufacturing and production operations.

Process reaction method. A method of determination of optimum controller settings when tuning a process-control loop. The method is based on the reaction of the open loop to an imposed disturbance.

Proportional band. The change in input of a proportional controller mode required to produce a full-scale change in output. Thus, if 10% change in error causes a 100% change in controller output, the *PB* is 10.

Proportional control mode. A controller mode in which the controller output is directly proportional to the controlled variable error.

Pyrometer. A temperature transducer that measures temperature by the EM radiation emitted by an object, which is a function of the temperature.

Quarter amplitude. A process-control tuning criteria where the amplitude of the deviation (error) of the controlled variable, following a disturbance, is cyclic so that the amplitude of each peak is one quarter of the previous peak.

Range. The region between the limits within which a variable is to be measured. Thus, a temperature is to be measured in the range of 20°C to 250°C defines the range. (See also *Span.*)

Rate action. Another name for the derivative control mode.

Reset action. Another name for the integral control mode.

Resolution. The minimum detectable change of some variable in a measurement system.

Rotameter. A flow measurement system that is based on the proportionality of the rise of a float in a tapered tube, arranged vertically, placed in the flow system.

RTD. A temperature transducer that provides temperature information as the change in resistance of a metal wire element, often platinum, as a function of temperature.

Self-regulation. The property of some variables in a process to adopt a stable value under given load conditions without regulation via a process-control loop.

Sensitivity. The ratio of the change in output magnitude to the change of input magnitude, under steady-state conditions, for a measurement device.

Setpoint. The desired value of a controlled variable in a process-control loop.

Software. When used in the context of computer software, refers to the programs that provide the instructions to the computer on operations and calculations to be performed.

Span. The algebraic difference between the upper range value and the lower range value. Thus, a temperature in the range of 20°C to 250°C has a span of 230°C. (See also *Range*.)

Strain gauge. A transducer that converts information about the deformation of solid objects, called the strain, into a change of resistance.

Supervisory control. A process-control installation for which a computer oversees the operation of the control systems and provides the setpoint for the process control loops, which are themselves still analog.

Thermistor. A temperature transducer constructed from semiconductor material and for which the temperature is converted into a resistance, usually with a negative slope and highly nonlinear.

Thermocouple. A temperature transducer in which a voltage is produced nearly linearly with the difference in temperature to be measured and a known reference.

Three-mode controller. Another name for a PID controller.

Time constant. A number characterizing the time required for the output of a device to reach approximately 63% of the final value following a step change of its input.

Transducer. A device that converts variation of one variable into variation of another variable (for example, temperature change into resistance change).

Transmitter. In process control, a device that converts a variable into a form suitable for transmission of information to another location (e.g., resistance changed to current that is propagated on wires to a control installation).

Transfer function. The response of an element of a process-control loop that specifies how the output of the device is determined by the input.

Ultimate cycle method. See *Ziegler–Nichols method*.

Vapor pressure thermometer. A temperature transducer for which the pressure of vapor in a closed system of gas and liquid is a function of temperature.

Venturi. A type of restriction used in flow systems to facilitate measurement of flow by the pressure drop across a restriction.

Ziegler–Nichols method. A method of determination of optimum controller settings when tuning a process-control loop (also called the ultimate cycle method). It is based on finding the proportional gain that causes instability in a closed loop.

SOLUTIONS TO THE ODD PROBLEMS

Following are the solutions to most of the odd problems at the end of each chapter. In many cases, more than one answer to a design question is correct, so the given answer is to be taken as typical.

Chapter 1

1.5 The area is estimated by rectangular areas; curve a is found to give the least area.

1.7 First peak error is 22.5°C, so the quarter amplitudes should be 5.6 and 1.8. Actual values are 7 and 2, so quarter amplitude is approximately satisfied.

1.9 1110_2

1.11 19°C, 19.3°C, 20.7°C, and 21°C

1.13 101379 Pa

1.15 $R = 397 \pm 7.5 \ \Omega$

1.17 Worse case $= \pm 4.2\%$, $RMS = \pm 2.5\%$

1.19 $T = 199°C$

1.21 See Figure S.1.

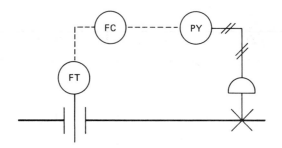

1.23 $\tau = 12$ s

1.25 $t = 1.99$ s

1.27 $Q_{ave} = 10.4$ gal/min, $\sigma = 1.24$ gal/min

Chapter 2

2.1   4.90 to 1.67 V, 46.2 to 1.10 mW

2.3 $R_4 = 2790$ Ω

2.5 $R_4 = 1285$ Ω

2.7 $R_3 = 600$ Ω, $R_3 = 600.64$ Ω for 0.25 mA

2.9 $C_4 = 0.21$ μF

2.11 $f_c = 15$ kHz, $R = 1060$ Ω, $C = 0.01$ μF, down 0.04% at 400 Hz

2.13 $f_c = 12$ kHz, $R = 1326$ Ω, $C = 0.01$ μF, down 7% at 30 kHz

2.15 See Figure S.2.

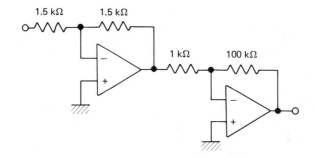

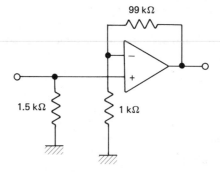

2.17 See Figure S.3.

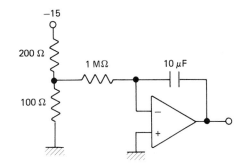

2.19 See Figure S.4.

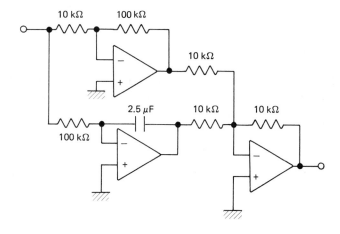

2.21 See Figure S.5, $P_{\text{rms}} = 0.38$ W.

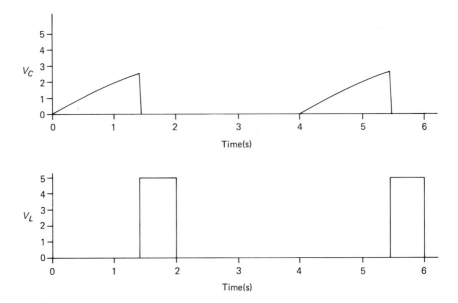

2.23 See Figure S.6 for the circuit and Figure S.7 for a plot of V_{out} vs. R_4. Nonlinearity is slight; least-squares-curve-fit straight line shows a 1% FS max deviation.

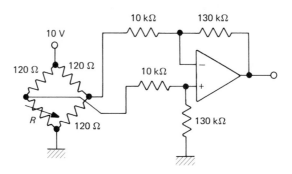

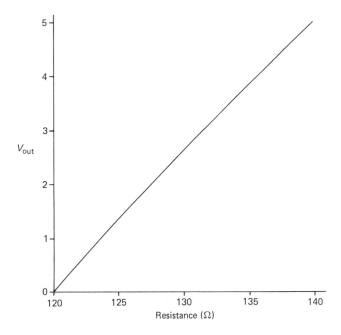

2.25 See Figure S.8.

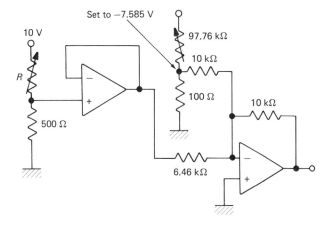

2.27 See Figure S.9.

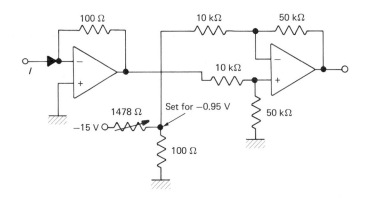

2.29 See Figure S.10.

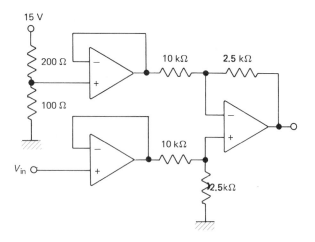

Chapter 3

3.1 (a) Ah, 12_8, 10_{10}
(b) 3Bh, 73_8, 59_{10}
(c) 16h, 26_8, 22_{10}

3.3 (a) 15h, 10101_2, 25_8
(b) 276h, 1001110101_2, 1165_8
(c) 1ABh, 110101011_2, 653_8

3.5 (a) 10101_2, (b) 101010100_2

3.9 See Figure S.11.

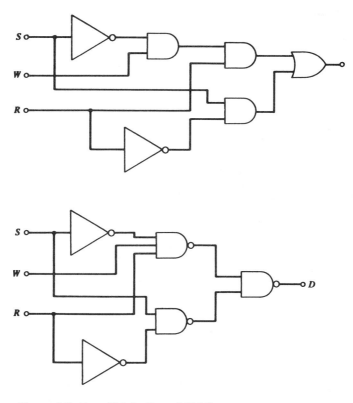

3.11 Refer to Figure 3.8, $R = 10 \text{ k}\Omega$, $R_f = 390 \text{ k}\Omega$.

3.13 $V_{\text{ref}} = 8.53$ V

3.15 Output = 61h or 01100001_2; input could be in the range of 2.382 to 2.422 V; input for 10110111_2 could be 7.148 to 7.187 V.

3.17 For -4.3 V, 47h; for -0.66 V, 1BCh; for 2.4 V, 2F5h; for 4.8 V, 3EBh. For 30Bh out, input is 2.607 to 2.617 V.

3.19 See Figure S.12. Resolution: 0.3125°C, 0.3867 psi, 0.2343 gal/min.

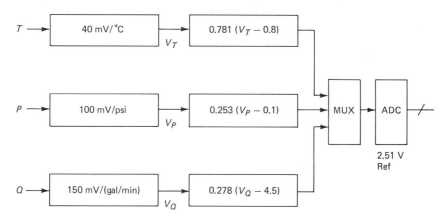

3.21 $\tau_s = 3.9$ ms

3.23 (a) 6Ch for 3.4 V and D6h for 6.7 V; (b) 5.71875 to 5.75 volts

3.25 See Figure S.13.

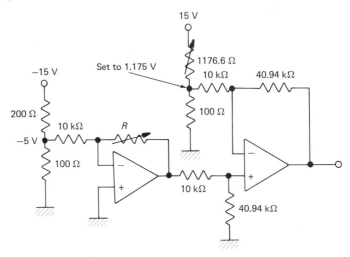

Chapter 4

4.1 251.7 K, $-21.43°C$, $-6°F$

4.3 403.1 K, $302°F$

4.5 3114 m/s, 10217 ft/s

4.7 $R = 108.1 \ \Omega$

4.9 $R(T) = 589.48[1 + 0.0018(T - 115)]$; $R(T) = 589.48[1 + 0.0018(T - 115) - 0.00000016(T - 115)^2]$. Linear has -0.89% error, quadratic has a $+0.016\%$ error.

4.11 Resolution required is 263 mV.

4.13 See Figure S.14, nonlinear, $0.4°C$ heating.

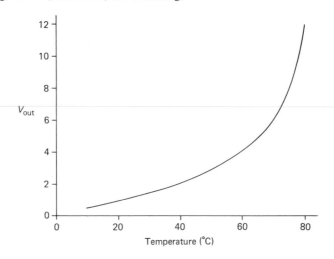

4.15 1225.25°C
4.17 (a) 12.21 mV, (b) 4990 Ω
4.19 0.22 mm
4.21 40 psi to 420 psi
4.23 See Figure S.15.

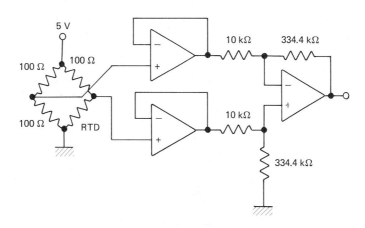

4.25 See Figure S.16.

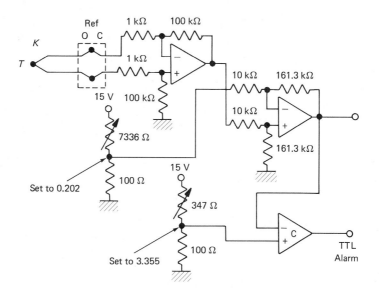

4.27 See Figure S.17.

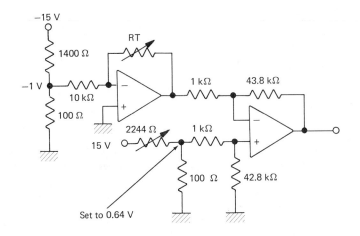

4.29 See Figure S.18.

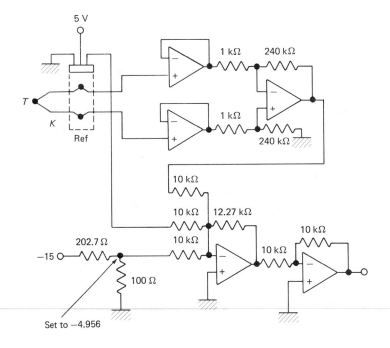

Chapter 5

5.1 0.6 mm, 250 Ω
5.3 862.75 pF to 897.96 pF, or approximately ±17.6 pF
5.5 See Figure S.19; resolution is 0.97 mm.

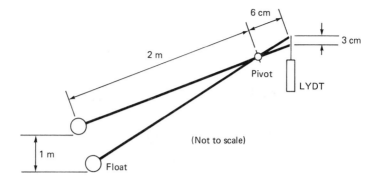

5.7 Refer to Figure 2.10; $R_1 = R_3 = 1 \text{ k}\Omega$, $R_2 = R_4 = 6.25 \text{ k}\Omega$, $C_3 = 0.02 \text{ μF}$. See Figure S.20 for plot.

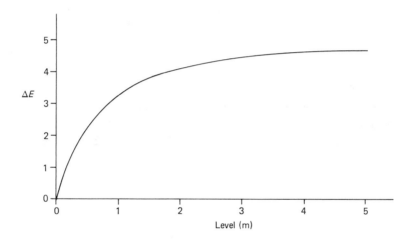

5.9 0.037 Ω

5.11 See Figure S.21 for circuit and plot. Linear.

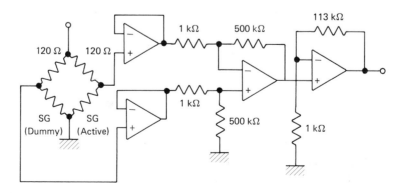

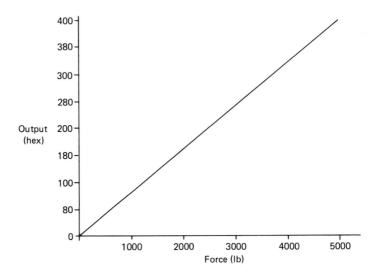

5.13 1047 rad/s

5.15 2.18 m/s^2

5.17 2.58 gs

5.19 $V_{\text{out}} = [0.0000646 \text{ V}/(\text{m/s}^2)]a$, $a_{\text{max}} = 96$ m/s^2, 11 Hz

5.21 1.2 mm

5.23 Use bridge with 120 Ω in each arm, 5-volt source. See Figure S.22 for the plot.

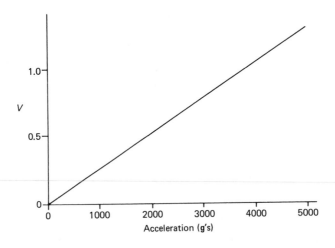

5.25 10.34 MPa, 102 atm

5.27 197.7 m/hr

5.29 See Figure S.23.

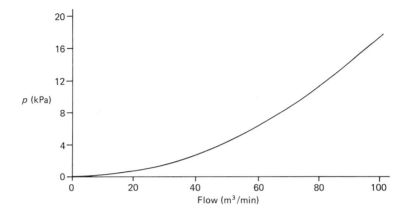

Chapter 6

6.1 10^{10} Hz, microwave

6.3 0.33 μs without liquid, 0.57 μs with liquid

6.5 5.64 m

6.7 See Figure S.24; 124.8 kΩ at 20 ms.

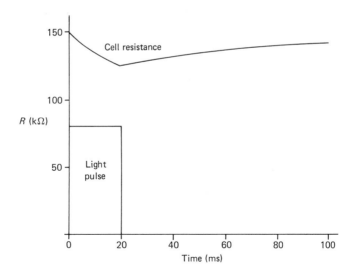

6.9 See Figure S.25; $V_{max} = 0.504$ V.

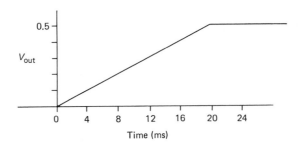

6.11 12.7 MW/m^2

6.13 0.72 m

Chapter 7

7.1 $V_{out} = 1250I + 15$

7.3 min: 01010_2, max: 10101_2, 93.75 rpm/LSB

7.5 2500 rpm

7.7 12.25 N

7.9 See Figure S.26.

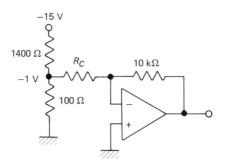

7.11 3.33 in^2

7.13 $Q(2/3) = 31.5$ m^3/hr, $Q(4/5) = 50$ m^3/hr

7.15 $Q_{min} = 11.9$ m^3/hr, $Q_{max} = 356$ m^3/hr, 3.13 cm for a flow of 100 m^3/hr.

7.17 **(a)** 8.33 psi or 44.4%; **(b)** 9.57 psi or 54.7%

Chapter 8

8.1 One possible solution is as follows:

Condition	Generator	Fan	Light	Door SW	PWR SW	Timer
1. Open door	OFF	OFF	ON	ON	OFF	OFF
2. Place food	OFF	OFF	ON	ON	OFF	OFF
3. Close door	OFF	OFF	OFF	OFF	OFF	OFF
4. Set timer	OFF	OFF	OFF	OFF	OFF	OFF
5. Power on	ON	ON	OFF	OFF	ON	ON
6. Open door	OFF	OFF	ON	ON	ON	OFF
7. Close door	ON	ON	OFF	OFF	ON	ON
8. Time up	OFF	OFF	OFF	OFF	OFF	OFF

8.3 See Figure S.27.

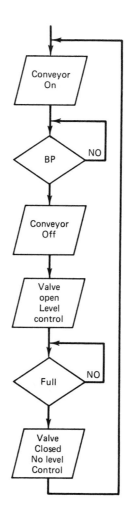

8.5 Assume a state is given by

$$(BP)(BF)(R)(LC)(V)(M)$$

State solution is

State		Output	
1. XX0000	→	000	Idle
2. XX1000	→	001	Conveyor on
3. 0X1000	→	001	Conveyor on
4. 0X1001	→	001	Waiting for BP
5. 1X1001	→	110	Conveyor off, start level and fill
6. 101110	→	110	Waiting for BF
7. 111110	→	000	All off
8. XXX000	go to 1		

8.7 See Figure S.28.

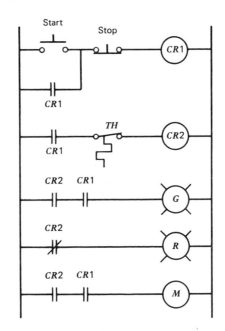

8.9 See Figure S.29.

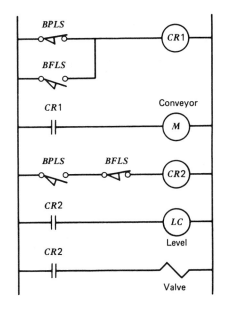

8.11 See Figure S.30.

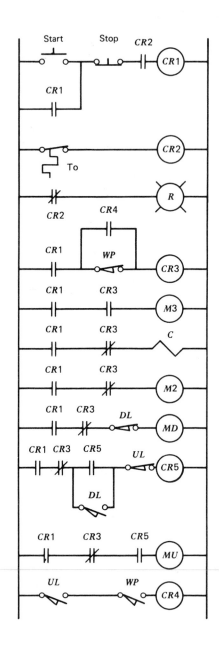

8.13 See Figure S.31.

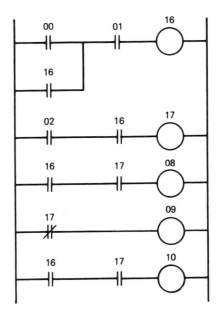

8.15 See Figure S.32.

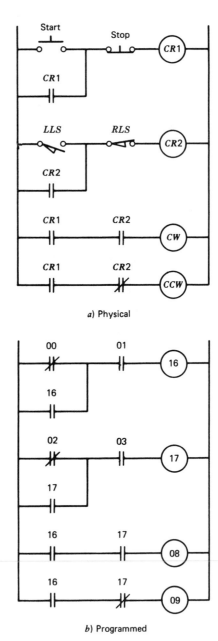

a) Physical

b) Programmed

Chapter 9

9.1 Q_A, T_0, T_A, Q_B

9.3 $e_p = 13.75\%$

9.5 Period = 39.3 minutes; see Figure S.33.

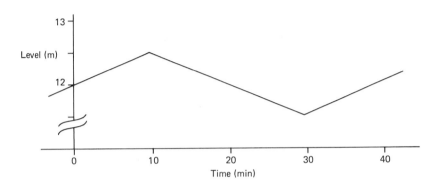

9.7 See Figure S.34.

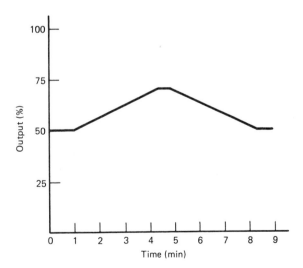

9.9 (a) 83.75%, (b) 0%, (c) 28.75% − 12.5t%

9.11 $P_D = 0.352 \cos(0.04t)$

9.13 See Figure S.35.

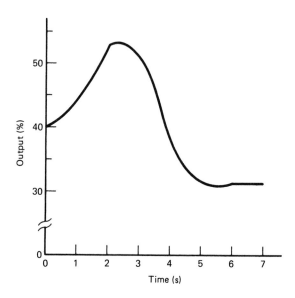

9.15 See Figure S.36.

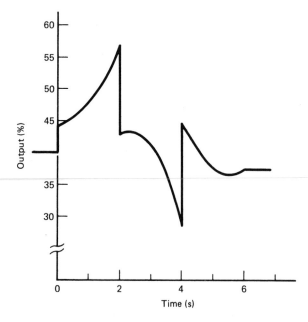

9.17 0.72 min

9.19 $p(t) = 13.5 \sin(\pi t) - 30[\cos(\pi t) - 1]$; phase shift $= 65.8°$ lag

Chapter 10

10.1 (a) 1.08 V, (b) -1.1 V to 0.5 V

10.3 Use Figure 10.5 with $R_1 = 2.2$ kΩ, $R_2 = 10$ kΩ, and V_{sp} set by a divider of the $+15$-volt supply to 5.4 V.

10.5 Use Figure 10.5 with V_{sp} selected from a divider of the 15-volt supply to be 1.356 V, $R_2 = 100$ kΩ, and $R_1 = 1.1$ kΩ.

10.7 See Figure S.37.

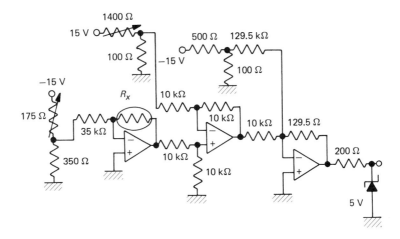

10.9 $G_I = 0.343$ s^{-1}, $R = 343$ kΩ, $C = 1$ μF

10.13 Use Figure 10.15 with $R_1 = 10$ kΩ, $R_2 = 62.5$ kΩ, $C_D = 10$ μF, $R_D = 937.5$ kΩ, $C_I = 1$ μF, $R_I = 576$ kΩ, $R_3 \ll 75$ kΩ.

10.15 8.426 psi and 5.666 psi

10.17 Replace single capacitor with a four-position switch with a capacitor for each required reset time.

10.19 See Figure S.38.

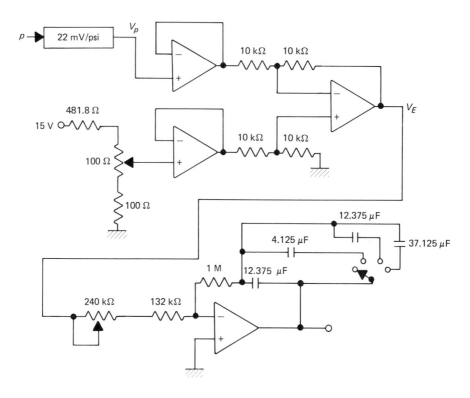

Chapter 11

11.1 Use a comparator with a trigger level of 2.31 volts.
11.3 Use the circuit of Figure 11.1 with dividers to produce 1.1 V and 0.935 V.
11.5 See Figure S.39.

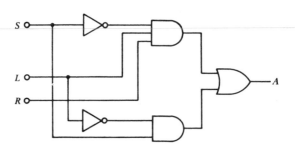

11.7 250 μs

11.9 See Figure S.40.

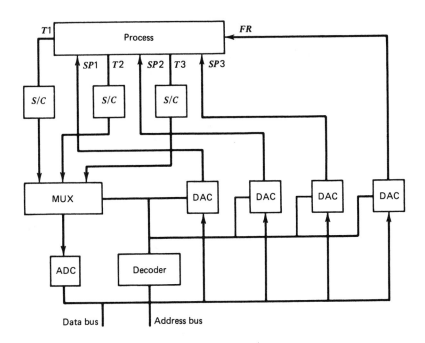

11.11 10 bits

11.13 See Figure S.41.

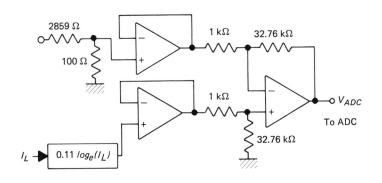

Computer equations:

$VADC = 5 * N/256; N = $ ADC input
$VL = $ VADC$/32.76 + 0.507$
$IL = EXP(VL/0.11)$

11.15 Equation is, $\Delta 1/1 = -0.9756 \Delta V/(5 + \Delta V)$. This is used directly in the higher-level language, as

$$STRAIN = -0.9756 * DELTAV/(5 + DELTAV)$$

$$MICROS = 1000000 * STRAIN$$

11.17 See Figure S.42.

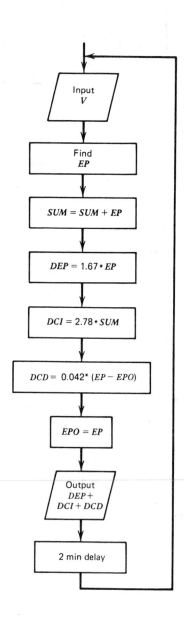

11.19 The appropriate equations, in program format, are as follows:

Assuming N = input from ADC

$$V = N * 5/256$$

$$L = 50 * LOG(2 * V + 1)$$

$$E = 100 * (50 - L)/119.7; \text{ Error}$$

$$SUM = SUM + E$$

$$P = 3.7 * E + 0.111 * SUM$$

$$OUT = P * 255$$

or, alternatively

$$P = P_0 + 3.811 * E - 3.7 * E0$$

$$P0 = P$$

$$E0 = E$$

$$OUT = P * 25$$

Chapter 12

12.1 See Figure S.43.

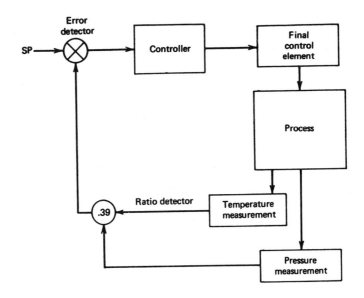

12.3   See Figure S.44.

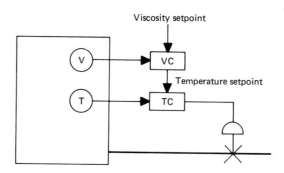

12.5   See Figure S.45.

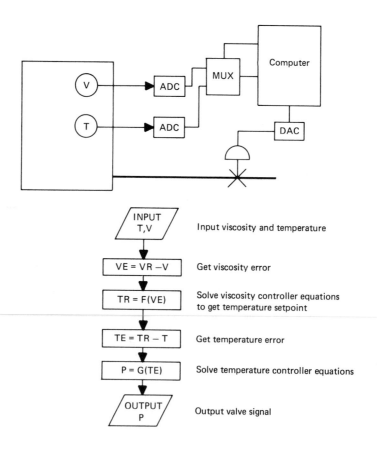

12.7 4.48; 105°

12.9 (a) $K_P = 2.1$, (b) $K_P = 1.89$, $T_I = 1.84$ min, (c) $K_P = 2.52$, $T_I = 1.11$ min, $T_D = 0.28$ min

12.11 $K_P = 4.95$

12.13 See Figure S.46.

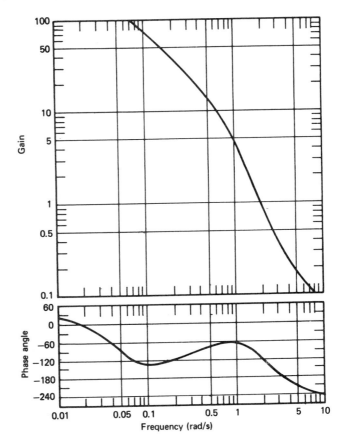

12.15 Instability for $K_P = 5$. Ultimate cycle gives $K_P = 2.5$, gain margin of 0.5 or about 6 dB, and a phase margin of 130°.

12.17 (a) At unity gain, the phase lag is about 170° and therefore unstable. At 180° the gain is about 0.9, greater than 0.56, and therefore unstable. (b) Reduce gain to about 2.5.

INDEX